DERATION
Terek
Samur
Alazani
Mingechaur Reservoir
AZERBAIJAN
Kura
Lake Sevan
ERBAIJAN
Aras
IRAN
Lake Urmia
CASPIAN SEA
Scale 1:4 740 000

RED LIST OF THE ENDEMIC PLANTS OF THE CAUCASUS

Armenia, Azerbaijan, Georgia, Iran, Russia, and Turkey

RED LIST OF THE ENDEMIC PLANTS OF THE CAUCASUS

Armenia, Azerbaijan, Georgia, Iran, Russia, and Turkey

James Solomon, Tatyana Shulkina, and George E. Schatz, Editors

Missouri Botanical Garden
IUCN
Critical Ecosystem Partnership Fund
WWF

ISBN: 978-0-915279-82-1
Library of Congress Control Number 2014930037
Monographs in Systematic Botany from the Missouri Botanical Garden Press, Volume 125

Scientific Editor and Head: Victoria C. Hollowell
Managing Editor: Allison M. Brock
Associate Editor: Tammy M. Charron
Associate Editor: Lisa Pepper
Associate Editor: Laura Slown
Press Coordinator: Cirri R. Moran
Editorial Assistant: Maryjane M. Hotaling

P.O. Box 299
St. Louis, Missouri 63166-0299, U.S.A.
www.mbgpress.info

Cover and book design by Design 5, LLC
www.design5online.com

This monograph was typeset 31 December 2013.

Cover photo:
Rhododendron caucasicum Pall. (Ericaceae) is common in alpine zones of the Caucasus Mountains.

Back cover photos:
Photo 1: *Lilium ledebourii* (Baker) Boiss. (Liliaceae).
Photo 2: *Iris iberica* Hoffm. (Iridaceae).
Photo 3: *Paeonia mlokosewitschii* Lomakin (Paeoniaceae).

Citation when citing the entire publication:

Solomon, J., T. Shulkina & G. E. Schatz (editors). 2013. Red List of the Endemic Plants of the Caucasus: Armenia, Azerbaijan, Georgia, Iran, Russia, and Turkey. Monographs in Systematic Botany from the Missouri Botanical Garden (MSB) 125. Missouri Botanical Garden Press, Saint Louis.

Citation when citing a single chapter:

Ekim, T., S. Terzioğlu, Ö. Emanağaoğlu & K. Coşkunçelebi. 2013. Turkey. *In*: J. Solomon, T. Shulkina & G. E. Schatz (editors), Red List of the Endemic Plants of the Caucasus: Armenia, Azerbaijan, Georgia, Iran, Russia, and Turkey. Monographs in Systematic Botany from the Missouri Botanical Garden (MSB) 125. Missouri Botanical Garden Press, Saint Louis.

Red List of the Endemic Plants of the Caucasus
Armenia, Azerbaijan, Georgia, Iran, Russia, and Turkey

James Solomon, Tatyana Shulkina, and George E. Schatz, Editors

REGIONAL COORDINATORS
Georgi Nakhutsrishvili – Regional Coordinator
Ketevan Batsatsashvili – Regional Coordinator Assistant

This book is a product of collaborative research among botanists from the countries of the Caucasus Region

ARMENIA
Kamilla Tamanian†, National Coordinator
Eleonora Gabrielyan, George Fayvush, Marine Hovhannesyan, Anush Nersesyan, Izabella Arevshatyan, Nazik Khandjyan, and Zhirajr Vardanyan

AZERBAIJAN
Valida Ali-zade, National Coordinator
Vahid Hajiev†, Vugar Kerimov, Salim Musayev†, Rena Abdiyeva, and Vahid Farzaliyev

GEORGIA
Georgi Nakhutsrishvili, National Coordinator
Ketevan Batsatsashvili, Revaz Gagnidze†, Zurab Manvelidze, Nino Memiadze, Shamil Shetekauri, and David Kharazishvili

IRAN
Jalil Noroozi, National Coordinator

RUSSIA
Dmitry Geltman, National Coordinator
Anatoly Mikheev†, Svetlana Litvinskaya, Ramazan Murtazaliev, Nikolay Portenier†, and Valeria Shvanova

TURKEY
Tuna Ekim, National Coordinator
Salih Terzioğlu, Özgür Emanağaoğlu, and Kamil Coşkunçelebi

IUCN Red List Coordinator
Craig Hilton-Taylor

Address for Correspondence
Missouri Botanical Garden
P.O. Box 299
St. Louis, MO 63166-0299
U.S.A.
www.missouribotanicalgarden.org

PARTICIPATING BOTANISTS

RENA ABDIYEVA – Institute of Botany, Azerbaijan National Academy of Sciences, Baku, Azerbaijan
VALIDA ALI-ZADE – Institute of Botany, Azerbaijan National Academy of Sciences, Baku, Azerbaijan
IZABELLA AREVSHATYAN – Institute of Botany, National Academy of Sciences, Yerevan, Armenia
KETEVAN BATSATSASHVILI – Ilia State University, Tbilisi, Georgia
KAMIL COŞKUNÇELEBI – Karadeniz Technical University, Trabzon, Turkey
TUNA EKIM – Istanbul University, Faculty of Biology, Istanbul, Turkey
ÖZGÜR EMANAĞAOĞLU – Karadeniz Technical University, Trabzon, Turkey
VAHID FARZALIYEV – Central Botanical Garden, Azerbaijan Academy of Sciences, Baku, Azerbaijan
GEORGE FAYVUSH – Institute of Botany, National Academy of Sciences, Yerevan, Armenia
ELEONORA GABRIELYAN – Institute of Botany, National Academy of Sciences, Yerevan, Armenia
REVAZ GAGNIDZE† – Ilia State University, Tbilisi, Georgia
DMITRY GELTMAN – Komarov Botanical Institute, Russian Academy of Sciences, St. Petersburg, Russia
VAHID HAJIEV† – Institute of Botany, Azerbaijan National Academy of Sciences, Baku, Azerbaijan
CRAIG HILTON-TAYLOR – IUCN Species Programme, Cambridge, United Kingdom
MARINE HOVHANNESYAN – Institute of Botany, National Academy of Sciences, Yerevan, Armenia
VUGAR KERIMOV – Institute of Botany, Azerbaijan National Academy, Baku, Azerbaijan
NAZIK KHANDJYAN – Institute of Botany, National Academy of Sciences, Yerevan, Armenia
DAVID KHARAZISHVILI – Batumi Botanical Garden, Batumi, Georgia
SVETLANA LITVINSKAYA – Kuban State University, Krasnodar, Russia
ZURAB MANVELIDZE – Shota Rusaveli Batumi State University, Batumi, Georgia
NINO MEMIADZE – Batumi Botanical Garden, Batumi, Georgia
ANATOLY MIKHEEV† – Komarov Botanical Institute, Russian Academy of Sciences, North Caucasian Ecological Station, Pyatigorsk, Russia
RAMAZAN MURTAZALIEV – Mountain Botanical Garden of the Dagestan Scientific Center, Makhachkala, Dagestan, Russia
SALIM MUSAYEV† – Institute of Botany, Azerbaijan National Academy of Sciences, Baku, Azerbaijan
GEORGI NAKHUTSRISHVILI – Ilia State University, Tbilisi, Georgia
ANUSH NERSESYAN – Institute of Botany, National Academy of Sciences, Yerevan, Armenia
JALIL NOROOZI – University of Vienna, Austria, and Plant Science Department, University of Tabriz, Tabriz, Iran
NIKOLAY PORTENIER† – Komarov Botanical Institute, Russian Academy of Sciences, St. Petersburg, Russia
GEORGE E. SCHATZ – Missouri Botanical Garden, St. Louis, Missouri, U.S.A.
SHAMIL SHETEKAURI – Ivan Javakhishvili Tbilisi State University, Tbilisi, Georgia
TATYANA SHULKINA – Missouri Botanical Garden, St. Louis, Missouri, U.S.A.
VALERIA SHVANOVA – Komarov Botanical Institute, Russian Academy of Sciences, St. Petersburg, Russia
JAMES SOLOMON – Missouri Botanical Garden, St. Louis, Missouri, U.S.A.
KAMILLA TAMANIAN† – Institute of Botany, National Academy of Sciences, Yerevan, Armenia
SALIH TERZIOĞLU – Karadeniz Technical University, Trabzon, Turkey
ZHIRAJR VARDANYAN – Institute of Botany, National Academy of Sciences, Yerevan, Armenia

CONTENTS

FOREWORD

Prior to the middle Miocene Period (15 mya), a fairly uniform warm-temperate forested region stretched across the entire north temperate region. Earlier fossil beds reflected this uniformity and a comparable biological richness in Europe, Asia, and North America at the appropriate latitudes. However, as the effects of continental glaciation began to be felt, first in Antarctica and later in the north, climates began to differentiate much more strongly than they had been previously from the Equator to the Poles. As this process continued through the Ice Ages of the Pleistocene, organisms became more restricted in distribution, and major disjunctions appeared in their ranges. The most important region of survival was in south-central China, in the region of the Hengduan and Qinling mountains, and the second in the southeastern United States. In western Eurasia lies the third area where considerable numbers of relicts persisted, notably in the Caucasus region.

In the Caucasus, as in the other two major centers of survival, there has also been a great deal of speciation, resulting in the evolution of many additional species in the respective mountain ranges. The Caucasus is situated in north-temperate Eurasia, a territory that has been studied in detail by botanists for many years. Most of the regions surrounding the Caucasus have been treated floristically in recent years, but not the Caucasus itself. The region lies between Europe and Asia, essentially at the southeastern corner of Europe, and it was included in modern European maps (Bohn et al., 2000–2003, 2007). However, its flora was not analyzed or presented in either *Flora Europaea*, which described all of the national floras of Europe, or in the Med-Checklist, which inventoried the vascular plants of the Mediterranean countries.

Because of the rich regional representation of endemic species, the Caucasus was identified as a biodiversity hotspot; for this reason it has attracted a considerable degree of international attention. The ecosystem profile of the Caucasus region, published by the Critical Ecosystem Partnership Fund (CEPF) in 2003 and updated in 2004, presented detailed studies of the mammals, birds, reptiles, fishes, and amphibians of the region, and even for some groups of insects. At least 153 species of mammals inhabit the Caucasus, one fifth of them endemic. More than 200 species of fishes inhabit the region, over one third of which are found nowhere else. Twenty-two of the 77 reptile species are endemic. As many as 400 species of birds inhabit or pass through the Caucasus, with millions of migrating birds using Caucasian stopover sites seasonally (Zazanashvili et al., 1999). In contrast to these reasonably complete reports, the CEPF analysis included only a single plant species, partly because of the lack of regional coordination in dealing with the available information.

The results included here comprise the first floristic analysis of the entire Caucasian region. This comprehensive analysis was initiated with the preparation of International Union for Conservation of Nature and Natural Resources (IUCN) Red List data for plant taxa of the Caucasus region, starting in 2003, when the Trust for Mutual Understanding approved a project submitted by the Missouri Botanical Garden (MO). The Trust awards grants to American non-profit organizations to support cultural and environmental programs conducted in partnership with institutions and individuals in countries of the former Soviet Union and Central and Eastern Europe. This support provided an extraordinary opportunity for Caucasian botanists to meet with each other after many years of isolation. This isolation had impeded developing a coordinated, regional approach to the study of the plants in the area, while such an approach was clearly necessary for a more complete understanding of each country's interrelated floras as a basis for conserving them.

The flora of the Caucasus region was little known to botanists from outside the former Soviet Union because of past restrictions on travel and the fact that most of the botanical literature was published either in Russian or a regional language. In order to promote the dissemination of this knowledge, the Missouri Botanical Garden started to work with Caucasian botanists in 1997 with financial support from the National Geographic Society and the Trust for Mutual Understanding. In 2001 the Missouri Botanical Garden put a List of Rare, Endangered, and Vulnerable Plants of the Republic of Georgia on its website (http://www.mobot.org/MOBOT/research/georgia/introduction.shtml) as a beginning point for presenting basic information on the plants of the region.

With generous support from the Trust for Mutual Understanding and Missouri Botanical Garden, the Georgian Institute of Botany convened a workshop in Tbilisi, Georgia, from 14–19 October 2003. The primary goal of this first workshop was to establish collaboration between Caucasian and American botanists and to develop a framework to assess the conservation status of endemic plants of the Caucasus region with the goal of producing a Red List for the Caucasus region. Among the participants were botanists from the major botanical institutions of Armenia, Azerbaijan, Georgia, and the Russian Federation (North Caucasus). Representatives from the World Wide Fund for Nature (WWF); the Government of Georgia; the IUCN; and CEPF also joined the workshop. From the Missouri Botanical Garden, Tatyana Shulkina, James Solomon, George E. Schatz, and Peter Jørgensen participated. The motivation of the participants in this meeting was to find ways to effectively preserve the extraordinary botanical diversity of the Caucasus, which is very much in keeping with the mission of the Missouri Botanical Garden: "to discover and share knowledge about plants and their environment, in order to preserve and enrich life."

During the workshop, American participants offered models for preparing the Red List of the Endemic Plants of the Caucasus based on their experience with similar projects in other countries. Wendy Strahm (IUCN) informed the audience about IUCN categories and criteria for the Red List. Dan Cao (CEPF) explained requirements of CEPF proposals. Representatives from each of the Caucasian countries presented analyses of their local situations, obstacles, and possibilities. During the course of the deliberations, it became clear that the conservation needs far outstripped the capacity of any one country to meet them. The participants were particularly interested in techniques new to them, such as geographic information systems (GIS) applications and developing databases. Tatyana Shulkina (MO) suggested that the Red List of the Endemic Plants of the Caucasus have a comprehensive introduction discussing the endangered plants of the region as a whole, but that in addition the section for each country would include an introduction to its own plants in its native language and in English. Georgi Nakhutsrishvili expressed the opinion of the Caucasian botanists, who suggested that the Red List should give each country an opportunity to have its own flora description and participate in the general overall list of plants. James Solomon (MO) was named as editor of the English language for this book.

A team was established to manage the development of a proposal to prepare the Red List. The Tbilisi Botanical Garden and Institute of Botany (TBI) was designated to coordinate all the activities in the Caucasus for this effort. Included were botanists from the major Caucasian institutions, along with botanists from the Komarov Botanical Institute, St. Petersburg, Russia, which holds the largest collection of Caucasian plants, as well as botanists from Turkey and Iran. The Turkish and Iranian parts of the Caucasus had not been included in earlier floristic studies of the region, making previous analyses incomplete.

Resulting from these efforts, a proposal entitled "Coordination and Development of Plant Red List Assessments for the Caucasian Biodiversity Hotspot" was submitted to the CEPF in 2004. The proposal stipulated the involvement of representatives from six countries: Armenia, Azerbaijan, Georgia, Russia, Turkey, and the United States. The Principal Investigator was George E. Schatz, the Missouri Botanical Garden (MO) and a member of the IUCN committee. Tatyana Shulkina (MO), who worked for many years at the Komarov Botanical Institute and subsequently at Missouri Botanical Garden, was designated the overall Coordinator. Georgi Nakhutsrishvili became the Regional Coordinator and Ketevan Batsatsashvili the Regional Coordinator Assistant (both Tbilisi, Georgia). The CEPF published a map of the Caucasus as a hotspot, a suitable target for investment. Conservation targets in this geographical region needed to be established and achieved in order to prevent biodiversity loss, and the most logical targets often do not coincide with floristic borders. Scientists at Conservation International's Center for Applied Biodiversity Science (CABS) produced an outline map to show the region for which they were responsible (see inside cover map).

Important input to the overall project was stimulated by a later workshop entitled "Globally Threatened Trees of the Caucasus" held in Tbilisi, Georgia, 26–28 September 2005. This workshop was organized by the Global Tree Campaign, a joint initiative of Fauna and Flora International and the United Nations Environment Monitoring Programme–World Conservation Monitoring Centre (UNEP–WCMC). The workshop leader, Sara Oldfield (Chair of the IUCN/Species Survival Commission, Global Tree Specialist Group, now Secretary General, Botanic Garden Conservation International, United Kingdom) brought together experts knowledgeable on Caucasian woody plants to assess the global conservation status of the tree species of the Caucasus. Among the 19 participants were representatives of all countries in the Caucasian region, as well as Germany, the United States, and the United Kingdom. Over the course of the workshop the participants reviewed 152 taxa of woody plants. The workshop was a great success and provided a very important step toward the Red List of Endemic Plants of the Caucasus. It contributed to the broader project by helping to train Caucasian botanists in IUCN evaluation techniques. At the same time, the results of the workshop demonstrated the need for regional coordination in an effort to treat the plants of the Caucasus adequately and to focus on their conservation needs. It also became clear that to achieve the aims of conservation in the Caucasus, regional integration, thorough literature and specimen reviews, and field studies were essential.

The next workshop, 5–9 June 2006, was held in Bakuriani, Georgia. It was organized by the Tbilisi Botanical Garden and Institute of Botany and the Bakuriani Botanical Garden. The IUCN representative was Craig Hilton-Taylor, Red List Programme Officer, United Kingdom. The main goal of the workshop was to train participating botanists in the preparation of National Red Lists and the application of IUCN criteria for evaluating the different categories of endangerment. Craig Hilton-Taylor and George E. Schatz (MO) guided the training program. Subsequent fieldwork added significantly to the quality of the data underlying the evaluations, verifying distributions, and adding some previously unknown subpopulations of Caucasian endemics.

On 2–8 October 2009, the Missouri Botanical Garden hosted the International Symposium for Plant Conservation in the Caucasus Region, with financial support from the Trust for Mutual Understanding. Representatives from each of the countries concerned gave presentations about their flora, vegetation, and endemic plants and provided an analysis of their five years of work on the Caucasian Red List. This meeting was particularly important because it was the first time that all of the participants from the different Caucasian countries were able to assemble at a single location at the same time to meet with each other and present their results. The symposium attracted media attention, and descriptions of the proceedings were published in newspapers and broadcast on the radio.

The present volume provides the first conservation analysis of the endemic plants of the Caucasus region and is a significant product of the collaborative integration made possible by botanists from the Caucasus region. We sincerely believe that it will be the basis for the development of future integrated studies and for planning effective conservation action for Caucasian plants on a regional basis. Continuing effort will be needed to keep the taxonomic and conservation status of Caucasian plants current. As a means of facilitating the conservation of what is truly a world treasure, the cooperative efforts among Armenian, Azerbaijani, Georgian, Iranian, Russian, and Turkish botanists in the preparation of this Caucasian Red List was unprecedented. We are grateful for the opportunity to work with our colleagues from different countries on this critically important project, and for the support of the Critical Ecosystem Partnership Fund, the Trust for Mutual Understanding, and many institutions in producing the Red List of the Endemic Plants of the Caucasus. We look forward confidently and enthusiastically to future conservation activities in the Caucasus.

Peter H. Raven
President Emeritus
Missouri Botanical Garden

PREFACE

The Caucasus region is home to a biodiverse treasure of global significance. Its flora, in particular, is among the richest on earth, but the region is now recognized as one of the world's 34 biodiversity hotspots areas whose biodiversity is facing significant threats. The region includes a fascinating and rich mixture of plants, ranging from alpine to temperate and subtropical species, found in the complex tapestry of habitats and ecosystems, most notably the important deciduous and coniferous forests, alpine meadows, and species-rich pastures throughout the region. For the people of the Caucasus, their flora is of immense importance because it is used for many purposes. For botanists and horticulturists, the diversity of trees, shrubs, perennials, annuals, and bulbous species found in the Caucasus are a source of endless interest and value. For economic botanists and ethnobotanists, the range of wild plants with actual and potential value to humanity, including crop wild relatives and wild medicinal plants as well as the traditional knowledge of their use, are resources of great significance. Given its importance, the region must be considered a resource requiring urgent and effective conservation measures.

While it is an honor and pleasure for me to have the opportunity of contributing a preface for this book, I am sad that this publication should be necessary to document the threats facing so many of the unique endemic plants of the region. The book is hugely important in that it includes the results of a conservation assessment of 1752 endemic plants, selected from a list of 2791 known endemics. Each of these species has been assessed according to the IUCN Red List categories. The first ever IUCN Plant Red Data Book (Lucas, 1978) included only one Caucasian plant, *Iris winogradowii* Fomin. The need for the preparation and publication of this special Red List for the Caucasus reflects not only our greatly increased knowledge of the status of plants in the region—for which we can be grateful—but also the continuing and relentless loss of plant diversity throughout the world. The information presented includes five species that are Extinct in the Wild (EW); 169 that are Critically Endangered (CR); 291 that are Endangered (EN); 272 that are Vulnerable (VU); 255 that are Near Threatened (NT); 242 that are of Least Concern (LC); and 518 species that are Data Deficient (DD). In total, 1752 species were assessed for the Red List of the Caucasus, which represents 62% of the 2791 endemics known for the region.

In 2002 the United Nations Convention on Biological Diversity adopted a Global Strategy for Plant Conservation (GSPC) to highlight the plight of tens of thousands of plant species threatened worldwide. The GSPC and its 16 targets (updated in 2010), to be achieved by 2020, provide a template to guide plant conservation actions worldwide. I consider this publication an important contribution towards GSPC Target 1 (documenting plant diversity) and Target 2 (conservation assessments).

I congratulate the editors and all of those who have contributed information for this important book. I am delighted that the Missouri Botanical Garden has been able to play a major role in its preparation. We celebrate our continued close collaboration with partner institutions, organizations, and individuals in the countries in the region and hope that our work together can continue to be further developed and enhanced in the future, to achieve many more successes.

The preparation of this book is very timely, not only to highlight and raise awareness of the urgent conservation needs for so many plant species, but also to help guide conservation efforts and the priorities needed to safeguard far too many unique plants. If we are to be successful in conserving the diversity of the all too numerous plant species included in this book, we need to do so on the basis of comprehensive, accurate, and up-to-date information. Let us use the book therefore as a call to action to put in place new, and support continuing, efforts to protect the remarkable flora of the Caucasus.

Peter Wyse Jackson
President
Missouri Botanical Garden

ACKNOWLEDGMENTS

The major support for this work was provided by Critical Ecosystem Partnership Fund (CEPF/ United States); International Union for Conservation of Nature and Natural Resources (IUCN); Trust for Mutual Understanding (TMU), United States; Missouri Botanical Garden (MO), United States; Tbilisi Botanical Garden and Institute of Botany (TBI), Georgia; Botanical Institute (ERE) of the National Academy of Sciences, Yerevan, Armenia; Botanical Institute (BAK) of the National Academy of Sciences, Baku, Azerbaijan; and the Komarov Botanical Institute (LE) of the Russian Academy of Sciences.

The line drawings of Caucasian endemics were prepared from living plants or herbarium specimens by Svetlana Litvinskaya, Kuban State University. Photographs were provided by Otar Abdaladze (Georgia), Shalva Sikharulidze (Georgia), Kamilla Tamanian (Armenia), George Fayvush (Armenia), Vugar Kerimov (Azerbaijan), Jalil Noroozi (Iran), Anatoly Mikheev (Russia, Stavropol), Ramazan Murtazaliev (Russia, Dagestan), and others.

Many people have helped us in numerous ways during the preparation of this manuscript. To all named, and any unintentionally omitted, we would like to express our sincere gratitude for their contribution to the project. A partial list of botanists who helped should begin with Armen Takhtajan, who supported the idea of the basic structure of the book, and Rudolf Kamelin, for helpful discussion and constructive comments and for very important corrections in the text. Significant progress in completing this book was made by members of the WWF Caucasus Programme Office, especially its director, Nugzar Zazanashvili, who provided material and documents on conservation within the entire Caucasian ecoregion. We are most grateful to Ihsan Al-Shehbaz and Michael Pimenov for checking the list of species that are included in the overall Red List. Many thanks to D. Shangelia for providing us with geological information included in the text. We also appreciate the contribution of the following botanists from the Caucasian countries and Russia: N. Adigüzel, Firuza Akhmed-zade, Vandika Avetisyan, S. Ayaz, Z. Aytaç, B. Bani, M. Bona, Aida Dadasheva, H. Duman, M. Ekici, A. Güner, F. Günes, N. Íkinci, Galina Kudrjashova, Helen Mordak, Estella Nazarova, N. Özhatay, Hagiaga Safarov, Eldar Sukurov, Ivan Tatanov, Nikolai Tzvelev, M. Vural, and S. Yüzbaşioğlu. We also wish to thank botanists who gifted their own personal records: A. Aksoy, E. Hamzaoğlu, and F. Karaer.

People from different institutions outside the Caucasus provided tremendous help, among them Craig Hilton-Taylor (IUCN), Cristopher Holtz (formerly CEPF), Sara Oldfield (BGCI), Jane Smart (IUCN), Wendy Strahm (formerly IUCN), Simon Stuart (IUCN), Helen Temple (formerly IUCN), Jack Tordoff (CEPF), and Jean-Cristophe Vie (IUCN). Officials from the Kuban State University have helped our work in the North Caucasus, Russia.

We are very grateful to Christopher Davidson and Sharon Davidson; Francis Cabot; and Mr. and Mrs. Constantine Sidamon-Eristoff for their financial support, advice, and friendship.

During the preparations for the field trip and writing manuscripts, Caucasian botanists worked at different herbaria, and we are grateful to the authorities of the following institutions: AEF, ANK, BAK, BATU, ERE, HUB, ISTE, ISTF, ISTO, KATO, KTUB, LE, MO, TB, TGM.

We would like to thank Victoria C. Hollowell, Allison M. Brock, Tammy M. Charron, Maryjane M. Hotaling, and Lisa Pepper (of Missouri Botanical Garden Press) and Fred Keusenkothen (of the Peter H. Raven Library), for their help in preparing this book.

This book was made possible only by the efforts of many people and we hope it will be useful for them, for other botanists, and the international community concerned about the world's biodiversity. The book is dedicated to the memory of our colleagues who started with us but passed away—Kamilla Tamanian (Armenia), Vahid Hadjiev (Azerbaijan), Salim Musayev (Azerbaijan), Anatoly Mikheev (Russia), and Nikolay Portenier (Russia).

REGIONAL OVERVIEW

Tatyana Shulkina and James Solomon, Missouri Botanical Garden

The Caucasus region is no doubt one of the most beautiful places in the world. This famous yet mysterious place was mentioned many times in the Bible and in Greek mythology. It is where biblical scholars believe Noah may have landed the Ark on Mount Ararat when the waters of the Flood subsided; where Prometheus from Greek legend was condemned by Zeus and chained to a cliff; and where the Argonauts journeyed in quest of the Golden Fleece.

The Caucasus today is one of 34 biodiversity hotspots around the world. The region was recently designated by Conservation International as a conservation priority, which indicates that the area has an exceptional concentration of endemic species and is also undergoing an exceptional loss of habitat (Mittermeier et al., 1999; Myers et al., 2000; Krever et al., 2001). The region's need for conservation investment is most urgent.

Geography and Geology

The Caucasus region is located between the Black and Caspian seas, ranging between 39°N and 47°N and 40°E and 50°E, which means the Caucasus region lies at latitudes similar to Italy and Spain. Because the Caucasus region is situated between Europe and Asia, its geographic affiliations have changed many times (Bohn et al., 2003, 2007).

There are two main mountain ranges, the Greater Caucasus and the Lesser Caucasus, which run parallel to one another, northwest to the southeast for over 1500 km. The Greater Caucasus Range stretches from the northeastern coast of the Black Sea and gradually diminishes in elevation toward the Caspian Sea at its easternmost extent. Throughout much of the Greater Caucasus Range, mountain peaks can reach 4000 m or more. The highest elevations form the watershed between the North and South Caucasus, and the scientific literature often divides the Caucasus into these two regions: North Caucasus and South Caucasus. The southern portion of the Caucasus is known as the Transcaucasus region, which extends through Georgia, Armenia, and Azerbaijan, between the two Caucasus ranges, and into bordering Iran and Turkey. Synclinal valleys lie south of the Greater Caucasus, stretching between the Black and Caspian seas. This depression is divided by the Surami Range into western and eastern lowlands, and the Surami Range serves as a significant barrier to the maritime influence of the Black Sea. The Surami Range meets the Lesser Caucasus in the south that, in turn, merge with the Armenian Plateau, which extends to the south into Turkey and Iran.

The Caucasus eco-region has been the subject of biodiversity hotspot studies and includes not only montane areas, but also steppes in the North Caucasus and areas adjacent to the mountains in the south. There were two reasons why the northern borders for the Caucasus biodiversity hotspot were deliberately expanded (see inside cover map). First, many of the most important protected areas are located along national boundaries that fall outside designated Caucasian Floristic Provinces (Takhtajan, 1986). Secondly, this expansion of the Caucasus hotspot includes territories important for wildlife migration between summer and winter feeding areas; thus, special biological corridors were created. As a result, the northern border of the Caucasus biodiversity hotspot does not correspond with floristic regions, although the southern border does run within the limits of the Armeno-Iranian Floristic Province. Takhtajan (1986) believed, however, that the hotspot borders in the east, south, and southeast were not clearly delineated and that additional information on endemic distributions was needed. Takhtajan suggested the inclusion of part of the Alborz Mountains in the Atropatenian Subprovince. Current studies of the distributions of endemic species in Iran have validated Takhtajan's insight. There is botanical evidence that the mountainous regions of northwestern Iran such as the inactive stratovolcanoes of Sabalan, Sahand, and Bozqush, as well as the high mountains of Misho-Dagh and Kiamaki have strong floristic relationships not only with each other, but also with the Caucasus territories. Accordingly, the northeastern part of the Pontic Mountains and Mount Ararat (Ağrı Dağı) in Turkey and the Alborz Mountains along the Caspian Sea in Iran are also included in the Caucasus biodiversity hotspot.

Much of the Caucasus region was part of the former Soviet Union, and previous floristic studies were usually restricted to these administrative regions. The Turkish and Iranian portions of the Caucasus region were never included in the Russian language floras published in the former Soviet Union (Grossheim, 1928–1934, 1939–1967; Takhtajan, 2003–2012). As a result, the present publication will be the first to cover the Caucasus region in its entirety. Politically, the North Caucasus region includes seven administrative regions or autonomous republics of the Russian Federation (Adygea, Chechnya, Dagestan, Ingushetia, Kabardino-Balkaria, Karachayevo-Circassia, and North Ossetia), as well as the territories of Krasnodar (Krasnodarsky Krai), Stavropol (Stavropolsky Krai), and part of the Rostov region. The South Caucasus region extends throughout the countries of Armenia, Azerbaijan, and Georgia, as well as northeastern Turkey and northwestern Iran. Significantly, the Caucasus region is a multinational region inhabited since ancient times by a diversity of culturally and linguistically distinct peoples.

The mountainous region of the Caucasus belongs to what is generally referred to as an alpine zone that stretches across the whole of Eurasia from the Pyrenees and the Atlas Mountains in western Europe and North Africa to the Malay Peninsula and Vietnam in eastern Asia. Orogenically the Caucasus system was the result of global tectonics, with its high mountains the result of millennia of collisions between the Arabian Nubia plate and its northward movement into the Eurasian plate.

The Caucasus region geologically represents the central segment of the Mediterranean mobile belt. This deformation zone is a remarkable mosaic of crustal microplates that form and move independently of the Nubia and Eurasian tectonic plates. The North Caucasus belongs to the Scythian platform, an epi-Variscan structure built from Precambrian and Paleozoic metamorphic rocks that create a stable substrate with overlying Upper Paleozoic molasse and Mesozoic-Cenozoic platform deposits (Gerasimov, 1966; Mrevlishvili, 1997). The fold nappe topography of the Greater Caucasus stretches for 1300 km from the Black Sea in the west, from the Taman Peninsula in Krasnodar Krai, in the Russian Federation, dropping gradually to the hilly terrain of the Absheron Peninsula in Azerbaijan on the Caspian Sea. At its northern extent, the Caucasus is represented by Precambrian and Paleozoic metamorphites, granite-metamorphic complexes, Middle Paleozoic island arc volcanites and carbonates, ophiolites, Upper Paleozoic molasses, and Jurassic, Cretaceous, and Paleogene clay-schists and sandstones. The southern portion, or the Main Range zone of the Caucasus, is built up from Precambrian and Paleozoic metamorphic complexes as well as Late Baikalian, Caledonian, and Variscan granitoids. Formations of the Main Range zone overthrust the intensely folded Upper and Middle Jurassic clay-slate formations of the southern slopes, which in turn border thick Upper Jurassic–Lower Paleogene terrigenous and carbonate flysch formations (Svanidze, 1972).

The Transcaucasus intermountain area is composed of Precambrian and Paleozoic gneiss-metamorphic complexes and granitoids (the Dzirula, Khrami, Loki, Akhumi, and Asrikchay Median massifs) and volcanites. Intense middle Jurassic volcanism continued into the middle Senonian series of the Late Cretaceous period, ca. 85 mya. In the Bathonian stage, accumulation of coal-bearing bands and intensive granitoid magmatism were recorded. Earlier, in the Late Jurassic period (164–145 mya), eruption of alkali basalts took place (Kolakovsky & Shakryl, 1978).

The Lesser Caucasus mountain range is the result of a complex tectonic fold thrust system. In its northwestern portion, the Adjara-Trialeti rift trough formed in the Middle Cretaceous period, which through the Cretaceous–Lower Paleogene epoch, 66–56 mya, was filled with volcanogenic carbonate formations. To the southeast of the Transcaucasus area, its analog is the Talysh zone in Azerbaijan. In the Paleozoic and almost during the entirety of the Mesozoic, the southern portions of the Lesser Caucasus were formed by the northern margin of the Iranian platform, where Baikalian and Grenvillian metamorphic basement rock represents the Arzakan massif (Armenia). Late Cretaceous–Early Paleogene deposits spread across the Yerevan-Ordubad trough, where intense volcanism occurred through the Eocene epoch, 56–33.9 mya. In the Early Miocene epoch, the formation of the Megri-Ordubad granitic batholith and its intrusion into Armenia terminated. It was during the Miocene epoch, 23–5.3 mya, that the present-day basin and plain of the rivers Aras, Ararat, and Nakhichevan formed, with their distinctive molasse depressions of conglomerate, sandstones, and shales. In the central part of the Lesser Caucasus, the smaller Sevan and Leninakan (Gyumri) depressions were also formed (Takhtajan, 1946). Intense surface volcanism covered a significant area of the central Lesser Caucasus region in the Late Miocene epoch, and this volcanic activity continued into the current Holocene epoch in Armenia and Georgia. Such volcanism also occurred in the Greater Caucasus region, where small Pliocene granitoid intrusions formed (Georgia) as were several older laccoliths in the area of Mineralnye Vody (Stavropol Krai, Russian Federation).

The Greater Caucasus Mountains usually are divided into four orographic zones: the Main or Watershed (Glavny), the Lateral (Bokovoi), the Rocky (Skalisty), and the Chalk (Melovoy). Each range consists of diverse rock types and varied fold and thrust systems and is characterized by Jurassic and Cretaceous rocks. The extreme western part of the Greater Caucasus Mountains is a small mountainous upheaval, consisting mostly of marls and limestone. In the central range, the height of the mountains increases considerably and is formed of crystalline rocks. Notable peaks in the central portion include Mount Elbrus (5633 m, the highest peak in Russia) and Dykh-Tau (or Dykhtau, Jagged Mount, 5205 m, the second highest in Russia, just north of the Georgian border). Also prominent are Mount Kazbek (5033 m, Kazbegi in Georgian) and Mount Bazardüzü (4480 m, Bazardüzü daği in Azerbaijani, the highest peak in Azerbaijan, on the border with Dagestan). The permanent snow cap of Mount Elbrus is the largest in the Caucasus, extending over 122 km^2. However, glaciers are found throughout the alpine zone, numbering more than 2000 and covering 1424 km^2. The Greater Caucasus Range is crossed by high passes, the two best known being the Dar'ial Gorge, in Georgia crossing the Bokovoi Range, and Mamison at 2911 m on the Georgian border with Russia. The eastern extent of the Greater Caucasus Range is much lower than the central portion and has mountains composed of sedimentary rocks. There are more than 470 karst caves in the southern slopes of the Greater Caucasus, mostly found in Georgia. One of the deepest, Snowy Cave, in the Bzyb Range, descends more than 1320 m and is covered by ice and snow. However, one of the longest, Novoafonskaya Cave, stretches for 50 km, and has many halls and tunnels. The Greater Caucasus region is also rich in minerals, non-ferrous metals, with mineral springs, especially in the North Caucasus, particularly in the region of Mineralnye Vody (Mineral Waters, Stavropol Krai) and near the cities of Sochi (Krasnodar Krai, Russian Federation) and Tskhaltubo, in the Imereti District of central Georgia. In the area surrounding the Georgian town of Borjomi are the mineral springs that are the source of the widely known mineral water of the same name.

In contrast, the Lesser Caucasus Mountains are largely of volcanic origin. There were extensive lava flows in the Tertiary period (65–1.8 mya) that covered most mountains, and only a few peaks remained uncovered. The Javakheti Volcanic Plateau in Georgia and the volcanic ranges that reach central Armenia and northeastern Turkey are still regularly subjected to earthquakes from this continued orogenic activity. There are active mud volcanoes in the eastern Caucasus, in the Kura–Aras lowland through which the Aras River (originating in Turkey) passes through Armenia, Azerbaijan, and Iran, and is thus variably known as the river Araks, Arax, or Araz. Mud volcanoes also occur on the Absheron Peninsula in Azerbaijan. In total, there are 320 mud volcanoes in the Caucasus, which is nearly 40% of the 800 known worldwide. Notably, terrains in Azerbaijan are rich in petroleum oil resources. The tallest peak in the south is Mount Ararat (Ağrı Dağı in Turkish) (over 5100 m), a dormant stratovolcano located in eastern Turkey, near the Iranian (16 km distant) and Armenian (30 km distant) borders. Since ancient times, Mount Ararat has been considered by Armenians as their holy place, home of the gods of the Armenian Pantheon, and is a national symbol of Armenia. In Armenia, the highest peak is the dormant stratovolcano Mount Aragats (Alagyaz in Armenian), which reaches 4090 m. In Iran, the two highest peaks are Mount Damavand (5671 m) and Mount Alamkuh (Alam Kūh, 4850 m) in the central portion of the Alborz Range. The Alborz Mountains arc around the southern Caspian Sea, extending from Armenia and Azerbaijan through northern Iran into Turkmenistan and Afghanistan; thus, these mountains may be locally known as the Alburz, Elburz, and Elborz. Mount Sabalan (4811 m), in Ardabil Province, is the third highest peak in Iran, and is known for the strikingly beautiful crater lake at its summit.

A wonderful historic place in the Lesser Caucasus is situated in Gemigaya, in the Ordubad region (3500–3700 m), in Nakhichevan, with unique petroglyphs that date from the fourth to the first century BC. This ethnographic complex is evidence of ancient civilization. Christianity also spread here beginning in the first century, when the Mother Albanian Church was built in Kish village, Sheki region, Azerbaijan. It is important to mention that Christianity became the official religion in Georgia as early as 337 A.D.

History of Vegetation

Extending from the early Jurassic period to the Oligocene epoch, most of the Caucasian territory was covered by water and there were just isolated emergent islands. By the mid-Miocene epoch, a major land bridge had appeared. Paleobotanical data show fossil plants are found in different regions in the Caucasus (Kolakovsky, 1974), with the oldest fossil discoveries referred to the Paleozoic period and identified as the scale trees *Lepidodendron* and *Sigillaria* (primitive relatives of modern Lycopsida or club mosses) found in Khrami, South Georgia (Shatilova & Ramishvili, 1990, 2011). More recent fossil finds in the East Caucasus (Darry-Dag) from the Lower Oligocene are similar to the modern palm *Sabal* Adans. (Arecaceae), and discoveries from Borjomi district (historic Tori region, Georgia) have been identified as *Hakea* Schrad., *Lomatia* R. Br., and *Roupala* Aubl. (all Proteaceae). Reports of fossil plants from Pliocene

(Pont) and Miocene rock strata in the central and western Caucasus (Kodori) contain leaves of many evergreen trees similar to *Magnolia* L. and *Persea* Mill. and other extant Lauraceae (Kolakovsky, 1964, 1974). In the northern part of the Lesser Caucasus (Dashkesan, Azerbaijan) fossilized leaves of mid-Jurassic pteridosperms or tree ferns such as *Cladophelbis*, *Marattiopsis*, *Ptilophyllum*, and *Sagenopteris* have been found. There was tropical or subtropical vegetation extending into the Caucasus, with many discoveries from the Tertiary flora of Abkhazia supporting relationships with tropical floras. Eight genera and 20 species of fossil Araliaceae have been discovered in Abkhazia (Kolakovsky, 1995), all of tropical East Asian origin. This region, known as the Colchis refuge, is also the richest known for fossil Lauraceae (Kolakovsky & Shakryl, 1976). There are two refugia known in the Caucasus region, Colchis in the western portion and Hyrcania in the east where representatives of these ancient floras survived during the cold and dry periods of the Neogene period (Rechinger, 1963–2012). The Colchis name is well known, but the lesser known Hyrcanian region was named after the ancient Hyrcanian Sea, a predecessor of the Caspian Sea. Hyrcanian National Park in Azerbaijan is included in the World Natural Heritage List.

Climate

Mountainous terrains produce a great variety of different climatic and vegetation zones today (Alisov, 1956; Grigoriev & Budyko, 1971). The Greater Caucasus Range acts as a significant barrier to horizontal air movement and protects the South Caucasus region from Arctic influence. Their high elevation also provides for altitudinal zonation of climate on its slopes. Permanent snow fields and glaciers exist at the highest elevations. Snow-capped peaks create spectacular landscapes and, near the snow line, magnificent alpine vegetation thrives. South of the Greater Caucasus Mountains lie a synclinal valley divided into western and eastern lowlands by the transverse Surami or Likhi Range in Georgia, and as a result the marine influences of the Black Sea do not penetrate deeply to the east.

The southwestern Caucasus includes parts of Georgia and Turkey and is characterized by a very mild and humid climate. Average air temperature of the coldest months is ca. 9°C, which means that the area falls within zones 8 or even 9, according to U.S. Department of Agriculture (USDA) climatic zones. This area is also remarkable for its high level of annual precipitation, which in some places reaches 4500 mm, as recorded for Mount Mtirala, Adjara, in southwestern Georgia. This type of climate also occurred in the Mediterranean region in the Lower Tertiary period but no longer exists there. The Black Sea coastal areas to the north are colder and dryer and are referable to USDA zone 7. There are also differences within the South Caucasus, with the climate changing from humid and warm in the east to drier and hotter to the west. In winter the eastern Caucasus is colder than the western region. Thus, Makhachkala (east Caucasus), the capital of Dagestan on the west coast of the Caspian Sea, and Sochi (west Caucasus) on the east coast of the Black Sea (Krasnodar Krai, Russia Federation, near the Georgian border), are at the same latitude; however, the average January temperature in Makhachkala is on average 10°C lower than in Sochi. By contrast, summers in the southeastern Caucasus region are hot and dry, with precipitation in some places only ca. 300 mm per year. While the western region of Georgia and northeastern Turkey have a floristic profile full of evergreen species and many relictual plant taxa, the eastern Caucasus is characterized by many steppes and semi-deserts in Azerbaijan and northeastern Iran. However, between these climatic extremes, many other vegetation types exist. Thus, around the town of Lenkoran in Azerbaijan, close to the Caspian Sea, the warm and relatively humid climate supports the growth of a relictual Hyrcanian forest, characterized by the endemic Persian ironwood, *Parrotia persica* (DC.) C. A. Mey. (Hamamelidaceae), a small tree related to modern witch hazels. This region can be referred to the USDA zone 7. The central portion of the South Caucasus is characterized by a continental type of climate. The Armenian region has cold winters with average temperature ca. –12°C, but quite hot summers with a mean temperature up to 43°C. Precipitation in the central plain ranges from ca. 180 to 300 mm, but in the montane areas it may exceed 1000 mm. This can be classified as USDA zone 5, however, in some places, can range from USDA zones 4 to 6. In the North Caucasus, the climate is also a continental one with hot summers. The coldest month, January, in Pyatigorsk (Stavropol Krai, Russian Federation), has an average temperature of –5°C while in Novorossiysk, a Russian port on the Black Sea (Krasnodar Krai, Russian Federation), the mean temperature is 1°C and can be referred to USDA zone 5, and in some places USDA zone 6. In general, the weather patterns in the Caucasus can be considered a Mediterranean climatic type, although there are exceptions. There are five regions in the world that can be characterized as Mediterranean climates supporting at least 25,000 plant species, of which more than half are endemic (Raven, 1976).

Flora and Vegetation

The flora of the Caucasus region consists of ca. 6300 species of vascular plants, according to Gagnidze (2002) or ca. 5000 species, according to the *Caucasian Flora Conspectus* (Takhtajan, 2003–2012). Botanical exploration of the Caucasus is known to have started as early as the beginning of the 17th century. Local botanists and amateurs not only undertook field trips to different areas of the Caucasus, but also collected seeds and propagules to create gardens in villages and towns, using local trees and herbs. The French Protestant, later royal jeweler, Jean-Baptiste Chardin (1686), described the gardens in Tiflis [Tbilisi], Georgia, as well as the surrounding areas on his 1672–1673 journey across the Transcaucasus, which also resulted in rich collections of local ornamental and medicinal plants. Another French visitor, Abbé Joseph Delaporte, came to Tiflis and published similar impressions in 1768, mentioning in particular the Garden at the Royal Palace (Gogolishvili & Skhiereli, 1986). It is this garden that would later become the Tbilisi Botanical Garden. Early scientific knowledge of the Caucasian vegetation came from the famous botanist, Joseph Pitton de Tournefort, who was among the pioneers to visit the Caucasus, with recollections of his journeys posthumously published in 1717. Tournefort amassed a large herbarium, now at the Natural History Museum in Paris, which was later studied by Carl Linnaeus, the father of modern binomial nomenclature, who treated most of the included taxa in his *Species Plantarum* (1753). Peter the Great, who created the Russian Academy of Sciences (RAS) in 1724, would send a diplomatic expedition to Constantinople, Turkey (1724–1727), accompanied by the first RAS academician, the physician and botanist Johann Christian Buxbaum. His travels across the Caucasus region were later documented in five volumes (Buxbaum [Gmelin], 1728–1740) and included the Southern Caucasus (Takhtajan, 2003–2012).

The first comprehensive flora of the Caucasus and Crimea was *Flora taurico-caucasica*..., published by Adolf Marschall von Bieberstein. Its three volumes (1808, 1819), included 2000 species and were based on Linnaeus' system of binomial nomenclature and the subsequent treatises of Karl Ludwig von Willdenow. Von Bieberstein authored many other works from 1798 to 1843 that helped to develop an understanding of the vegetation in these regions. His herbarium collections may be found in St. Petersburg (LE), Moscow (MW), Tartu (TU), and elsewhere. The period from the second half of the 18th through the beginning of the 19th century saw many collectors active in the region, including both amateur and professional botanists such as M. Adams, E. Eichveld, I. Falk, S. Gmelin, I. Güldenstädt, F. Hohenaker, F. Kolenati, C. A. Meyer, A. Nordman, P. Pallas, and C. Steven. Taxonomic monographs and revisions began to appear more frequently, and several authors presented treatments on specific groups of plants, such as Christian von Steven's monographs of *Saxifraga* L. (Saxifragaceae, 1813), *Pedicularis* L. (Scrophulariaceae, 1823), *Pinus* L. (Pinaceae, 1838), and the family Boraginaceae (1851). Botanists widely exchanged their findings and descriptions, and Caucasian plants were included in de Candolle's *Prodromus* (1824–1873), Ledebour's *Flora Rossica* (1841–1853), and Boissier's *Flora Orientalis* (1867–1884). Vladimir Lipsky summarized all prior work on the Caucasus flora, with over 30 treatises to his name, and published a comprehensive list of 4430 species in his *Flora Kavkaza* (Lipsky, 1899), arranged according to the Boissier system. During the 1860s and 1870s, the vegetation of the Caucasus was further studied by A. Bekker, A. Overin, G. Radde, F. Ruprecht, and P. Trautvetter, and some of these botanists worked in the Caucasus for extended periods. Gustav Radde, who spent more than 40 years there, created the Caucasian Museum in Tbilisi, now known as The Museum of Georgia, noted for its rich herbarium collections. The list of botanists who worked at the end of the 19th and beginning of the 20th centuries is very long, but a few of the better known include N. Albov, I. Akinfiev, F. Alexeenko, N. Busch, B. Fedtschenko, A. Krasnov, N. Kuznetzov, V. Markovich, I. Schmalhausen, and G. Woronow, as well as the non-Russian botanists N. Déchy, E. Levier, and C. Sommier. By the beginning of the 20th century, floristic work had turned to the monographic style; taxonomy had broadened into systematic and experimental taxonomy; and even minutia such as the rules of nomenclature would change and evolve. A team of enthusiastic botanists that included N. Busch, A. Fomin, and N. Kuznetzov began a comprehensive flora, the *Flora Caucasica Critica* (Kuznetzov et al., 1901–1913). There were 45 issues published before World War I, but the Russian Revolution stopped the work. Numerous studies of the Caucasian vegetation were made by Alexander Grossheim, who published the most complete *Flora Kavkaza* (Flora of the Caucasus, 1928–1934). He even started a new, second edition in 1939 which was interrupted by his death in 1948. Although the second edition was never completed, seven volumes were published, the last in 1967. His *Manual of the Caucasian Plants* (Grossheim, 1949) was comprehensive, well done, and remains in use today.

The next step in botanical studies that treated the Caucasus flora was initiated in 1931 with the production of the *Flora URSS*. This continent-spanning project was handled by the Botanical Institute in Leningrad, later in 1940, the Komarov Botanical Institute, St. Petersburg. The program head was the respected academician Vladimir Komarov and among its many outstanding authors were E. Bobrov, E. Busch, N. Busch, S. Czerepanov, A. Fedorov, A. Fomin, I. Mandenova, A. Pojarkova, Y. Prokhanov, B. Schischkin, S. Uzepchuk, I. Vasil'chenko, and N. Zvelev. The *Flora URSS* project was interrupted by World War II, but finished in 1960 (Komarov et al., 1934–1960). After this, local floras for all the southern Caucasian republics and other regional floras followed, with most of these written in Russian, but some were in local languages. As an example, two editions of *Flora of Georgia* were published in the Georgian language, with the most recent in 2007 (Akademiia nauk Gruzinskoi SSR, 1971–2007). Its authors included botanists such as I. Mandenova, A. Makashvili, L. Kemularia-Nathadze, A. Charadze, and D. Sosnovsky.

Armen Takhtajan initiated the preparation of the *Caucasian Flora Conspectus* in 1985 that included all Former Soviet Union (FSU) territories in the Caucasus. Among the participants at this time were L. Menitsky, T. Popova, N. Portnier, H. Mordak, N. Zvelev, and other Russian and Caucasian botanists. The first volume of this floristic conspectus was published in 2003 and the most recent in 2012 (Takhtajan, 2003–2012). However, until the current publication, there was no single source that attempted to treat even a portion of the rich flora of the Caucasus region. The International Union for Conservation of Nature and Natural Resources (IUCN) includes both the North Caucasus (Russia) and the South Caucasus (Armenia, Azerbaijan, Georgia, Iran, and Turkey) in the Caucasus biodiversity hotspot.

Following the floristic divisions presented by Takhtajan (1986), the Caucasus region belongs to the Holarctic and Boreal Subkingdoms and can be further subdivided. Within the Circumboreal Floristic Region, the Euxine Floristic Province is botanically the richest, encompassing the western Caucasus from Anatolia in Turkey east around the Black Sea to the coastal area of Krasnodar Krai, Russian Federation. The predominant vegetation are forests with evergreen elements, but along the seashore, swamps and other estuarine lowlands are found. The mountainous systems of the Greater and the Lesser Caucasus belong to the Caucasian Floristic Province, including the central South Caucasus. In the north, this province extends to the southern part of the Kuban lowlands, the Stavropol Hills and plains in Dagestan. Vegetation in this floristic region has much in common with that of the Euxine Floristic Province. Montane broad-leaved forests are characteristic of the Caucasian Province. In the upper montane zone, coniferous forests occur, and, at the lowest elevations, riparian and marsh forests occupy large areas, along with forest steppes and meadow steppes. The Caucasus biodiversity hotspot, as defined by the Critical Ecosystem Partnership Fund's ecosystem profile (CEPF, 2003), also covers the southern extent of the Eastern European Floristic Province and includes the valley of the Kuban River and the Stavropolsky Hills in Russia. The vegetation in the north is mostly meadows and steppes that have been extensively converted to agricultural uses. The northern portion of the western Caucasus, along the Black Sea between the towns of Anapa and Tuapse, belongs to the Crimea-Novorossiysk Floristic Province, part of the Mediterranean Floristic Region.

The southern portion of the Caucasus biodiversity hotspot falls into the Armeno-Iranian Floristic Province, part of the Irano-Turanian Floristic Region (Takhtajan, 1986). This extends to northeastern Turkey, most of the Armenian Highlands, and the dry region of the southern Caucasus, Azerbaijani Zuvand and adjacent Iran. The Armeno-Iranian Floristic Province is heterogeneous but is characterized by various kinds of steppes, including grass steppes and steppes with cushion-like or dwarf shrubs. In some areas there are remnants of oak forests, usually degraded and park-like. The eastern and southeastern slopes of Mount Aragats in Armenia are included by some botanists in the Atropatenian Floristic Subprovince. Its vegetation is mostly xerophytic and consists of sparse phryganoid formations, steppes and open woodlands. Hyrcania refers classically to the ancient Persian satrapy or lands bordering the southern coastline of the Caspian Sea. Relictual Hyrcanian forest extends from the southwestern Caspian coastal plains of Lenkoran, in Azerbaijan, and across northern Iran, through the Iranian provinces of Gilan, Mazanderan, and Golestan, and into adjacent Turkmenistan. The Hyrcanian Floristic Province differs significantly from the floras of neighboring regions and can be described as an enclave of an ancient relictual mesophilous forest from the Tertiary flora. The last region included in the Caucasus biodiversity hotspot are the territories of the eastern portion of the South Caucasus and includes the Precaspian lowland. This flatland lies at the northern edge of the Caspian Sea in Kazakhstan and Russia. The Precaspian Depression and the lower Terek River, flowing into the Caspian Sea with its broad alluvial fan, both fall into the Turanian or Aralo-Caspian Floristic Province. Landforms consist mostly of deserts, semi-deserts, and steppe formations.

Human impacts have led to significant changes in the diverse landscapes that form the Caucasus region. During the past 100 years, the area under cultivation has increased five-fold, and in some places 10-fold. The overgrazing of montane vegetation has led to the disappearance of local plant species, replaced by invasive weedy taxa. This process has been slowed in some areas where livestock pressures have been reduced. Coastal areas have become recreational zones where towns and villages stretch almost continuously for hundreds of kilometers along the Black Sea coast. Vegetation here has been supplanted by introduced ornamental plants such as the North American southern magnolia, *Magnolia grandiflora* L. (Magnoliaceae), cedar of Lebanon, *Cedrus libani* A. Rich. (Pinaceae), *Eucalyptus* L'Hér. from Tasmania and southeastern Australia (mostly *Eucalyptus viminalis* Labill., Myrtaceae), as well as many non-native palms and ornamental shrubs. Today only ca. 50,000 km^2, or ca. 10% of the Caucasus, is estimated to remain in pristine condition. Along with land conversion to agriculture, overgrazing and urbanization, large-scale development projects for the petroleum industry, and related road construction projects should be mentioned as significant threats to biodiversity. To preserve and manage these areas and their plant species has great scientific and economic importance.

The Caucasus is one of eight global centers of domestication identified for economically important plants, with many food plants having their origins in the southern Caucasus. The ancient state of Chaldea, or Urartu to the Assyrians, was located in southwestern Asia and, in the ninth century BC, included the southern part of the Caucasus. Chaldea was famed for its advanced agriculture, known in particular as the first country where wheat was known to have been cultivated, and where its progenitors were widespread in the wild (Zhukovsky, 1964; Vavilov, 1992). Cereals, pears, pomegranates, and grapes can be found growing wild and may have been the ancestors or close relatives of economically important plants grown today. These wild relatives constitute a very important source of genetic diversity for future breeding and hybridization. Current molecular studies conducted by Barbara Schaal, Tengiz Beridze, and James Beck have led to the conclusion that the Caucasus region is the possible center of origin for the cultivated grape, *Vitis vinifera* L. subsp. *sylvestris* (C. C. Gmel.) Hegi (Vitaceae) (Beridze, pers. comm.). Modern ethnobotanical studies continue to a limited extent in the Caucasus right now, and the region has been known since the time of the ancient Greeks as a rich source for medicinal, dye, and ornamental plants. Even now it is possible to find new medicinal, food, and other economic plants here, as well as time-honored traditions for baking bread and preparing wine that survive still in rural areas. Limited studies showed that new drug plants (Miller et al., 2004) and ornamentals (Shulkina, 2004) can still be found in this very rich region for cultivation.

The Caucasus region is a meeting place for many different floras with connections to the Mediterranean, European, and West Asian floras. The three families most abundant in species are the Asteraceae, Poaceae, and Fabaceae, followed by the Apiaceae, Brassicaceae, Caryophyllaceae, Cyperaceae, Lamiaceae, Rosaceae, and Scrophulariaceae. These plant families are of cosmopolitan distribution, and there is no endemism at the family level, suggesting that the Caucasus region was never completely isolated from Europe or Asia. Even so, the taxonomic diversity of the Caucasus flora is remarkable. Plant families with more than 100 species, with their most speciose genus (in parentheses), include the Apiaceae (*Heracleum* L.), Asteraceae

(*Centaurea* L.), Brassicaceae (*Erysimum* L.), Campanulaceae (*Campanula* L.), Caryophyllaceae (*Dianthus* L.), Fabaceae (*Astragalus* L.), Poaceae (*Festuca* L.), and Rosaceae (*Alchemilla* L., *Potentilla* L.). Coincidentally, all of these genera were described by Linnaeus and are well-known and speciose taxa either Eurasian or worldwide in distribution. Plant families with ca. 50 or more species in the Caucasus region are the Boraginaceae (*Symphytum* L.), Lamiaceae (*Thymus* L.), and Ranunculaceae (*Ranunculus* L.), and families with 30 or more species include the Alliaceae (*Allium* L.), Iridaceae (*Iris* L.), Liliaceae (*Gagea* Raddi, *Lilium* L.), Primulaceae (*Primula* L.), and Rubiaceae (*Asperula* L.). In addition, there are a few notable species that represent groups primarily distributed in the tropics. These include yams, *Dioscorea caucasica* Lipsky (Dioscoreaceae), parasitic plants such as *Cytinus ruber* (Fourr.) Fritsch (Rafflesiaceae), and *Rhamphicarpa medwedewii* Albov (Scrophulariaceae). The familiar fig tree of domestication, *Ficus carica* L. (Moraceae), grows in the Caucasus and in the Mediterranean area.

Anthemis saportana Albov, endemic in Georgia and Russia, from the western Caucasus region.

Potentilla camillae Kolak., an endemic species in the western Caucasus (Georgia, Russia).

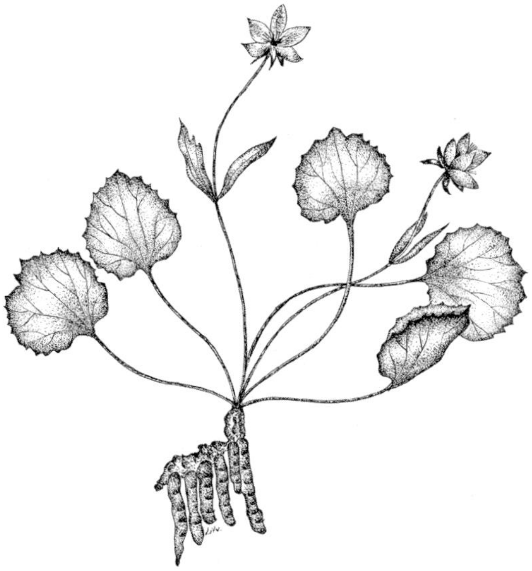

Ranunculus helenae Albov, an endemic species considered vulnerable to threat in Georgia and Russia, from the western Caucasus.

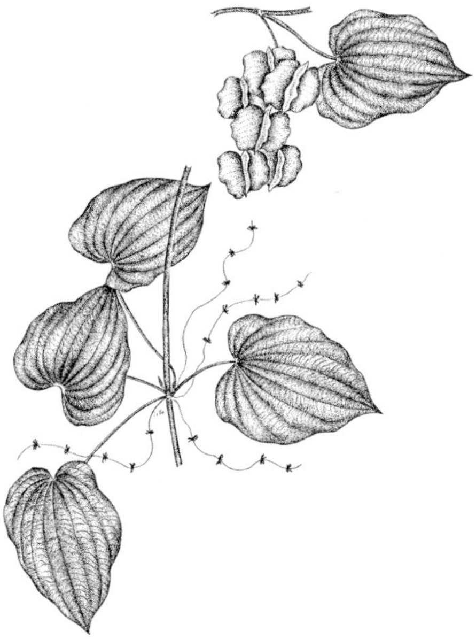

Dioscorea caucasica Lipsky, an endangered endemic species under threat in Georgia and Russia, from the western Caucasus.

Analysis of the Caucasian genera shows their relationship with plants from Europe and Asia. In the Orobanchaceae, *Leptorhabdos* Schrenk ex Fisch. & C. A. Mey. has the one species, *Leptorhabdos parviflora* Schrenk, distributed in the eastern Caucasus, while the second species is quite disjunct and reported from the Himalayas. Therefore, the presence of *Leptorhabdos* in the Caucasus clearly suggests a floristic affinity between these two regions. Other genera such as *Pyrus* L. or pears (Rosaceae) also support a floristic relationship between the two regions.

The Caucasian flora is unique in many respects, and the region is a natural refuge for plant species, especially for taxa that are not found or occur very rarely in the Mediterranean area, but having been common there in the past (Grossheim, 1936; Wulff, 1943). Many examples could be mentioned, such as the box-tree, *Buxus colchica* Pojark. (Buxaceae), holly, *Ilex colchica* Pojark. (Aquifoliaceae), rhododendron, *Rhododendron ponticum* L., and tree heath, *Erica arborea* L., both in the Ericaceae. These woody shrubs and trees are mesophilic relicts that show floristic affinity with the Mediterranean (Kolakovsky, 1961) and are found in Euxine Colchic deciduous forests along the southern Black Sea coast. Representatives of the Tertiary flora persist in Caucasus wetlands, and many are aquatic plants such as water chestnut, *Trapa colchica* Albov (Trapaceae), yellow water lily, *Nuphar lutea* (L.) Sm. (Nymphaeaceae), and lotus, *Nelumbo caspica* (DC.) Fisch. (Nelumbonaceae). The Chakva valley is a particularly humid locale in Colchis where ferns are well represented. Remarkable among them are a species of filmy fern, *Hymenophyllum tunbrigense* (L.) Sm. (Hymenophyllaceae), whose fronds lack stomata and are obligately restricted to very humid habitats. In the Caucasus region are remnants of the flora that existed in the Mediterranean during the Upper Pliocene epoch (5.3–2.6 mya). This relictual flora exists now only in a series of enclave communities along the Black Sea (Davis,

1965–1988). Floodplain forests in Colchis support plants of moist, warm habitats such as *Kosteletzkya pentacarpa* (L.) Ledeb. (Malvaceae), *Osmunda regalis* L. (Osmundaceae), *Hibiscus ponticus* Rupr. (Malvaceae), and *Rhamphicarpa medwedewii* Albov (Scrophulariaceae). The presence of these plants in Colchis supports the subtropical relationship of these plant communities.

Presently the number of endemic taxa known for the Caucasus region is 2791. Analysis of the list has helped to elucidate the number of endemic species for each country included in the Caucasus region (Armenia, Azerbaijan, Georgia, and the Caucasian portions of Iran, the Russian Federation, and Turkey). In Table 1, three figures are given for each of the six countries: (1) the number of species for that country's portion of the Caucasian flora, (2) the number of endemic species for the Caucasus portion, and (3) the number of national endemic species that occur in each individual country.

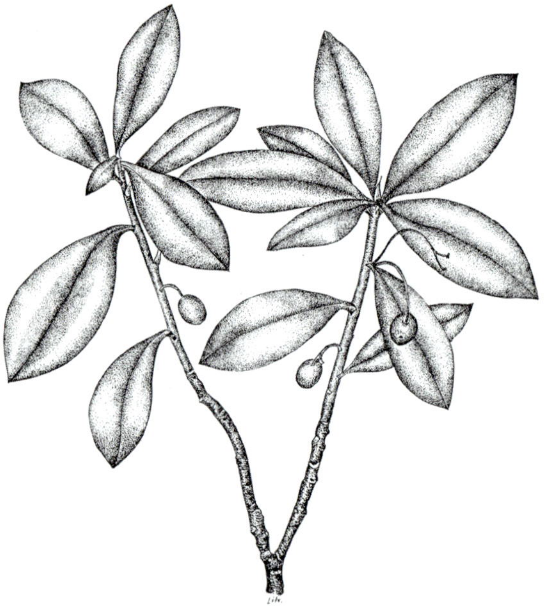

Daphne albowiana Woronow ex Pobed., an endangered Colchic endemic in southwestern Caucasus, with scattered populations restricted to limestone substrates in Georgia and Russia.

Table 1. Distribution of endemic species of the Caucasus biodiversity hotspot within the six countries.

COUNTRY	TOTAL SPECIES	CAUCASIAN ENDEMICS	NATIONAL ENDEMICS
Armenia	3600	850	126
Azerbaijan	4500	950	200
Georgia	4100	1320	276
Iran/Caucasus	3300	420	178
Russia/Caucasus	3700	1300	280
Turkey/Caucasus	4000	400	210

Taxonomic analysis indicates that there are 21 genera of vascular plants endemic across the Caucasus region, belonging to a number of different families, and that most are monotypic or oligotypic and some have unclear relationships. Thus, *Alboviodoxa elegans* (Albov) Woronow ex Grossh. (Asteraceae) has no close relatives in the Caucasus and is related to species of *Amphoricarpos* Vis. that occur in the Aegean Islands of the Mediterranean. *Muehlbergella oweriniana* (Rupr.) Feer (Campanulaceae) is related to species of *Edraianthus* A. DC. that occur in the Balkan Peninsula and the Apennines, yet *Muehlbergella* Feer differs by characters that include its dense cushion life form and single flowers. Represented by the single endemic species *Sredinskya grandis* (Trautv.) Fed. (Primulaceae), *Sredinskya* (Stein) Fed. can be compared with *Kaufmannia* Regel and *Cortusa* L., and it has characters in common with both of these genera from Central Asia. The taxonomic relationships of *Symphyoloma* C. A. Mey. (Apiaceae) and *Pseudobetckea* (Höck) Lincz. (Valerianaceae) are unclear and usually such isolated genera may be considered of ancient origins.

Table 2. Endemic genera in the Caucasus biodiversity hotspot.

FAMILY	GENERA	REGION
Apiaceae	*Agasyllis* Spreng. (1 species)	western, central, and southern Caucasus
	Arafoe Pimenov & Lavrova (1 sp.)	western Greater Caucasus
	Chymsydia Albov (2 spp.)	southern Greater Caucasus
	Macrosciadium V. N. Tikhom. & Lavrova (2 spp.)	western Caucasus
	Mandenovia Alava (1 sp.)	northeastern Greater Caucasus
	Symphyoloma C. A. Mey. (1 sp.)	central and eastern Greater Caucasus
Asteraceae	*Alboviodoxa* Woronow ex Grossh. (1 sp.)	western Greater Caucasus
	Cladochaeta DC. (2 spp.)	central and eastern Caucasus
	Grossheimia Sosn. & Takht. (6 spp.)	Caucasus
	Kemulariella Tamamsch. (5 spp.)	western, central, and eastern Greater Caucasus
Boraginaceae	*Trigonocaryum* Trautv. (1 sp.)	central and eastern Greater Caucasus
Brassicaceae	*Pseudovesicaria* Rupr. (1 sp.)	central and eastern Greater Caucasus
	Zuvanda (Dvorak) Askerova (1 sp.)	eastern Greater Caucasus
Campanulaceae	*Gadellia* Shulkina (1 sp.)	Caucasus and northern Asia Minor
	Muehlbergella Feer (1 sp.)	eastern Caucasus
Caryophyllaceae	*Charesia* E. A. Busch (1 sp.)	central Greater Caucasus
	Petrocoma Rupr. (1 sp.)	central Caucasus
Primulaceae	*Sredinskya* (Stein) Fed. (1 sp.)	western Greater Caucasus
Rosaceae	*Woronowia* Juz. (1 sp.)	western Greater Caucasus
Scrophulariaceae	*Paederotella* (E. Wulff) Kem.-Nath. (3 spp.)	Caucasus
Valerianaceae	*Pseudobetckea* (Höck) Lincz. (1 sp.)	eastern Caucasus

Endemic genera occur in different parts of the Caucasus, mainly in its western extent where their species grow mostly in subalpine and alpine zones of the Greater Caucasus Range, on rocky, mostly calcareous soils, or in meadows. They tend to be herbaceous perennials of mesophytic nature and indigenous origin. The most speciose endemic genus is *Kemulariella*

Tamamsch. (Asteraceae), with five species distributed from the western to eastern Greater Caucasus.

Fewer endemic genera can be referred to xerophilous conditions at mid-elevations in mountain communities or sometimes to phrygana communities. Rarely do species of endemic genera occur at low elevations. There are also genera that were thought to be endemic but have now been discovered outside of the Caucasus region as defined in this publication. One example is *Vavilovia* Fed. (Fabaceae), a genus that was described by A. Fedorov (1952) and has subsequently been found in Iraq, Lebanon, and Syria (Mikić et al., 2011).

It should also be mentioned that there is no agreement between Russian taxonomists on the treatment of the endemic genera. Thus, Czerepanov (1995) did not recognize the generic concepts of *Alboviodoxa* Woronow (Asteraceae) or *Charesia* E. A. Busch (Caryophyllaceae). He treated *Alboviodoxa elegans* (Albov) Woronow ex Grossh. within the Mediterranean genus *Amphoricarpos* Vis., with its three species occurring in the Aegean Islands, and *Charesia akinfievii* (Schmalh.) E. A. Busch within the larger generic concept of the intercontinental *Silene* L. On the other hand, endemic genera included by the Russian botanist Khokhrjakov (1991) more broadly included those that were found not only in the Caucasus, but also in surrounding territories, e.g., *Malvalthaea* Iljin (Malvaceae) from Central Asia. He also recognized the genus *Annaea* Kolak. (Campanulaceae), which was not accepted by most contemporary botanists. As further examples, Georgian botanists (Gagnidze et al., 2002) did not recognize the genera *Arafoe* Pimenov & Lavrova and *Macrosciadium* V. N. Tikhom. & Lavrova (both Apiaceae) nor *Muehlbergella* Feer (Campanulaceae). These varying taxonomic circumscriptions all have a subjective component to them, which in turn has a direct influence on the names that are used in species lists. This will continue to change in the future as taxonomists undertake additional studies and revise their taxonomic perspectives.

Campanula ciliata Steven, endemic to higher elevations throughout the Caucasus region, from Azerbaijan to Georgia and Russia.

Campanula circassica Fomin, endemic to the western Caucasus, known from Georgia and Russia.

Campanula saxifraga M. Bieb., endemic to the central Caucasus region, from Georgia and Russia.

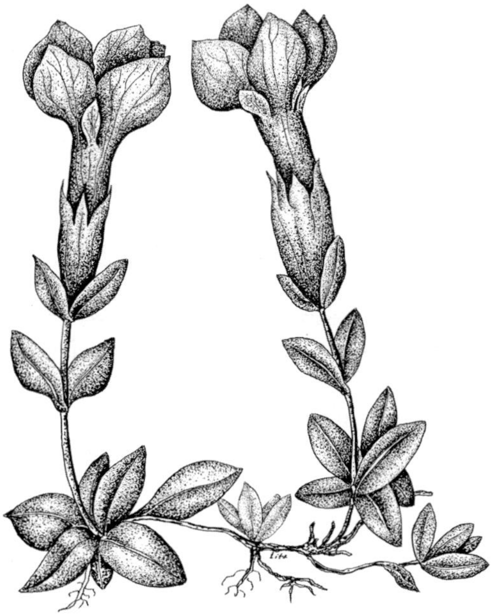

Gentiana oschtenica (Kusn.) Woronow, endemic to the Greater Caucasus region and listed from Georgia and Russia.

The number of endemic taxa for the Caucasus biodiversity hotspot is remarkably high. As has been compiled by Caucasian botanists in this volume, 2791 endemic taxa occur in this region and nowhere else. Endemic species and genera occur in every type of vegetation, but herbaceous plants predominate in subalpine and alpine zones whereas woody plants occur more so in forests. It is essential to mention that most species that occur in alpine and subalpine meadows and rocky places do not occur in the Arctic zone, as the Caucasus region does not have Arctic-alpine species (Fedorov, 1952).

The genus *Campanula* L. (Campanulaceae) can serve as a good example of alpine plants that do not extend to the Arctic. There are 108 species of the genus in the Caucasus of which 72 are endemic. Many occur in mountainous habitats, but none of them extend to the Arctic. A similar pattern can be seen with *Primula* L. (Primulaceae) in the Caucasus and even with the classical Arctic-alpine genus *Saxifraga* L. (Saxifragaceae). The latter genus is represented in the Caucasus with many endemic species, such as *Saxifraga abchasica* Oett., *Saxifraga caucasica* Sommier & Levier, and *Saxifraga colchica* Albov. Conversely, other species of *Saxifraga* that occur in the Arctic zone and in the European mountains are practically absent here, nor do they have close relatives among the Caucasian species. Thus, *Saxifraga oppositifolia* L. and *Saxifraga aizoides* L., two widespread Arctic-alpine species, are not found in the Caucasus. *Oxyria digyna* (L.) Hill (Polygonaceae) is one of the rare Arctic-alpine species that does occur in the Caucasus. Such distributions support the hypothesis that Caucasian endemics are of local origin and most are relictual taxa from the Tertiary period (Fedorov, 1952; Kamelin, 2006). Analysis of the Caucasian flora led Kamelin (2006) to the conclusion that development of the Caucasian flora was quite isolated from the surrounding areas. He listed 150 species that were common in adjacent countries but completely absent in the Caucasus. Neither Eastern European elements nor Ural-Siberian ones appear to have penetrated into the Caucasus during the Pleistocene epoch (2.6 mya to 11,700 yr bp). Western Asian elements are thought to have migrated through the Kopet-Dag Mountain Range, which lies along the border of Turkmenistan and northern Iran, influencing the southern Caucasian steppes, presumably during the Akchagyl/Apsheron boundary (from 1.2 mya to as much as 2.6 mya ago). This was an interval of significant rejuvenation for the geomorphology of the eastern Caucasus and the Caspian Sea basin, during the Late Pleistocene epoch (Adamia et al., 2010). The evolution of the Caucasian flora was very much isolated, and its pockets of endemism can be seen as analogous to the development of floras on oceanic islands. The complex geological history of the Caucasus region underlies its high level of plant endemism. Nevertheless, some endemic species do indicate a floristic connection that spanned the continents of the Northern Hemisphere. *Erythronium* L. (Liliaceae) is represented by the single endemic species *Erythronium caucasicum* Woronow but has representatives also in North America, southern Europe, Siberia, the Russian Far East, and Japan.

High biodiversity in the Caucasus is supported by the wide variety of vegetation types. A large portion of the region is covered by broad-leaved and coniferous forests (Galushko, 1978–1980; Doluchanov, 1989; Nakhutsrishvili, 1999; Zazanashvili et al., 1999, 2000). In the northern Caucasus, forests are concentrated in the western part, whereas toward the east these woody assemblages gradually reduce to bands along streams, and in the east, in Dagestan, steppe and shrub formations are more prevalent. The territorial extent for forests in the North Caucasus is estimated at ca. 41,800 km^2 (Litvinskaya & Murtazaliev, 2009). In other parts of the Caucasus, forests were also predominant. However, because of extensive recent deforestation in the post-Soviet period, particularly during the 1990s, this forest cover has been significantly reduced. The forest cover has been estimated at only 36.7% in Georgia (Doluchanov, 1989), while Gadzhiev (2003, pers. comm.) indicated that forest cover has diminished from 35% to 11% in Azerbaijan. Overgrazing by goats and sheep has further reduced this coverage, and the timberline has been observed to retreat altitudinally and upslope.

What most characterizes Caucasian forests is their relative taxonomic richness. Broad-leaved and coniferous forests in the Caucasus are typically composed of many tree species, with two or three species as canopy dominants. Rarely do Caucasian forests consist of a single woody species, although stands of the relictual Persian ironwood, *Parrotia persica* (DC.) C. A. Mey. (Hamamelidaceae) in the southeast Caucasus are exceptional. Almost half of the forested territory in the Greater Caucasus Range is covered by *Fagus orientalis* Lipsky (Fagaceae), but this characteristic beech is associated with other trees in different environmental conditions, forming different communities and vegetation types. Other tree species that predominate in Caucasian forests include Nordmann's fir, *Abies nordmanniana* (Steven) Spach, and the eastern spruce, *Picea orientalis* (L.) Peterm. (both Pinaceae), and both contribute to the darkened aspect of the understory of these coniferous forests. Neither conifer is considered to be closely related to more northern species of *Abies* Mill. and *Picea* A. Dietr. However, among the herbaceous plants of these forests are taiga elements that migrated to the south. These include the alpine enchanter's nightshade *Circaea alpina* L. (Onagraceae), the twinflower *Linnaea borealis* L. (Caprifoliaceae), and the one-flowered wintergreen *Moneses uniflora* (L.) A. Gray (Ericaceae), but not the Arctic starflower *Trientalis europaea* L. (Primulaceae).

Different oaks (*Quercus* L., Fagaceae) play important roles in climax vegetation formations, often as canopy dominants in Caucasus forests (Menitzky, 2005). The chestnut-leaved oak, *Quercus castaneifolia* C. A. Mey., and the Georgian oak, *Quercus iberica* Steven, are familiar species distributed across the entire Caucasus region, while the white oak, *Quercus hartwissiana* Steven, is restricted to Colchis along the eastern margins of the Black Sea. The foothills of the North Caucasus ranges are covered with dense oak forests that first appear as tree islands among expanses of open steppe on slopes, mountain summits, and level areas along ridge lines. The oaks of such forest margins include the widespread European species *Quercus robur* L. (English oak, French oak), which is often found as an indicator of these first forest islands. In more humid areas, there are associations of oak-maple-ash, while in drier areas, maples drop out and oak-ash predominate. This forest type may range up to 500–700 m in elevation and in the Central Caucasus may follow valleys to elevations of 1100–1800(–2000) m. Forests dominated by the European sessile oak, *Quercus petraea* L. ex Liebl., along with *Quercus robur* L. are usually confined to the upper portions of slopes. In the northwestern portion of the Caucasus region, especially along the Black Sea, the downy oak, *Quercus pubescens* Willd., widespread in Europe, is the major forest-forming species. The coastal area in the southwest used to have the Imeretine oak, *Quercus imeretina* Steven ex Woronow, which grew at low elevations as a canopy dominant in forests there, but populations of this endemic oak are now considered as vulnerable to threat. These forests now consist of the Oriental beech, *Fagus orientalis* Lipsky, of eastern European distribution, and the sweet chestnut, *Castanea sativa* Mill. (both Fagaceae), widespread elsewhere in Europe. Other woody species here include *Quercus hartwissiana* Steven and a Georgian oak *Quercus dshorochensis* K. Koch (considered by some to be a segregate of the European sessile oak, *Quercus petraea*). In some places, the Caucasian elm *Zelkova carpinifolia* (Pall.) K. Koch is seen; in swampy areas, the black alder *Alnus barbata* C. A. Mey. may be common, with the endemic *Quercus imeretina* Steven ex Woronow occurring sporadically.

Under current threat today, the Euxine Colchis broadleaf forests along the southern Black Sea coast represented refugia for floristic elements of temperate eastern Europe during the Pleistocene epoch. The predominant plant formations in this ancient Euxine flora were forests of mainly oaks, beech (*Fagus orientalis* Lipsky, Fagaceae), and mixed forests of beech–chestnut. Evergreen relictual forest elements are well represented in the Caucasus while elsewhere almost extinct, as in the Mediterranean. In southwestern Caucasus, in some areas along the Black Sea, the coastal forests start at sea level, with lowland mixed and alder forests. Representative woody taxa are the black alder *Alnus barbata* C. A. Mey, and the linden *Tilia begoniifolia* Steven (Tiliaceae). Comparison between the Colchis Caucasian forests and mid-European forests led Rikli (1943) to conclude that the number of species in the ancient and relictual Colchis forests were almost twice as many as those in European forests. Forests in the Colchis typically begin at 400–600 m, and the timberline is reached at altitudes of 2200–2800 m, with subalpine and alpine meadows appearing.

The Colchis forests are dominated by relictual endemics, such as the Caucasian birch or gold bark, *Betula medwedewii* Regel (Betulaceae), and the closely related *Betula megrelica* Sosn., both under current threat. Other endemics include the Pontine or Armenian oak, *Quercus pontica* K. Koch, the Georgian buckthorn, *Rhamnus imeretina* Booth, Petz. & Kirchn. (Rhamnaceae), the Colchis rowan, *Sorbus subfusca* (Ledeb.) Boiss. (Rosaceae), and the Colchis hazelnut, *Corylus colchica* Albov (Betulaceae). The understory in the Colchis forest is also rich in endemic and rare plants (Adzinba, 1987; Doluchanov, 1989, 2010). Shrubs are represented by rhododendrons such as *Rhododendron ponticum* L., *Rhododendron smirnowii* Trautv. ex Regel, and *Rhododendron ungernii* Trautv. ex Regel. Other evergreen taxa include the common cherry laurel, *Lauro-cerasus officinalis* M. Roem. (Rosaceae), the Colchis or Persian ivy, *Hedera colchica* (K. Koch) K. Koch (Araliaceae), the Colchis holly, *Ilex colchica* Pojark., and the Colchis butcher's broom, *Ruscus colchicus* P. F. Yeo (Ruscaceae). The Colchis forest assemblages in the Caucasus are not particularly rich in herbaceous species, largely due to shaded cover of the evergreen shrubby cover. However, a few shade-tolerant herbs can grow, and among these are birthwort, *Aristolochia pontica* Lam. (Aristolochiaceae), sowbread, *Cyclamen adzharicum* Pobed. (Primulaceae), and navelwort, *Omphalodes cappadocica* (Willd.) DC. (Boraginaceae).

One striking feature of the pine forests in the Caucasus is that the number of *Pinus* species is small. There are only three species that develop forest, the Turkish pine, *Pinus pityusa* Steven (= *Pinus brutia* Ten. subsp. *pityusa* (Steven) Nahal), the Pallas pine *Pinus pallasiana* D. Don, and most common, the Caucasian pine, *Pinus kochiana* Klotzsch ex K. Koch (Pinaceae). *Pinus eldarica* Medw. is a relictual endemic species and occurs in a very limited area in Azerbaijan. Only *Pinus kochiana* develops true pine forest stands on the northern slope of the Greater Caucasus, in Dagestan, and in the southern Caucasus. Here, the shrub layer consists of *Cistus ponticus* Juz. (Cistaceae), Eurasian smoketree, *Cotinus coggygria* Scop. (Anacardiaceae), and European privet, *Ligustrum vulgare* L. (Oleaceae), among others. Some species that are common in the European pine forest are absent here, for example, heather *Calluna vulgaris* (L.) Hull (Ericaceae), dyer's broom *Genista tinctoria* L. (Fabaceae), and the cutleaf anemone *Pulsatilla patens* L. (Ranunculaceae).

A number of species from genera endemic to the Caucasus region can be found in the forest formations, such as the vulnerable species *Kemulariella abchasica* (Kem.-Nath.) Tamamsch. and *Kemulariella tugana* (Albov) Tamamsch. (Asteraceae) that occur in western Caucasian forests, in rocky places, in clearings, and also in subalpine meadows. Among the more interesting and valuable endemic species are the peonies (*Paeonia* L., Paeoniaceae). Four yellow-flowered species should be noted: the pale yellow peony, *Paeonia wittmanniana* Hartw., occurs in forests in western and central Caucasus; a lemon yellow peony, *Paeonia macrophylla* (Albov) Lomakin, is recorded in Adzharia, southwestern Georgia, *Paeonia steveniana* Kem.-Nath. is found mostly along forest borders; and a bright yellow peony, *Paeonia mlokosewitschii* Lomakin, is known from the forests of the central Caucasus region, from Lagodekhi, Georgia, at the foot of the Greater Caucasus Range.

In the Lesser Caucasus, in Armenia, southern Azerbaijan, and eastern Georgia, mixed forests occur from 550 to 2300 m. Species of oaks such as *Quercus iberica* Steven and *Quercus petraea* L. ex Liebl. (both Fagaceae) are common and occur with hornbeam, *Carpinus betulus* L. (Betulaceae), species of apple, *Malus* L. and numerous species of pears, *Pyrus* L. (both Rosaceae). The Caucasian oak, *Quercus macranthera* Fisch. & C. A. Mey., is very common from 1800 to 2600 m in elevation on southern slopes, while the Oriental beech, *Fagus orientalis* Lipsky (both Fagaceae), occurs on northern slopes. Closer to the timberline one can see semi-arid ecosystems and mountain steppe vegetation. There is an interesting assemblage of woody plants in southeastern Azerbaijan, in the Talysh Mountains, surrounded on the north and west by steppes, which have been sometimes called wooded islands (Karjagin, 1950–1961; Gadzhiev et al., 1979). Precipitation in the Talysh region reaches 1900 mm in the lowlands decreasing with altitude where a Mediterranean-type vegetation with drought-resistant thermophilic shrubs and small trees known as shibliak appears. The Talysh lowland of Hyrcania (Nakhutsrishvili et al., 2010) has many woody plants that are relicts of the Tertiary period. There are many rare and beautiful trees in this Hyrcanian flora such as the silk tree, *Albizia julibrissin* Durazz. (Fabaceae). Another Hyrcanian endemic is the honey locust, *Gleditsia caspia* Desf. (Fabaceae), whose closest relatives are found far from the Caucasus region, disjunct in eastern Asia and North America. However, the gem of the Talysh forest is the Persian ironwood, *Parrotia persica* (DC.) C. A. Mey. (Hamamelidaceae) and the only witch hazel known from the Caucasus. This taxon was once widespread throughout Europe as indicated by the fossil record (Probst, 1981; Sales & Hedges, 1996; Akhani et al., 2010). *Parrotia persica* is now confined to Talysh in Azerbaijan and northeastern Iran, as well as one other small enclave in the Alazan Valley in western Azerbaijan. Trunks and branches of these trees fuse and grow together, making an unusual picture of a fantastic forest. Other species present in the Hyrcanian forest but absent in Colchis are *Acer velutinum* Boiss. (Aceraceae), *Amoria bobrovii* (Khalilov) Roskov and *Astragalus bakuensis* Bunge (both Fabaceae), *Danae racemosa* (L.) Moench (Ruscaceae), *Phuopsis stylosa* (Trin.) Benth. & Hook. f. ex B. D. Jacks. (Rubiaceae), *Pyrus grossheimii* Fed. and *Rosa alexeenkoi* Crép. ex Juz. and *Rubus juzepzukii* Sanadze (all Rosaceae), as well as other monocots and herbs. In contrast to the Euxine Floristic Province (Takhtajan, 1986), the Hyrcanian Floristic Province is not distinguished by species of *Rhododendron* L., *Abies* Mill., or *Picea* A. Dietr. This supports the hypothesis that these two centers have developed almost independently. At the same time there are some relictual taxa common to both regions such as *Acer laetum* C. A. Mey. (Aceraceae), *Pterocarya pterocarpa* (Michx.) Kunth ex I. Ilinsk. (Juglandaceae), *Ruscus colchicus* P. F. Yeo (Ruscaceae), and *Zelkova carpinifolia* (Pall.) K. Koch (Ulmaceae).

For other vegetation types, there are also open forests or woodlands with steppe-like understory in the Caucasus region, which are especially common in the southern regions (Gabrielyan & Fragman-Sapir, 2008). The dominant genus in such a typical woodland is oak, *Quercus* L. (Fagaceae), mostly the Caucasian oak, *Quercus macranthera* Fisch. & C. A. Mey. ex Hohen. In Armenia there also occurs the Cyprus oak, *Quercus boissieri* Reut., with pomegranate, *Punica granatum* L. (Lythraceae), and hop hornbeam, *Carpinus schuschaensis* H. J. P. Winkl. (Betulaceae). In eastern Georgia, southern Armenia, and Azerbaijan, junipers, especially *Juniperus foetidissima* Willd. and *Juniperus oblonga* M. Bieb. (Cupressaceae), are an important component of the vegetation. In Turkey, individual thickets of the Pontic oak, *Quercus pontica* K. Koch, are found. The forb and grass cover includes species of *Bromopsis* (Dumont.) Fourr. (Poaceae) and *Carex* L. (Cyperaceae). The other type of open forest is characterized by pistachio, *Pistacia mutica* Fisch. & C. A. Mey. (Anacardiaceae), and Cyprus oak, *Quercus boissieri* Reut., as the predominant trees. Accompanying these are other woody taxa, among them are the European smoke tree, *Cotinus coggygria* Scop. (Anacardiaceae), and several species of hackberry, *Celtis* L. (Ulmaceae), and mountain ash, *Sorbus* L. (Rosaceae).

Tall herbaceous subalpine vegetation is quite characteristic of the Caucasus region and exists along with crook-stem forest that is formed by beech at the upper altitudinal range of their distribution on slopes (Nakhutsrishvili, 2013). Herbaceous plants in these subalpine environments are remarkably tall, with leafy stems reaching 1.5–2 m in height, and sometimes to 3–4 m. This is usually confined to the lower extent of the subalpine zone. In contrast, there is no equivalent vegetation type in the Alps of Europe or in the Rocky Mountains of North America. This gigantic growth is supported by favorable environmental conditions such as rich soil, high humidity, and optimal air temperatures throughout the year. Such tall herbaceous communities are situated in the Greater Caucasus from 1800 to 2200 m and from 2700 to 2800 m in the Lesser Caucasus where the climate is drier. In the southeastern Caucasus, in Iran, this vegetation type reaches to 3500 m (Noroozi, pers. comm.). Herbaceous plants in the subalpine zone in the Lesser Caucasus do not exhibit such gigantism in growth, and heights are usually less than 1.5 m. Thus, plant assemblages in the western Caucasus differ slightly from those known for the eastern Caucasus. Approximately 100 species have been listed for tall herbaceous plant communities (Gagnidze, 1974). A conspicuous element are species of

Apiaceae, such as cow parsnip, *Heracleum* L., especially *Heracleum mantegazzianum* Sommier & Levier, which can reach 4 m in height. Other tall umbelliferous genera such as *Prangos* Lindl., *Ferula* L., and *Leutea* Pimenov are quite conspicuous in subalpine communities in the southeastern Caucasus in Iran. Other tall plants are from the Asteraceae (*Cicerbita* Wallr., *Inula* L., *Senecio* L., and *Cirsium* Mill.), Ranunculaceae (*Aconitum* L. and *Delphinium* L.), Dipsacaceae (*Cephalaria* Schrad. ex Schult. & Schult. f.), Polygonaceae (*Rheum* L. and *Rumex* L.), Papaveraceae (*Papaver* L.), and Campanulaceae (*Gadellia* Shulkina) among others. Petaloid monocots are also sometimes present with species of *Lilium* L. and *Veratrum* L., but for the most part, this tall herbaceous community consists mostly of dicots. Among endemic genera that occur in these subalpine meadows is the monotypic *Agasyllis* Spreng. with *Agasyllis latifolia* (M. Bieb.) Boiss. (Apiaceae), which inhabits meadows throughout the Caucasus region. Other endemic genera include *Gadellia* Shulkina (Campanulaceae) with the single species *Gadellia lactiflora* (Boiss.) Shulkina and *Grossheimia* Sosn. & Takht. (Asteraceae) with six species. *Gadellia* was segregated from the larger taxon *Campanula* L., as was *Grossheimia* from the speciose *Centaurea* L., based on distinctive morphological features that included both vegetative and reproductive features, as well as chromosome numbers. *Gadellia* can be as tall as 1.5–2 m, occurring mostly in meadows and among tall grasses in the Greater Caucasus, in its western portion and also in Asia Minor. *Grossheimia* species occur in both the Greater and Lesser Caucasus and also in northeastern Turkey. *Grossheimia macrocephala* (Muss.-Puschk. ex Willd.) Sosn. & Takht. is known to reach 1.5 m. Turkish botanists include this species within the larger taxon *Centaurea.*

Subalpine meadows and especially rocky places are very rich in endemic species. In the western Caucasus, *Woronowia speciosa* (Albov) Juz. represents an endemic monotypic genus in the Rosaceae. It has large, strikingly ornamental leaves and occurs in mountain meadows on limestone soil. *Woronowia* Juz. is taxonomically very close to *Geum* L., but differs by its habit and inflorescence structure. *Woronowia* grows with other endemic species, in what is considered a relictual plant community, whose composition includes *Carex pontica* Albov (Cyperaceae), a species that is isolated taxonomically and geographically, having close relatives only in the Himalayas and southwestern China, Japan, and New Guinea. Other endemic members of this relictual plant community are the genera *Chymsydia* Albov (Apiaceae) and *Sredinskya* Fed. (Primulaceae). *Chymsydia* has two species, one of which, *Chymsydia agasylloides* Albov, grows on calcareous rocky soil in subalpine zones in Abkhazia, along the eastern coast of the Black Sea. The other, *Chymsydia colchica* (Albov) Woronow, occurs to the south of this area in the Samegrelo Mountains. Both locations are part of the Euxine Floristic Province (Takhtajan, 1986). Due to overgrazing, the relictual plant community characterized by the showy *Woronowia speciosa* (Albov) Juz. has disappeared in many parts of western Georgia to be replaced by grass communities dominated by the bunchgrass *Nardus* L. (Poaceae), unpalatable to grazing animals. *Sredinskya grandis* (Trautv.) Fed. (Primulaceae) occurs here and higher, in subalpine and alpine meadows in the western Caucasus. This species appears similar to *Cortusa* L., also in the Primulaceae, but absent in the Caucasus region. *Sredinskya* is taxonomically separated from *Primula* L. based on distinctions in the flower and inflorescence structure. Also found in the same area, growing in rocky places, are beautiful, pink-flowered plants of *Alboviodoxa elegans* (Albov) Woronow ex Grossh. in the monotypic *Alboviodoxa* Woronow ex Grossh. (Asteraceae). This genus was taxonomically segregated from *Amphoricarpos* Vis., which has three species occurring in the Aegean Islands. Based on differences in morphological characters and the discontinuity of this geographical distribution, recognition of the segregate genus *Alboviodoxa* seems supported.

Petrocoma hoefftiana (Fisch. ex C. A. Mey.) Rupr. is in the monotypic genus *Petrocoma* Rupr. The species is endemic to the Greater Caucasus region and is considered vulnerable to threat in Russia.

Several endemic genera are found in the central and eastern portions of the Greater Caucasus, within the Caucasian Floristic Province (Takhtajan, 1986), and occur mostly in alpine and subnival vegetation. The alpine zone in the Caucasus region usually extends from elevations of 2400–3000 m, while in the southeastern Caucasus this zone is higher, from 3200 to 3800 m. Carpet-like meadows with short grasses are common. The small rosette-like plants of the Valerianaceae, such as *Pseudobetckea caucasica* (Boiss.) Lincz., can be seen in the alpine zone, sometimes close to snow patches or directly growing on talus. Two other endemic genera, *Petrocoma* Rupr. and *Charesia* E. A. Busch (Caryophyllaceae), each with a single species, were included by western botanists in the related genus *Silene* L.; however, Caucasian botanists treat them separately in segregate genera. *Petrocoma hoefftiana* (Fisch. ex C. A. Mey.) Rupr. grows on cliffs and slopes at mid and high elevations, while *Charesia akinfievii* (Schmalh.) E. A. Busch grows on moraines and in areas with shallow soil. In the alpine and subnival zones can be seen *Pseudovesicaria digitata* (C. A. Mey.) Rupr. (Brassicaceae), another small, rosette-forming plant. *Trigonocaryum involucratum* (Steven) Medw. (Boraginaceae) belongs to the oligotypic and threatened *Trigonocaryum* Trautv., which occurs not only in the alpine zone but also on gravels and at sites with shallow soil in the mid-mountain zone. Another rosette plant, *Symphyoloma graveolens* C. A. Mey. (Apiaceae) is the sole species of this endemic genus found on rocky talus slopes of the alpine zone and is critically endangered. Yet another endemic and oligotypic genus, *Paederotella* (E. Wulff) Kem.-Nath. (Scrophulariaceae), includes just three species, *Paederotella pontica* (Rupr. ex Boiss.) Kem.-Nath. (western Caucasus), *Paederotella teberdensis* Kem.-Nath. (northern Caucasus), and *Paederotella dagestanica* (Trautv.) Kem.-Nath. (eastern Caucasus). All three species grow on cliffs in the subalpine and alpine zones, with *Paederotella pontica* and *Paederotella dagestanica* considered vulnerable taxa.

Steppes cover large areas in the northern Caucasus region and mountain steppes are also numerous in the southern and especially southeastern portions, where they usually include many arid elements. Steppes are the youngest plant communities in the Caucasus because these areas are relatively recent, geologically speaking. Steppes are composed of a large number of grass species mostly from *Stipa* L., *Festuca* L., *Bothriochloa* Kuntze, and other Poaceae. However, even in the steppe vegetation some species are absent when this flora is compared to that of the neighboring floristic provinces. In dry places where the soil has been damaged by erosion with denudation of the steppe plants, a sparse phryganoid type of vegetation develops with scattered short trees and cushion-forming plants, both shrubs and perennial herbs. Many endemic species are restricted to this vegetation type especially in the southeastern Caucasus region (see Iran chapter). Among the woody plants are almond, *Amygdalus fenzliana* (Fritsch) Lipsky, cherry plum, *Cerasus incana* (Pall.) Spach, and pears, *Pyrus salicifolia* Pall. (all Rosaceae), buckthorn, *Rhamnus pallasii* Fisch. & C. A. Mey. (Rhamnaceae), and the gymnosperm shrub, *Ephedra procera* Fisch. & C. A. Mey. (Ephedraceae). One of the special types of phrygana is tragacanth vegetation that consists mostly of species of *Astragalus* L. and *Onobrychis* Mill. (both Fabaceae) along with *Acantholimon* Boiss. (Plumbaginaceae) and *Gypsophila* L. (Caryophyllaceae). Tragacanth communities take their name from the tough, thorn-shrubby *Astragalus* L. (Fabaceae) taxa, which are such a characteristic feature in this dry, sparse landscape. In the Iranian part of the Caucasus this type of vegetation is rich in species of *Astragalus* L. and *Cousinia* Cass. (Asteraceae), and these are mostly endemic species with narrow distributions. These slopes are very picturesque in the spring because they include numerous ephemeral annual and ephemeroid perennial species, mostly monocots. Among these are species of *Allium* L. (Alliaceae), *Bellevalia* Lapeyr., *Fritillaria* L., *Gagea* Salisb., and *Tulipa* L. (these four Liliaceae), *Gladiolus* L. and *Iris* L. (Iridaceae), and *Muscari* Mill. (Asparagaceae). Many of these flowering monocots are generally endemic to the Caucasus region and in particular to Armenia, Azerbaijan, and Iran. In spring, the lowland steppes (ca. 500–600 m) are resplendent in grasses (*Bothriochloa* Lem. and *Bromus* L., Poaceae) and blooming cresses (*Arabidopsis* (DC.) Heynh. and *Crucianella* Joss. [Brassicaceae] and other widely distributed genera). One shrub of interest is the Christ's thorn, *Paliurus spina-christi* Mill. (Rhamnaceae), related to the buckthorn, *Rhamnus pallasii* Fisch. & C. A. Mey., both of wide distribution.

In the southern Caucasus region there are semi-deserts on different soil types, ranging from saline to gypseous or clay and sandy soil profiles. Sagebrush communities that are characterized by *Artemisia* L. (Asteraceae) are discussed in the Armenian and Azerbaijani chapters. An endemic genus found in the phryganoid and xerophytic assemblages is *Mandenovia* Alava (Apiaceae), with one species, *Mandenovia komarovii* (Manden.) Alava, which is vulnerable and under threat. This species is closely related to the Mediterranean *Tordilium* Neck. and until recently was often treated within this genus. *Alococarpum erianthum* (DC.) Riedl & Kuber represents a monotypic genus in the Apiaceae that is distributed in the dry steppe vegetation, endemic to northwestern Iran. From the Asteraceae, an endemic genus that grows in dry places is *Cladochaeta* DC. In this publication all populations are treated as *Cladochaeta candidissima* DC., although *Cladochaeta caspica* Sosn. & Grossh., considered distinct by some botanists, is restricted to sandy soil along the Caspian Sea coast on the Absheron Peninsula (Azerbaijan), while *Cladochaeta candidissima* DC. grows in gravelly soil in mountainous areas. This genus was segregated from similar plants of *Helichrysum* Mill. and recognized for its differences in flower structure. Within the Campanulaceae, the monotypic genus *Muehlbergella* Feer has the single species *Muehlbergella oweriniana* (Rupr.) Feer, a critically endangered cushion plant with small bellflowers that is native to rocky slopes in Dagestan (Yarovenko & Murtazaliev, 2009). It is closely related to the more speciose *Edraianthus* (A. DC.) A. DC., with species occurring in the western Balkan Peninsula.

The largest number of endemics are confined to the Greater Caucasus Range, occurring mostly at high elevations, in subalpine and alpine zones, in forests, and often on rocky slopes, where such plant communities have been maintained continuously with relictual elements. Most comprise perennial herbs, and just a few are trees. Typical herbaceous examples are an alpine lily of the valley, *Convallaria transcaucasica* Utkin ex Grossh. (Asparagaceae), an alpine mustard, *Sobolewskia caucasica* (Rupr.) N. Busch (Brassicaceae), and a Caucasus tulip found in temperate mixed forest, *Tulipa lipskyi* Grossh. (Liliaceae). Woody examples include the Caucasian birch, *Betula raddeana* Trautv. (Betulaceae), and *Quercus imeretina* Steven ex Woronow (Fagaceae), endemic to Kokhida, western Georgia, and along the Black Sea coast, extending slightly into Russia.

Another large grouping of endemic species is found in the eastern Caucasus region, in Dagestan and surrounding territories. These plants are mostly montane xerophytes and almost all are perennial taxa. Representatives include *Astragalus daghestanicus* Grossh. and *Astragalus inoblitus* Maassoumi (Fabaceae), *Acantholimon tragacanthinum* (Jaub. & Spach) Boiss. (Plumbaginaceae), *Helianthemum dagestanicum* Rupr. (Cistaceae), and *Matthiola daghestanica* (Conti) N. Busch (Brassicaceae).

Yet a third cluster of endemic plant species occur on the southern slopes of the Greater Caucasus. These consist of both herbaceous and woody species. Examples include the showy Caucasian snowdrop, *Galanthus caucasicus* (Baker) Grossh. (Amaryllidaceae), *Iris iberica* Hoffm. (Iridaceae), the Georgian peach, *Amygdalus georgica* Desf. (Rosaceae), and the Eldar pine, *Pinus eldarica* Medw. (Pinaceae).

A fourth endemic group includes plant species that grow in the southeastern Caucasus and are characterized as mostly xerophytes found in lower- and mid-montane zones. Among the diversity are a cushion prickly thrift, *Acantholimon schemachense* Grossh. (Plumbaginaceae), the hybrid hornbeam tree, *Carpinus schuschaensis* H. J. P. Winkl. (Betulaceae), and Grossheim's pink, *Silene grossheimii* Schischk. (Caryophyllaceae). This heterogeneity of endemic species throughout the Caucasus infers that the plant taxa are of different origins. Mesophytes are usually paleoendemics, whereas xerophytes may be regarded as mostly neoendemics.

There are some genera that are especially numerous in the Caucasus, among them *Campanula* L. (Campanulaceae) with ca. 70 endemic species, *Silene* L. (Caryophyllaceae) with 70 species, *Veronica* L. (Scrophulariaceae) with 66 species, and *Heracleum* L. (Apiaceae) with 30 species. Similarly, *Astragalus* L. (Fabaceae), which is one of the largest genera among flowering plants with ca. 3000 species, has 189 endemic species in the Caucasus.

THE IUCN PROTECTED AREAS SYSTEM IN THE CAUCASUS

Text prepared by Nugzar Zazanashvili, World Wildlife Fund

Areas with high concentrations of endemic species have been declared as protected areas throughout the Caucasus where they have played an important role in biodiversity conservation for nearly a century. The development of new protected areas in the Caucasus has become dramatically apparent during the last two decades, especially in Azerbaijan, Georgia, and the Turkish part of the Caucasus where a number of new national parks have been established and the number of protected areas expanded. Management effectiveness of these areas has also increased. Several different categories of protected areas exist in the Caucasus region and can be classified as strict nature reserves, national parks, sanctuaries, nature parks, natural monuments, and other protected areas.

Strict Nature Reserves. These reserves correspond to category Ia of the International Union for Conservation of Nature and Natural Resources (IUCN)'s widely recognized protected area management categories (Dudley, 2008). The first strict nature reserve in the region was created in 1912 in the Lagodekhi Gorge on the southeastern slopes of the Greater Caucasus Range in Georgia. Today, there are 35 strict nature reserves in the Caucasus Ecoregion covering approximately 2% of its area. Strict nature reserves have traditionally been afforded a high level of protection, generally enforced by an armed ranger service. Public access is not allowed in these areas, though scientific research and supervised educational activities are permitted. In recent years, however, limited access to tourists and other visitors has been allowed in an effort to supplement dwindling financial support from government sources, particularly in the former Soviet republics.

National Parks. These are the second highest category (II) of IUCN protection in the Caucasus region (Dudley, 2008). Many national parks have a strictly protected core, where scientists study natural processes in unaltered habitats. Some national parks allow traditional natural resource use in designated areas. Though national parks are some of the most common types of protected areas in the world, they are relatively new to the former Soviet republics. The Borjomi-Kharagauli National Park, created in 1999, was the first to be established in the South Caucasus, according to international standards of zoning and management planning. Today, there are 29 national parks in the Caucasus, covering more than 2% of the region.

Other Protected Areas. There are a number of other areas that may be less well protected (IUCN categories III-VI). These include sanctuaries, wildlife refuges, nature parks, and protected landscapes, covering about 6% of the Caucasus, with all protected areas occupying approximately 10% of the region.

The existing system of strict nature reserves, national parks, sanctuaries, and other types of protected areas provides a significant foundation for conservation of biodiversity in the Caucasus Ecoregion. The nature reserves, some decades old, serve as a basis by which to assess the impact on biodiversity in areas that are not protected.

Yet many deficiencies and gaps exist within the current system. In particular, an initial comparison of the geographical distribution of the main nature reserves and national parks of the Caucasus with the Caucasus portion of the Map of Natural Vegetation in Europe (Bohn et al., 2000–2003) indicated what was particularly alarming was the indicator for vegetation of flood plains (formation U), which is distinguished by its unique biodiversity and very important ecological functions. Almost totally critical are index values of oak forests, with the exception of a few sub-formations. Among the forest formations, oak forests are characterized by the highest levels of species diversity. Conservation indicators are not balanced even in those formations whose average index is relatively high. Out of four alpine units, indices of three are critical, and of seven subalpine units, four are critical. Special attention should be focused on wetlands of formation R (tall reed vegetation and tall sedge swamps, and aquatic vegetation). It is also clear that the formations of the Caucasus steppes and deserts deserve more attention. It is obviously desirable to elaborate separate strategies for conservation of the Caucasus arid and semi-arid ecosystems, vegetation, and associated species diversity, which could be implemented under the umbrella of the United Nations Convention to Combat Desertification (UNCCD), to which the Caucasus countries are signed parties. Special analysis does not reveal a satisfactory picture, which calls for perfection to overcome asymmetry, fill gaps, and create connectivity in the system of protected areas of the Caucasus (Zazanashvili, 2005; Bohn et al., 2007).

Nature Conservation Legislation

Armenia, Azerbaijan, Georgia, and Russia began to adopt new environmental legislation following the dissolution of the Soviet Union in 1991. Legislation was enacted that focused on environmental protection, protected areas, wildlife management, and forestry. Other laws on air pollution, water, land use, and environmental impact assessments were also enacted. For the conservation of plants and vegetation, the following legal acts have particular importance. In Armenia, the Law on Specially Protected Nature Territories (2006), preceded by the Law on Flora (1999), was enacted. In Azerbaijan, the Law on Specially Protected Natural Areas and Monuments was established in 2000. In Georgia, the Law on the System of Protected Areas (1996) preceded the more recent Law on Red List and Red Book (2003). In Russia, significant legislation included the Federal Law on Protection of Natural Environment (2002) and the Federal Law on Specially Protected Areas (1995, with changes in 2001–2008). Other normative acts to provide recognition of National Red Lists/Books exist in these four countries.

In Turkey, articles in the 1982 Constitution have guaranteed the right to a clean environment and lay out the principles for the protection of cultural and natural areas. A number of other laws on allocation of forests for protection, hunting and fishing, water use, tourism, coastal areas, export of animal species, and national parks have come into force in the past two decades, in particular, the National Parks Law of 10 August 1983.

Iran's Constitution proclaimed the need to prevent pollution and environmental degradation. Laws governing management of game, forest, and rangeland resources have been in effect since 1967. Laws and acts dealing with environmental protection, air pollution, and water use were put in place, beginning in the 1970s and 1980s.

All six countries (Armenia, Azerbaijan, Georgia, Iran, Russia, and Turkey) have signed the majority of international conventions, including the Convention on Biological Diversity (CBD, 2002), the Ramsar Convention on Wetlands of international importance (RCW, 1971), and the Convention Concerning the Protection of the World Cultural and Natural Heritage (UNESCO, 1972). Most of the countries have adopted the Convention on International Trade in Endangered Species of Wild Fauna and Flora (CITES, 2003) as well. Countries are also implementing other multilateral strategies and programs (Williams et al., 2006). A great deal of conservation work has been done by the World Wildlife Fund (WWF) through its headquarters in Tbilisi.

EX SITU CONSERVATION: SEED STORAGE

In 2001 the National Botanical Garden of Georgia (formerly the Tbilisi Botanical Garden), in collaboration with the Missouri Botanical Garden in St. Louis, started the Caucasus Regional Seed Bank (CRSB) in Tbilisi, to preserve seeds from native plants of the Caucasus region. The seed bank was financially supported by the U.S. Civilian Research and Development Foundation (CRDF), and in 2005 the program joined the Millennium Seed Bank Partnership, an international conservation project coordinated by the Royal Botanic Gardens, Kew. Georgian botanists have been collecting seeds for ex situ seed storage in the CRSB at the Tbilisi facility, and duplicates of all material, both seeds and herbarium specimens, have been sent to the Millennium Seed Bank Partnership, to the Kew repository. This multilateral research has been supported worldwide in order to advance the conservation of wild plant species. Other Caucasian countries such as Armenia and Azerbaijan have joined the Millennium Seed Bank Partnership.

In 2011 the Caucasus Regional Seed Bank (CRSB) in Tbilisi held more than 850 species, with 180 of these being Caucasian endemics. Documentation for each taxonomic sample includes information on locality, ecological and habitat descriptions, population size, species distribution, threats, plant description, local vernacular name(s), ethnobotanical data, and collector information. All data are kept electronically using the Botanical Research And Herbarium Management System (BRAHMS), which is a flexible database management system that is globally accessible to herbaria and botanists. At both Kew and Tbilisi, CRSB seeds are preserved in special facilities, according to protocols accepted by the Millennium Seed Bank Partnership. Along with cold-storage facilities, collections of living plants are also established at both locations.

INVASIVE PLANTS VERSUS ENDEMICS

Invasive plant species are a serious threat for Caucasian regional endemics. Although the expansion of alien plants into native floras is a process thousands of years old, invasive plants have become a serious problem only in the last century. This plant interchange has extremely undesirable impacts for natural ecosystems, especially for those as unique and rich in endemics as the Caucasus. Many invasive plant species are less competitive in their native environments but can competitively eradicate their new neighbors in introduced settings. A total of no less than 400 invasive taxa in ca. 80 families are recorded for the Caucasus region. Many invasives, over 30%, came from Europe, including the Mediterranean. Among these are two legume species (Fabaceae) native elsewhere in Europe: Spanish broom, *Spartium junceum* L., an escape from cultivation, and gorse or whin, *Ulex europaeus* L. *Spartium junceum* is particularly aggressive and reduces local biodiversity, even replacing natural deciduous forests by forming pure stands of densely packed broom. *Ulex europaeus* threatens sensitive psamophyllous (sand-loving) vegetation along the coastal shores of the Black Sea and displaces the almost extinct sea daffodil, *Pancratium maritimum* L. (Amaryllidaceae), which impacts its obligate pollinator, the Convolvulus hawkmoth, *Agrius convolvuli* L. Also invasive is Johnson grass, *Sorghum halepense* (L.) Pers. (Poaceae). A similar 30% of invasives originate from Asia, and these include the Tree of Heaven, *Ailanthus altissima* (Mill.) Swingle, from northeastern and central China, and the Indian nettle or three-seeded mercury, *Acalypha indica* L. (Euphorbiaceae). Still other invasive plants came from North America. The black locust, *Robinia pseudoacacia* L. (Fabaceae), is native to the southeastern United States but is now a weedy tree in the North Caucasus that at lower elevations displaces native woody and herbaceous species. With even worse impact, the North American ragweed, *Ambrosia artemisiifolia* L. (Asteraceae), is invasive in the Caucasus wherever vegetation is disturbed and can be a significant weed in agricultural cropland, causing reduced productivity. On the other hand, a number of Caucasian plants have become invasive in many other countries. Some of these plants were intentionally introduced and have become aggressive weeds, e.g., *Heracleum mantegazzianum* Sommier & Levier (Apiaceae).

Robinia pseudoacacia L. (black locust). Photo by D. Kikodze.

Ambrosia artemisiifolia L. (ragweed). Photo by D. Kikodze.

IDENTIFICATION OF IMPORTANT PLANT AREAS IN GEORGIA

The flora of Georgia numbers ca. 4100 species of vascular plants of which 276 are endemic to the country with ca. 60% of these under threat, assessed as IUCN categories CR, EN, and VU. Data on Georgia's endemics were used for preliminary identification of Important Plant Areas (IPAs) as a basis for the establishment of new protected areas in the country. Up to 20% of Georgia's endemic plant species are calciphilous lithophytes occurring on the western limestone mountain ridges and massifs (Gagra, Bzipi, Egrisi, Askhi, Okriba, Khvamli, and from the Racha ridge to Rikoti pass) that extend almost parallel to the main watershed range of the Greater Caucasus in Colchis. Species of this ecological grouping also occur in the Lesser Caucasus, with patchy distributions in the Greater Caucasus and elsewhere in Georgia (Kakheti, Meskheti, Djavakheti, Trialeti). Approximately 80% of the endemics confined to these limestone rock and scree habitats are assessed as threatened (by overgrazing, infrastructure development, tourism and recreation, global climate change). By comparison, ca. 40% of the threatened endemics considered in the IPA identification process are confined to limestone habitats. Geographic Information System (GIS) analysis of the distribution of the endemic calciphytes has allowed the preliminary identification of IPAs on limestone ridges. This has led to the establishment of borders for new protected areas in Abkhazeti, Samegrelo, Racha-Lechkhumi—territories in northwestern and western Georgia noted as priority conservation areas, according to *An Ecoregional Conservation Plan for the Caucasus* (Williams et al., 2006). This report was based on the principles of IPA identification described in the Guide to Implementing Target 5 of the Global Strategy for Plant Conservation, "At least 75 per cent of the most important areas for plant diversity of each ecological region protected with effective management in place for conserving plants and their genetic diversity" (Convention on Biological Diversity, 2011). In addition, Identifying and Protecting the World's Most Important Plant Areas by Plantlife International (2004) and the recommendations by Anderson (2002) were also taken into account. The available data allowed testing of the International Union for Conservation of Nature and Natural Resources (IUCN, 2001) criterion A (estimate for the decline in the number of mature individuals over time) toward the identification of priority areas to target plant conservation at a site-based level by locating subpopulations of the most threatened plant species. The grid-based diversity mapping method in DIVA-GIS (Hijmans et al., 2005) located those areas with the greatest number of the target species and selected the least number of grid cells to capture all species. These biodiversity hotspots (high richness values) were mainly found on the limestone ridges of the Greater Caucasus in Abkhazeti, Samegrelo, Racha-Lechkhumi in northwestern and western Georgia, with nine to 14 species per grid cell, and also on the limestone habitats of the Lesser Caucasus in Trialeti, with seven or eight species in a grid-cell. Another hotspot was found in semi-arid areas in the vicinity of Mtskheta, Kartli Province, in eastern Georgia, with seven or eight species in a grid cell. The areas with high values of threatened endemic species richness did partly overlap with the existing Tbilisi National Park or the planned Samegrelo Protected Areas. The IPA identification process will continue to include Caucasian plant endemics of which 80% are known from rock and scree communities on Colchic limestone ridges in the Greater Caucasus.

EX SITU CONSERVATION: LIVING PLANT COLLECTIONS

Although in situ habitat protection is the top priority for conservation, this by itself is not sufficient to preserve every species. Caucasian botanical gardens serve an essential role as ex situ conservation organizations. Not only do these gardens keep appropriate plant stock for restoration, but they provide botanical and horticultural expertise. Many endemic and rare Caucasian plants are grown in diverse institutions, both within the Caucasus and outside this region.

The majority of living accessions of endemic plants are hosted at the National Botanical Garden in Tbilisi, Georgia. Two individuals, Drs. Boris Serdjukov and Lamara Asieshvili, have collected ornamental plants from all over the Caucasus for many years. This collection is diverse and rich, featuring many endemic and rare plant species, mostly perennials. One does not need to go to southern Georgia and climb dry slopes to see *Scorzonera ketzkhowelii* Sosn. ex Grossh. (Asteraceae), a wonderful perennial with gray leaves. We cannot easily go to the high Zha-zharo Pass to take a picture of *Scilla rosenii* K. Koch (Asparagaceae), the alpine gem of the genus with flowers that are similar to *Cyclamen* L. (Primulaceae). However, it is possible to visit the Tbilisi garden and see both endemic beauties. Visitors can see as many as ten endemic peonies from the Caucasus at the Garden in Tbilisi. Every year the National Botanical Garden sends seeds collected in the Garden and information about each taxon's biology upon request. The Garden staff study phenology and general behavior of plants in cultivation and in natural populations. This is information that is significant to the investigation of climate change and its impact on such endemic and rare plants. In addition to the Tbilisi garden, collections of native plants can be seen in other Georgian botanical gardens, e.g., Batumi and Kutaisi, as well as at the Yerevan Botanical Garden in Armenia and the Baku Botanical Garden in Azerbaijan. In the northern Caucasus living collections are maintained at the ecobiological station at Pyatigorsk and near Mineralnye Vody (Stravropol Krai, Russian Federation). Also noteworthy in the North Caucasus are the alpine gardens at Makhachkala (Dagestan), located along the western shore of the Caspian Sea.

ALPINE GARDENS

There are two remarkable places in the Caucasus where one can see collections of mountain plants. First established in 1910, the Bakuriani Alpine Botanical Garden was founded by the Georgian Academy of Sciences. Bakuriani is located in the Borjomi district in central-southern Georgia at 1700 m, surrounded by spruce forests and known for its snowy winters. The Bakuriani alpine gardens are home to ca. 500 species from the Caucasus region, 75 of which are trees and shrubs. Many of these taxa are included in Red Lists for countries within the Caucasus area, in particular for Georgia. The Bakuriani alpine gardens are central to the ex situ cultivation of these alpine plants, so many of which are assessed as under threat. One of most striking examples is *Iris winogradowii* Fomin (Iridaceae), which was the only plant mentioned from the Soviet Union in the IUCN Plant Red Data Book (Lucas, 1978) and was thought to be close to extinction. Originally collected from Mount Lomtismta in the Adzharo-Imeretinsky Range near Bakuriani, *Iris winogradowii* Fomin was propagated at the alpine garden and at the Komarov Botanical Institute in St. Petersburg. This vulnerable, yellow reticulate iris has now been re-established in its native range and horticulturally introduced globally. Its conservation status is now regarded as vulnerable. The Bakuriani Alpine Botanical Garden has many species of Caucasian *Lilium* L. (Liliaceae) and *Campanula* L. (Campanulaceae), as well as those promising for horticulture such as the rosid *Woronowia speciosa* (Albov) Juz., with its showy leaves and yellow flowers, the campanuloid *Gadellia lactiflora* (M. Bieb.) Shulkina, and the critically endangered aster *Scorzonera ketzkhovelii* Sosn. ex Grossh., with its striking silver-gray foliage. The pride of the Bakuriani collections are the wild Caucasian *Paeonia* L. (Paeoniaceae), four of which are yellow-flowered.

The Gunib Alpine Botanical Garden of the Dagestan Scientific Center was first established as an experimental station in 1972 and is now affiliated with the Russian Academy of Sciences. The Gunib garden is located on the limestone plateau of the eastern Greater Caucasus, and this native flora is represented by ca. 120 species, with two-thirds of these noted as rare endemic taxa under threat in the wild. Examples include the endangered *Iris acutiloba* C. A. Mey. (Iridaceae) and *Primula juliae* Kusn. (Primulaceae), and the critically endangered *Muehlbergella oweriniana* (Rupr.) Feer (Campanulaceae). The Gunib Alpine Botanical Garden has many experimental programs underway, including studies of plant populations in the wild as well as their genetics and ecophysiology.

Both alpine botanical gardens share a similar climate and are situated in subalpine zones, ranging from 1500 m to 2300 m, which supports the cultivation of alpine and subalpine plants from throughout the Caucasus region. Collections include trees and shrubs but consist mostly of perennial herbs. The major goal of the collections is to preserve rare and endemic plants from different regions of the Caucasus Mountains.

IMPORTANT PLANT AREAS IN ARMENIA

Current studies on the endemic plant taxa in the Caucasus region have given Armenian botanists the taxonomic support needed to establish national areas that best reflect maximum concentration of endemic plants, which include both Caucasian regional endemics and national endemics for Armenia. The key strategy for protecting this biodiversity is the establishment and maintenance of these Important Protected Areas (IPAs). Floristic analysis indicates that there are 29 localities with such rich biodiversity. Comparison of these places with current protected areas in Armenia, including national parks and natural reserves, reveals that 21 of the 29 biodiverse territories (ca. 73%) are situated nearby or entirely within protected areas. The remaining eight fall outside any current protection and these merit special attention by the Armenian government.

Botanists have studied different landscapes and plant communities, and those that are of particular botanical interest are also listed among protected territories. Very rich collections of Caucasian endemic plants can be seen at the Yerevan Botanical Garden. This garden was established in 1935 and is affiliated with the Armenian National Academy of Sciences. According to J. Akopian, the collection contains about 1000 species of Armenian flora. The garden itself is located on 80 hectares of semi-desert land at the northeastern edge of the capital city, with 16 hectares devoted to the Caucasus flora, with an estimated over 200 endemics. Plants of different life forms are included in the collection, which was known for its conifers (Cupressaceae and Pinaceae). One can see herbs, semi-shrubs, shrubs, and trees representative of the Caucasus such as *Rhododendron caucasicum* Pall. (Ericaceae) and *Parrotia persica* C. A. Mey. (Hamamelidaceae). They belong to different bio-ecological groups such as geophytes, succulents, lianas, and aquatic plants. Priority is given to rare and threatened species, endemics, insufficiently studied species, and species of educational, ornamental, or economic value. The wild relatives of agricultural crops have been a particular focus of the collections at the Yerevan Botanical Garden.

Many protected areas have historical sites within their boundaries, such as Tatevi Anapat, a 17th-century Armenian monastery.

BIBLIOGRAPHY

Adamia, S., V. Alania, A. Chabukiani, G. Chichua, O. Enukidze & N. Sadradze. 2010. Evolution of the Late Cenozoic Basins of Georgia (SW Caucasus): A Review. Pp. 239–260 *in* M. Sosson, N. Kaymakci, R. A. Stephenson, F. Bergerat & V. Starostenko (editors), Sedimentary Basin Tectonics from the Black Sea and Caucasus to the Arabian Platform. Geological Society, Special Publication 340. Bath, United Kingdom

Adzinba, Z. I. 1987. Endemy Flory Abkhazii. [Endemic Flora of Abkhazia.] Metzniereba, Tbilisi [in Russian].

Akademiia nauk Gruzinskoi SSR. 1971–2007. Flora Gruzii. [Flora of Georgia.] 15 Vols. Metzniereba, Tbilisi [in Georgian].

Akhani, H., M. Djamali, A. Ghorbanalizadeh & E. Ramezani. 2010. Plant Biodiversity of Hyrcanian Relict Forests, Iran: An Overview of the Flora, Vegetation, Palaeoecology and Conservation. Pakistan J. Bot. 42: 231–258.

Alisov, B. P. 1956. Klimaticheskie Oblasti i Raiony SSSR. [Climate Regions in the USSR.] Moscow University, Moscow [in Russian].

Anderson, S. 2002. Identifying Important Plant Areas: A Site Selection Manual for Europe, and a Basis for Developing Guidelines for Other Regions of the World. Plantlife International, London.

Bohn, U. & R. Neuhäusl, unter Mitarbeit von G. Gollub, C. Hettwer, Z. Neuhäuslova, H. Schlűter & H. Weber. 2000–2003. Karte der Natűrlichen Vegetation Europas. [Map of the Natural Vegetation of Europe.] Scale 1:2.500.000. Teil I, Erlauterungatext, 655 pp.; Teil II: Legende 153 pp.; Teil 3, Karten. Landwirtschaftsverlag, Munster.

Bohn, U., N. Zazanashvili & G. Nakhutsrishvili. 2007. The map of the natural Vegetation of Europe and its application in the Caucasus ecoregion. Bull. Georgian Acad. Sci. 175(1): 112–121.

Boissier, E. 1867–1884. Flora Orientalis: sive, Enumeratio plantarum in Oriente a Graecia et Aegypto ad Indiae fines hucusque observatarum, Vols. 1–5, Suppl. 6 [1888]. H. Georg., Basil.

Buxbaum, J. C. 1728–1740. Plantarum minus cognitarum centuria I. [-V.] complectens plantas circa Byzantium & in oriente observatas, Vols. 1–5. Ex typographia Academiae, Petropoli. [Posthumously published in part by J. G. Gmelin.]

Candolle, A. P. de. 1824–1873. Prodromus Systematis Naturalis Regni Vegetabilis, Vol. 17. Sumptibus Sociorum Treuttel et Würtz, Paris.

Chardin, J. 1686. Journal de Voyage de Chevalier Jean Chardin en Perse. Daniel Horthemels, Paris.

Convention on Biological Diversity (CBD). 2002. Conference of the Parties, COP 6 Decision VI/9. Global Strategy for Plant Conservation. 2002–2010. Sixth Ordinary Meeting of the Conference of the Parties to the Convention on Biological Diversity (COP 6), The Hague, Netherlands. <http://www.cbd.int/decision/cop/?id=7183>, accessed 11 September 2013.

Convention on Biological Diversity (CBD). 2011. Global Strategy for Plant Conservation: The targets 2011–2020. <http://www.cbd.int/gspc/targets.shtml>, accessed 11 September 2013.

Convention on International Trade in Endangered Species of Wild Fauna and Flora (CITES). 2003. Global Strategy for Plant Conservation (CBD): analysis and links with CBD [decisions 12.12 & 10.86]. PC13 WG5 Doc. 1. Thirteenth meeting of the Plants Committee, Geneva, Switzerland, 12–15 August 2003. <http://www.cites.org/eng com/pc/13/E-Minutes%20 PC13%20-%20Annex%203. pdf>, accessed 11 September 2013.

Critical Ecosystem Partnership Fund (CEPF). 2003. Ecosystem profile: Caucasus biodiversity hotspot [Final version, July 2003; updated September 2004]. <http://www.cepf.net/where_we work/ regions/europe_central_asia/caucasus/Pages/default.aspx>, accessed 11 September 2013.

Czerepanov, S. K. 1995. Plantae Vasculares Rossicae et Civitatum Collimitanearum. [Vascular Plants of Russia and Adjacent Countries.] Mir i Semja, St. Petersburg [in Russian].

Davis, P. H. (editor). 1965–1988. Flora of Turkey and East Aegean Islands, Vols.1–10. University Press, Edinburgh.

Delaporte, Abbé J. 1768. La Sibérie – la Nouvelle Zemble –la Russie, Vol. VII *in* Le Voyageur Francois ou la Connoissance de l'Ancien et du Nouveau Monde. L. Cellot, Paris.

Doluchanov, A. 1989. Rastitel'nost' Gruzii, Lesa. [The Vegetation of Georgia Forests.] Metzniereba, Tbilisi [in Russian].

Dudley, N. (editor). 2008. Guidelines for Applying Protected Area Management Categories. International Union for Conservation of Nature and Natural Resources (IUCN), Gland, Switzerland.<http://data.iucn.org/dbtw-wpd/edocs/paps-016.pdf>, accessed 13 September 2013.

Fedorov, A. A. 1952. Istoria vysokogornoi flory Kavkaza v chetvertichnoe vremja kak primer avtokhtonnogo razvitia tretichnoi floristicheskoi osnovy. [History of high mountain Caucasian flora in quaternary time as an example of autochthonous development tertiary floristic basis.] Mater. Istorii Fl. Rastitel'n. S.S.S.R. 3: 49–84 [in Russian].

Gabrielyan, E. T. & O. Fragman-Sapir. 2008. Flowers of the Transcaucasus and Adjacent Areas. Ganter Verlag, Ruggele, Konigstein, Germany.

Gadzhiev, V. D., K. G. Kulieva & Z. V. Vagabov. 1979. Rastitel'nost' Vysokogorii Talysha. [Flora and Vegetation of the Talysh Mountains.] Elm, Baku [in Russian].

Gagnidze, R. 1974. Botaniko-Geograficheskiy Analiz Florotcenoticheskogo Kompleksa Subal'Piyskogo Vysokotrav'ja Kavkaza. [Botanical-geographical Analysis of the Floral Coenotic Complex of Subalpine Tall Herbaceous Vegetation of the Caucasus.] Metzniereba, Tbilisi [in Russian].

Gagnidze, R. I., T. Gviniashvili, S. Shetekauri & N. Margalitadze. 2002. Endemic genera of the Caucasian flora. Feddes Repert. 113: 616–630.

Galushko, A. I. 1978–1981. Flora Severnogo Kavkaza: Opredelitel', 1–3. [Flora of the North Caucasus: A Guide, Vols. 1–3.] Rostov University, Rostov-on-Don [in Russian].

Gerasimov, I. (editor). 1966. Kavkaz. [The Caucasus.] Nauka, Moscow.

Gerasimov, I., et al. (editors). 1971. Natural Resources of the Soviet Union: Their Usage and Renewal. W. H. Freeman & Co., San Francisco.

Gogolishvili, M. A. & V. S. Skhiereli. 1986. Tsentralnyi Botanicheskii Sad Akademii Nauk Gruzinskoi SSR. Kratkaia Istoria. [The Central Botanical Garden of the Academy of Sciences of the Georgian SSR: A Brief History.] Metzniereba, Tbilisi [in Russian].

Grigoriev, A. A. & M. I. Budyko. 1971. Schematic maps of climatic types. Pp. 34–50 *in* Gerasimov et al. (editors), Natural Resources of the Soviet Union: Their Usage and Renewal. W. H. Freeman & Co., San Francisco.

Grossheim, A. A. 1928–1934. Flora Kavkaza. [Flora of the Caucasus.] Vols. I–IV. Narodnyi Komissariat Zemledeliya SSR Armenii, Baku, Izd. Az. Otd. Zak. Fil. Ak. Nauk, Tiflis [in Russian].

Grossheim, A. A. 1939–1967. Flora Kavkaza. [Flora of the Caucasus.] A. A. Fedorov (editor), Second edition, tomes I–VII. Moskva, 19. Nauka, Baku [in Russian].

Grossheim, A. A. 1949. Opredelitel' rasteniy Kavkaza. [Manual of the Caucasian Plants.] Sovetskaia Nauka, Moscow [in Russian].

Hijmans, R. J., L. Guarino, A. Jarvis, R. O'Brien, P. Mathur, C. Bussink, M. Cruz, I. Barrantes & E. Rojas. 2005. DIVA-GIS, Vers. 5.2, Manual. <http://www.diva-gis.org/docs/DIVA-GIS5 manual.pdf>, accessed 11 September 2013.

International Union for Conservation of Nature and Natural Resources (IUCN). 2001. IUCN Red List Categories and Criteria, Version 3.1. Prepared by the IUCN Species Survival Commission. IUCN, Gland, Switzerland, and Cambridge, United Kingdom.

Kamelin, R. V. 2006. Materialy k analisu flory Kavkaza. O nekotorykx osobennostiakx sostava flory Kavkaza i ikx znachenii dlia ponimania istorii flory etoi strany. [Data on the analysis of the Caucasian flora. Some peculiarities in the composition of the Caucasian flora and their importance for the understanding of the history of the country.] Bot. Zhurn. 91: 649–673.

Karjagin, I. I. (editor). 1950–1961. Flora Azerbaijana. Vols. 1–8. Nauka, Baku [in Russian].

Khokhrjakov, A. P. 1991. Rodovoi endemism v svjazi s problemoi floristicheskogo raionirovania na primere Kavkaza i sopredel'nykx territoryi. [The generic endemism in the connections with problem of floristical regionality for the Caucasus and neighboring territories.] Byull. Moskovsk. Obshch. Isp. Prir. Otd. Biol. 96: 96–106 [in Russian].

Kolakovsky, A. A. 1961. Rastitel' nyi mir Kolkhidy. [Flora and Vegetation of Colchis.] Moscow University, Moscow [in Russian].

Kolakovsky, A. A. 1964. Pliotzenovaia Flora Kodora. [Pliocene Flora of the Kodori River.] Metzniereba, Sukhumi [in Russian].

Kolakovsky, A. A. 1974. Nekotorye Dannye po Paleogeographii Kavkaza v Svjazi s Razvitiem Ego Flory. [Some Data on Paleogeography of the Caucasus in Connection with Flora Formation.] Sukhumi Botanical Garden. Metzniereba, Tbilisi [in Russian].

Kolakovsky, A. A. 1995. Sredizemnogornaia Oblast': Arena Evolutzii Zvetkovyx v Severnom Polusharii. [Mediterranean Mountain Region: Arena of Evolution for Flowering Plants in the Northern Hemisphere.] Soobshch. Akad. Nauk Gruzinsk. S.S.R. [in Russian].

Kolakovsky, A. A. & A. K. Shakryl. 1976. Sarmatskaya Flora v Abkhazii. Sarmatian Flora of Abkhazia.] Trudy Sukhumsk. Bot. Sada 22: 98–148.

Kolakovsky, A. A. & A. K. Shakryl. 1978. Kimmeriiskaya flora Gul'ripshi. [Kimmerian flora of Gulripsha.] Trudy Sukhumsk. Bot. Sada 24: 134–156.

Komarov, V. L., B. K. Shishkin & E. G. Bobrov (editors). 1934–1960. Flora URSS, Vols. 1–30. Akademii Nauk SSSR, Leningrad. [in Russian].

Krever, V., N. Zazanashvili, H. Jungius, L. Williams & D. Petelin (editors). 2001. Biodiversity of the Caucasus Ecoregion. World Wide Fund for Nature. Signar, Moscow.

Kuznetzov, N. I., N. A. Bush & A. V. Fomin. 1901–1913. Flora Caucasica Critica... [Materialy dlia Flory Kavkaza: Kritischeskoe Sistematichesko-geograficheskoe Izsliedovanie]. K. Mattisena, IUrev.

Ledebour, C. 1841–1853. Flora Rossica, Vols. I–IV. E. Schweizerbart, Stuttgart.

Linnaeus, C. 1753. Species Plantarum, Vols. I, II. Laurentii Salvii, Stockholm.

Lipsky, V. I. 1899. Flora Kavkaza. Svod svedenii o flore Kavkaza za dvukhsotletnii period ee issledovaniya, nachinaya ot Turnefora i konchaya XIX vekom. [Flora of Caucasus. A Corpus of Information about Caucasian Flora for Bicentennial Period of its Research, starting from Tournefort and ending by the 19th Century.] Tipo-Litografiia Gerolda, St. Petersburg [Suppl. I, 1902].

Litvinskaya, S. & R. Murtazaliev. 2009. Kavkazski Element vo Flore Rossiskogo Kavkaza: Geographia, Sozologia, Ekologia. [The Caucasian Elements within the Flora of the Russian Caucasus: Geography, Sociology, Ecology.] Kuban State University, Krasnodar [in Russian].

Lucas, G. 1978. The IUCN Plant Red Data Book: Comprising Red Data Sheets on Two Hundred-Fifty Selected Plants Threatened on a World Scale. International Union for Conservation of Nature and Natural Resources, Morges, Switzerland [H. Synge, compiler.]

Marschall von Bieberstein, F. A. 1808–1819. Flora taurico-caucasica exhibens stirpes phaenogamas in Chersoneso Taurica et regionibus caucasicis sponte crescentes..., Vols. 1 & 2 [1808], Vol. 3 [1819]. Typis Academicis, Charkouiae [Kharkov].

Menitzky, Y. L. 2005. Oaks of Asia. Science Publishers, Inc. Enfield, New Hampshire.

Mikić, A., N. Sarukhanyan, J. Akopian, A. Vanyan, I. Gabrielyan, V. Torchyan, S. Sarukhanyan, B. Cupina, T. Zeremski-Škorić, Đ. Krstić, P. Smýkal, G. Kenicer & M. Ambrose. 2011. Conservation strategies of *Vavilovia formosa* (syn. *Pisium formosum*), a high-mountainous pea relative, in Armenia. Proc. 16th Symp. Europ. Grass. Fed.: 371–373.

Miller, J. S. K. McCue, T. Consiglio & J. Stone. 2004. Endemic Medicinal Plants of Georgia (Caucasus). Institute of Botany, Georgian Academy of Sciences, Tbilisi.

Mrevlishvili, N. 1997. Geologia Gruzii. [Geology of Georgia.] Tbilisi Universitet, Tbilisi.

Myers, N., R. A. Mittermeier, C. G. Mittermeier, G. A. B. da Fonseca & J. Kent. 2000. Biodiversity hotspots for conservation priorities. Nature 403: 853–845.

Nakhutsrishvili, G. S. 1999. The Vegetation of Georgia (Caucasus). Braun-Blanquetia, Vol. 15. Dipartimento di Botanica ed Ecologia dell'Università, Camerino.

Nakhutsrishvili, G. S. 2013. On the altitudinal differentiation of the vegetation of Georgia. *In* The Vegetation of Georgia (South Caucasus). Geobotany Studies, Springer-Verlag, Berlin.

Nakhutsrishvili, G. S., N. Zazanashvili & K. Batsatsashvili. 2010. Regional profile: Colchic and Hyrcanic temperate rainforests of the Western Eurasian Caucasus. Pp. 214–221 *in* D. DellaSala (editor), Temperate and Boreal Rainforests of the World. Ecology and Conservation. Island Press, Washington, D.C.

Probst, W. 1981. Zur Vegetationsgeschichte und Klimaentwicklung des Südkaspischen Waldgebietes (Noriran). *In* W. Frey & H.-P. H. Uerpman (editors), Beiträge zur Umweltgeschichte des Vorderen Orients. Beihefte zum Tübinger Atlas des Vorderen Orients, Reihe A (Naturwiss.) 8: 26–39. Dr. Ludwig Reichert, Weisbaden.

Ramsar Convention on Wetlands (RCW). 1971 [2 Feb.]. Convention on Wetlands of International Importance especially as Waterfowl Habitat. Ramsar (Iran). UN Treaty Series No. 14583. As amended by the Paris Protocol, 3 December 1982, and Regina Amendments, 28 May 1987. <http://www.ramsar.org/cda/en/ramsar-about-about-ramsar/main/ramsar/1-36%5E7687_4000_0__>, accessed 11 September 2013.

Raven, P. H. 1976. Ethics and attitudes. Pp. 155–179 *in* J. B. Simmons (editor), Conservation of Threatened Plants. Plenum Press, New York.

Rechinger, K. H. (editor). 1963–2012. Flora Iranica. Akademische Druck-u. Verlagsanstalt und Naturhistorisches Museum Wien, Graz and Wien.

Rikli, M. 1943. Das Pflanzenkleid der Mittelmeerland. Bd. 1. H. Huber, Bern.

Sales, F. & I. C. Hedges. 1996. Biogeographic aspects of selected SW Asiatic woody taxa. Ann. Naturhist. Mus. Wien. 98(B Suppl.): 149–161.

Shatilova, I. & S. Ramishvili. 1990. Materialy po Istorii Flory i Rastitel'nosti Gruzii. [Material on History of Flora and Vegetation of Georgia.] Metzniereba, Tbilisi [in Russian].

Shatilova, I., N. Mchedlishvili, L. Rukhadze & E. Kvavadze. 2011. Istoria Flory i Rastitel'nosti Gruzii. [The History of the Flora and Vegetation of Georgia, South Caucasus.] Georgian National Museum, Tbilisi [in Russian].

Shulkina, T. 2004. Ornamental Plants from Russia and Adjacent States of the Former Soviet Union. Russian Academy of Sciences, Rostok, St. Petersburg.

Steven, C. 1813. Observationes in *Saxifraga*. Mém. Soc. Imp. Naturalistes Moscou 4: 73–82.

Steven, C. 1823. Monographia *Pedicularis*. Mém. Soc. Imp. Naturalistes Moscou 6: 1–60.

Steven, C. 1838. De Pinibus Taurico-caucasicis. Bull. Soc. Imp. Naturalistes Moscou 1838(1): 43–53.

Steven, C. 1851. Observationes in Asperifolias Taurico-caucasicas. Bull. Soc. Imp. Naturalistes Moscou 24(2): 558–609.

Svanidze, T. I. 1972. Yrskaia Flora Gruzii. [Jurassic Flora in Georgia.] Ph.D. Dissertation. Tbilisi [in Russian].

Takhtajan, A. L. 1946. Istorii razvitia rastitel'nosti Armenii. [To the history of vegetation development in Armenia.] Trudy Bot. Inst. Akad. Nauk Armyansk. S. S. R. 4: 51–107 [in Russian].

Takhtajan, A. L. (editor). 1954–2010. Flora Armenii. [Flora of Armenia.] Vols. 1–10. Nauka, Moscow–Leningrad, Koeltz Scientific Books, Germany [in Russian].

Takhtajan, A. L. 1986. Floristic Regions of the World. University of California Press, Los Angeles.

Takhtajan, A. L. (editor). 2003–2012. Caucasian Flora Conspectus, Vols. 1–3. St. Petersburg University Press, St. Petersburg [in Russian].

Tournefort, J. P. de. 1717. Relation d'un voyage du Levant, fait par ordre du roy : contenant l'histoire ancienne & moderne de plusieurs isles de l'archipel, de Constantinople, des côtes de la Mer Noire, de l'Armenie, de la Georgie, des frontieres de Perse & de l'Asie Mineure ..., Tome III. Chez Anisson et Poseul, A. Lyon.

United Nations. n.d. United Nations Convention to Combat Desertification (UNCCD). <http://www.unccdint/en/Pages/default.aspx>, accessed 11 September 2013.

United Nations Educational, Scientific and Cultural Organization (UNESCO). 1972 [16 Nov.]. Convention Concerning the Protection of the World Cultural and Natural Heritage. Report of the General Conference, 17th Session. World Heritage, Paris. <http://whc.unesco.org/archive/convention-en.pdf>, accessed 11 September 2013.

Vavilov, N. I. 1992. Origin and Geography of Cultivated Plants. Translated by D. Löve, Cambridge University Press, Cambridge. Originally published as Proiskhozhdenie I geografiía kul'turnykh

rastenii (Nauka, Leningrad, 1987).
Williams, L., N. Zazanashvili, G. Sanadiradze & A. Kandaurov (editors). 2006. An Ecoregional Conservation Plan for the Caucasus. World Wildlife Fund, Second ed. Contour Ltd., Tbilisi. <http://assets.panda.org/downloads/ecp__second_edition.pdf>, accessed 27 August 2013.
World Wildlife Fund (WWF). [No date.] <http://worldwildlife.org/>, accessed 11 September 2013.
Wulff, E. V. 1943. An Introduction to Historical Plant Geography. Chronica Botanica Co., Waltham, Massachusetts.
Yarovenko, Y. A. & R. A. Murtazaliev. 2009. Unique World of Dagestan's Flora and Fauna [Unikalnyi Mir Flory i Fauny Dagestana]. Publisher House Epokha, Makhachkala.
Zazanishvili, N. 2005. Application of the map of the natural vegetation of Europe in developing a protected areas network in the Caucasus Ecoregion. Pp. 251–262 *in* U. Bohn, C. Hettwer & G. Gollub (editors), Application and Analysis of the Map of the Natural Vegetation of Europe [Anwendung und Auswertung der Karte der Natürlichen Vegetation Europas]. Buhdesamt für Naturschutz (BfN) Skripten 156, Bonn.
Zazanishvili, N., G. Sanadiradze & A. Bukhnikashvili. 1999. Caucasus. Pp. 269–273 *in* R. A. Mittermeier, N. Myers, P. G. Gil & C. G. Mittermeier (editors), Hotspots: Earth's Biologically Richest and Most Endangered Terrestrial Ecoregions. Conservation International, New York.
Zazanishvili, N., R. Gagnidze & G. Nakhutsrishvili. 2000. Main type of vegetation zonation on the mountains of the Caucasus. Acta Phytogeogr. Suec. 85: 7–16.
Zhukovsky, P. M. 1964. Kul'turnye rastenia I ikx sorodichi. [Cultivated Plants and Their Relatives.] Kolos, Leningrad [in Russian].

ARMENIA

Հայաստանի Հանրապետություն

Կ.Գ. Թամանյան, Գ.Մ. Ֆայվուշ, Է.Ց. Գաբրիելյան, Ժ.Հ. Վարդանյան, Մ.Է. Հովհաննիսյան, Ա.Հ. Ներսեսյան, Ի.Գ. Արևշատյան, Ն.Մ. Խանջյան

Վայոց Ձորի մարզ

Տաթև վանք

Կովկասի տարածաշրջանի ֆլորան, որն ընդգրկում է հինգ երկրների տարածք, ներառում է մոտ 6300 տեսակ անոթավոր բույսեր: Ծրագիրը, որի արդյունքները ձևակերպված են տվյալ գրքում, նվիրված է ողջ տարածաշրջանի էնդեմների գնահատականին ըստ IUCN-ի կողմից ընդունված միջազգային համակարգի չափանիշների: Հայաստանի տարածքում հանդիպում են Կովկասյան էնդեմների մոտ 850 տեսակներ, որոնցից 123-ը հանդիսանում են մեր երկրի նեղտեղային էնդեմներ: Ըստ գոյության վտանգվածության աստիճանի 24 տեսակներ համապատասխանում են "կրիտիկական վիճակում գտնվողներ" (CR) չափանիշին, 33-ը` "վտանգավոր վիճակում գտնվողներ" (EN) և 9-ը` "խոցելի" (VU) չափանիշներին: Մնացած տեսակները դեռևս չեն գտնվում անմիջական վտանգվածության տակ:

Տվյալ գիրքը հանդիսանում է Կովկասի ֆլորայի պահպանությանը նվիրված անգլերեն լեզվով առաջին հրատարակությունը և պատրաստված է այս տարածաշրջանի բոլոր երկրների մասնագետների մասնակցությամբ:

Հայաստանի Հանրապետությունը (տարածքը` 29.743 կմ^{2}) գտնվում է Կովկասյան տարածաշրջանի հարավային մասում և սահմանակից է Վրաստանին, Ադրբեջանին, Թուրքիային և Իրանին:

Հայաստանը լեռնային երկիր է, ամենաբարձրադիր կետը` Արագած լեռան գագաթն է` 4095 մ, իսկ ամենացածրը` Դեբեդ գետի ավազանում է` 400 մ-ից ցածր: Հայաստանի տարածքում է գտնվում Կովկասյան խոշորագույն քաղցրահամ բնական ջրամբարը` Սևանա բնական լիճը (1360 քառ. կմ):

Հայաստանը հնագույն մշակույթի կենտրոն է, հայտնի Քրիստոնեական ժամանակաշրջանից դեռ շատ առաջ: Շնորհիվ բնակլիմայական պայմանների բազմազանության և հանրապետության երկու ֆլորիստիկ ենթաթագավորությունների (Բորեալ և Հինմիջերկրածովյան) շփման վայրում գտնվելու, երկրի բուսականությունը և ֆլորան շատ հարուստ են և բազմազան: Այստեղ ներկայացված է բուսականության Կովկասյան տիպերի ճնշող մեծամասնությունը:

Անտառային բուսականությունը զբաղեցնում է Հայաստանի ընդամենը 10-11%-ը: Որպես հիմնական անտառակազմող տեսակներ հանդես են գալիս կաղնիները (*Quercus iberica* և *Quercus macranthera*), հաճարենին (*Fagus orientalis*), բոխին ու դաժին (*Carpinus betulus* և *Carpinus orientalis*) և սոճին (*Pinus kochiana*), որոնք զբաղեցնում են անտառային ընդհանուր տարածքների մոտ 90%-ը: Իրենց կազմով չափազանց հետաքրքիր և յուրահատուկ են գիհուտային և արիդային նոսրանտառները, որոնցում գերակայում են գիհու չորադիմացկուն տեսակները (*Juniperus excelsa*, *Juniperus foetidissima*), արաքսյան կաղնին (*Quercus boissieri*), պիստակենին (*Pistacia mutica*), վայրի նշենին (*Amygdalus fenzliana*), տանձենիների վայրի տեսակները: Բուսաբանական և բնապահպանական առումներով շատ կարևոր են սոսու (*Platanus orientalis*) և կենու (*Taxus baccata*) ռելիկտային անտառպուրակները:

Հայաստանի ստորին լեռնային գոտում լայնորեն են ներկայացված կիսաանապատները, որոնք շատ բազմազան են ֆլորիստիկ կազմով և բուսական համակեցություններով: Հայաստանում ներկայացված է նաև իսկական ավազային անապատ: Դեպի վեր՝ միջին լեռնային գոտում, լայնորեն են ներկայացված տափաստանները, որոնք նմանվում են Եվրասիական տափաստաններին: Այս գոտում համեմատաբար մեծ մակերեսներ են զբաղեցնում տրագականտները, (*Astragalus*-ի տեսակները, *Onobrychis cornuta*-ն)՝ որոնք չափազանց գեղատեսիլ են ծաղկման շրջանում և լեռնալանջերին տալիս են յուրահատուկ գունային երանգավորում:

Վերին լեռնային գոտում գտնվում են մերձալպյան և ալպյան մարգագետիններն ու գորգերը: Դրանք հանդիսանում են հանրապետության կարևորագույն կերային հանդակները: Հատկապես գեղեցիկ են ալպիական գորգերը, որոնցում գերակշռում են տարախոտերի վառ ներկայացուցիչները՝ զանգակները, խատուտիկները, բոցերը և այլն:

Ողջ Հայաստանով մեկ լայնորեն են ներկայացված արտազոտային տեղավայրերը՝ ժայռերը, թափվածքները, քարացրոնները և գետերը, լճերը, ճահիճները: Զբաղեցնելով ոչ մեծ տարածքներ, դրանք ևս աչքի են ընկնում ֆլորայի ու բուսականության բազմազանությամբ:

Չնայած հանրապետության փոքր տարածքին, նրա ֆլորան ներկայացված է մոտ 3600 տեսակի անոթավոր բույսերով:

Հանրապետության բուսական բազմազանության վրա ազդող կարևորագույն վտանգը կայանում է միջավայրի կորուստը և փոփոխությունը, (օգտակար հանածոների արդյունահանումը, "անտառահատումը„ ենթակառուցվածքների զարգացումը և արոտավայրերի ծանրաբեռնվածության անհավասար տեղաբաշխումը): Բացի այդ, մեծ վտանգ է ներկայացնում կլիմայի գլոբալ փոփոխությունը:

Ներկա ժամանակահատվածում երկրի առաջատար բուսաբանական հաստատությունը Գիտությունների Ազգային ակադեմիայի Բուսաբանության ինստիտուտն է, որում հերբարիումային հարստագույն ստեղծված հավաքածու է (մոտ 300.000 նմուշ) և որին կից գործում է Երևանի բուսաբանական այգին իր երկու մասնաճյուղերով՝ Վանաձորում և Սևանում: Բացի այդ, բուսաբանական հետազոտություններ են տարվում Երևանի պետական համալսարանում և Հայաստանի պետական ագրարային համալսարանում:

Անկախություն ձեռք բերելուց հետո Հայաստանը ստորագրել է մի շարք միջազգային կոնվենցիաներ և ընդունել է ՀՀ օրենքներ՝ "Բուսական աշխարհի մասին" (1999), "Հատուկ պահպանվող բնական տարածքների մասին" (2006), "Սևանա լճի մասին" (2001), "Անտառային օրենսգիրք" (2005) և այլն:

Ներկայումս Հայաստանում գոյություն ունեն 3 ազգային պարկեր ("Սևան", "Դիլիջան", "Արփի լիճ"), 3 պետական արգելոցներ (Շիկահողի, Էրեբունու և "Խոսրովի անտառ"), 26 արգելավայրեր և 4 դենդրոպարկներ: Բացի այդ, հաստատվել է բնության հուշարձանների ցուցակը, որն ընդգրկում է 230 օբյեկտներ:

ARMENIA

Kamilla Tamanian†, Eleonora Gabrielyan, George Fayvush, Marine Hovhannesyan, Anush Nersesyan, Izabella Arevshatyan, Nazik Khandjyan, and Zhirajr Vardanyan

CAPITAL

Yerevan, the largest city.

OTHER CITIES

Gyumri, Vanadzor, Kapan.

POPULATION

According to CIA World Factbook, 2,974,184 (August 2013 estimate).

LANGUAGE

Armenian; 70% of the population speak Russian; 20%, English.

FLORA

3600 (total plant species); 123 (total national endemics); 850 (Caucasian region endemics).

The Second Red Book of Armenia (Tamanian et al., 2010) enumerated 452 plant species at risk.

BOTANICAL INSTITUTIONS

The Botanical Institute of the National Academy of Sciences houses the national herbarium with about 400,000 specimens of which 200,000 represent native Armenian plants. The Institute also has collections of living plants in the Yerevan Botanical Garden, located on 80 hectares of semi-arid land to the northeast of the capital city. Two field stations are located in the regions of Vanadzor and Lake Sevan. Address: Institute of Botany, Yerevan 0063, Armenia; Tel.: (+374 10) 621781; FAX: (+374 10) 628211; e-mail: botanyinst@sci.am.

Other botanical institutions and gardens include the Department of Botany at Yerevan State University. Address: Charents str. 8, Yerevan, Armenia.

There are four state arboreta (Dendroparks): (1) Arboretum Soranner located in the town of Berd, Tavush Province (marz) in northern Armenia, bordering Georgia and Azerbaijan; (2) Arboretum Idjevan in the town of Idjevan, Tavush region; (3) Arboretum Sochut in the town of Stepanavan, Lori Province (marz) in northern Armenia, bordering Georgia; and (4) Arboretum Vanadzor in the city of Vanadzor, Lori region. The Vanadzor arboretum was established in 1936, and has a nice collection of Armenian plants.

PROTECTED AREAS

There are four national parks (Sevan, in the Lake Area; Arpi-lich, surroundings of the Lake Arpi; Arevik in southern Armenia; and Dilizhan in Tavush Province, northern Armenia). Three state reserves include Khosrovsky Les, Shikakhokh, and Erebuni. There are 27 additional protected areas, all of which are managed by the national government.

Armenia is an ancient country known long before the Christian era as part of Chaldea or Urartu by the Assyrians. Mount Ararat is revered as a holy place for all Armenians, and the mountains of Ararat were mentioned in the Bible as the place where Noah's Ark came to rest when the waters of the Flood subsided. The higher of the two peaks of Ararat, at 5137 m, is quite visible across the Armenian plateau. It currently lies 32 km south of the Armenian border in Turkey.

Armenia occupies the southern portion of the Caucasus region and is bounded to the north by Georgia, to the east by Azerbaijan, to the south by Iran, and to the west by Turkey. It is a landlocked country lying at a distance of about 145 km from the Black Sea to its west,

175 km from the Caspian Sea to its east, and 750 km from the Mediterranean Sea. Armenia lies between 38°50′–41°18′N and 43°27′–46°37′E and spans 400 km in width. The country is situated in the Lesser Caucasus, whose mountain ranges of complex configuration run in different directions and alternate with plains. Armenia is generally a mountainous country, with its lowest point at 400 m along the Debed River and the highest peak, Mount Aragats, at 4090 m. The average altitude across the country is ca. 1850 m. Lake Sevan, with a surface of 1360 km^2, lies at an elevation of ca. 1900 m and is located to the northeast of Yerevan in the central province of Gegharkunik. Lake Sevan is one of the largest freshwater lakes at high altitude in the world.

CLIMATE

The climate in Armenia is continental in this south-central area of the Transcaucasus. Air temperature varies by elevation. Thus, at lower elevations, in the Ararat valley, temperature can drop in winter down to –32°C (–26°F) and reach a maximum in summer of 43°C (110°F). In mountainous regions, the January average is –12°C (10°F), and the mean temperature during the summer is 18°C (65°F). The average annual precipitation is about 150–300 mm (6–12 inches) in the plains, and 300–800 mm (12–31 inches) at higher elevations, exceeding 1000 mm in some areas. The rainfall reaches its annual maximum in the spring time. Long-lasting snow cover exists in the mountains above 1300 m, attaining 2 m in depth, while at lower elevations, the annual snow accumulation is 0.5 m or less. Arid conditions prevail throughout the country, even in mountainous regions. The amount of sunshine is high, 2600–2800 hours per year. In general, the climate of Armenia is best characterized as dry continental, with an annual precipitation pattern similar to that of the Mediterranean.

FLORA AND VEGETATION

According to Takhtajan (1978, 1986), the Armenian flora can be divided into the Caucasian and Armeno-Iranian Floristic Provinces. N. I. Vavilov (1926, 1992) included Armenia in the Near Asian center of origin of cultivated plants (bread and hard wheats, peas, lentils, grapes, etc.). Centuries of artificial selection by farmers has resulted in a significant diversity of grapes, apricots, and peaches under cultivation. Along with food plants, Armenians continue to use medicinal plants and mushrooms that are collected from the wild. There remain many promising plants for foodstuffs in the Armenian flora, including the wild relatives of cultivated plants such as cereal grasses, wild fruits, berries, vegetables, fodder, and aromatic plants and herbs.

The diversity of landscapes and the mountain topography are significant determinants of Armenia's rich floristic biodiversity and contribute to a high degree of endemism. The 10 most abundant plant families in Armenia are the Asteraceae (429 species), Fabaceae (346 species), Poaceae (335 species), Rosaceae (210 species), Brassicaceae (203 species), Caryophyllaceae (183 species), Lamiaceae (153 species), Scrophulariaceae (142 species), Apiaceae (132 species), and Cyperaceae (98 species).

Due to orographic diversity and history, many different types of vegetation have developed in Armenia. For example, the number of described types of forest communities is over 200. The forested areas collectively have 274 known species of trees and shrubs. Six principal tree species cover 89% of the total forest area, represented by the oriental beech (*Fagus orientalis* Lipsky, Fagaceae), the Iberian oak (*Quercus iberica* Steven ex M. Bieb., Fagaceae), the macranthera oak (*Quercus macranthera* Fisch. & C. A. Mey. ex Hohen.), the hornbeams (*Carpinus betulus* L. and *Carpinus orientalis*

Mill., Betulaceae), and Koch's pine (*Pinus kochiana* Klotzsch. ex K. Koch, Pinaceae). The remaining 11% include broadleaf deciduous species from *Acer* L. (maple, Aceraceae), *Betula* L. and *Corylus* L. (birch, hazelnut, Betulaceae), *Fraxinus* L. (ash, Oleaceae), *Juglans* L. (walnut, Juglandaceae), *Malus* Mill. and *Pyrus* L. (apples and pears, Rosaceae), *Platanus* L. (sycamore or oriental plane tree, Platanaceae), *Tilia* L. (linden, Tiliaceae), *Ulmus* L. (elms, Ulmaceae), as well as the conifer *Taxus* L. (Taxaceae). In northern Armenia, humid forests occupy large areas where slopes may be covered with beech (*Fagus orientalis* Lipsky), maple (*Acer trautvetteri* Medw.), and common mountain ash (*Sorbus aucuparia* L.). In more arid areas, the drier forest associations include Iberian oaks (*Quercus iberica* Steven ex M. Bieb.), ash (*Fraxinus excelsior* L.), and Hyrcanian maple (*Acer hyrcanum* Fisch. & C. A. Mey.) and are characterized by a well-developed herb cover, with elements of mountain steppe vegetation. Beech forests occur at elevations from 800 to 2000 m, usually on north-facing slopes. Monospecific stands of beech occur between 1000 and 1800 m. In the mixed beech forest types, the accompanying trees are often oak, ash, elm, linden, hornbeam, or maple. Oak woods are characterized by complex and varied typological composition. Oak forests occupy slopes with southern aspects and at elevations from 600 to 2200 m. Pine forests usually also have hornbeam, oak, and rarely beech. At elevations from 1900 to 2300 m, mostly mixed, sparse, low-density subalpine forests are present, with the most common species being *Betula litwinowii* Doluch., *Acer trautvetteri* Medw., and *Sorbus aucuparia* L. There are natural groves of oriental plane trees (*Platanus orientalis* L.) in southern Armenia that extend into Azerbaijan. Oriental plane trees mix with walnuts along small mountain rivers forming narrow borders that create a particular park-like, open landscape. In drier places away from the rivers this forest contains juniper (mostly *Juniperus excelsa* M. Bieb.), pistachio (*Pistacia mutica* Fisch. & C. A. Mey.), oak (*Quercus boissieri* Reut.), and wild pear (*Pyrus* species). A humid subalpine forest does not exist in the southern part of the country due to moisture deficits. Instead, the vegetation consists of tragacanth phryganoid formations and mountain steppes. In many areas that were forested until recently, the remaining secondary vegetation is now highly degraded. In the Lake Sevan basin at 1900 m, fragments of forest still exist in the mountains above the Gyunei shoreline. Broad-leaved deciduous forests of open aspect are common at 900–1200 m. They are characterized by pistachio (*Pistacia mutica* Fisch. & C. A. Mey., Anacardiaceae), Georgian maple (*Acer ibericum* M. Bieb., Aceraceae), almond (*Amygdalus fenzliana* (Fritsch) Lipsky, Rosaceae), pomegranate (*Punica granatum* L., Lythraceae), and jasmine (*Jasminum fruticans* L., Oleaceae). Other associated woody species are the elm-leaved sumac, *Rhus coriaria* L., the Eurasian smoke tree, *Cotinus coggygria* Scop. (both Anacardiaceae), and the oriental hackberry, *Celtis tournefortii* Lam. (Ulmaceae).

There are many other plant formations on the Armenian plains. The lower mountain zones (480–1200 m) and hills are covered by semi-desert (phryganoid) formations, most often gypsophilous vegetation. Specialized edaphic conditions occur in salt marshes and sand deserts. Middle and upper mountain zones (1200–2200 m) are characterized by various kinds of steppes, meadow-steppes, and shrubby steppes, as well as cushion (thorny tragacanth) vegetation. The subalpine and alpine zones exist at elevations from 2200 to 4000 m and feature tall grasses and meadows with cushion plants and tussocks.

True sand deserts cover only a small area in the Ararat valley in Armenia. The main components include phog, *Calligonum*

Campanula oblongifolia (K. Koch) Charadze, endemic to the southern and eastern Caucasus.

Pyrus takhtadzhianii Fed., endemic to the southern Caucasus.

polygonoides L. (Polygonaceae), milfoil, *Achillea tenuifolia* Lam. (Asteraceae), the small desert ephemeral, *Ceratocarpus arenarius* L. (Amaranthaceae), and the regional endemics *Astragalus paradoxus* Bunge (Fabaceae) and *Allium materculae* Bordz. (Alliaceae), all of which have an Irano-Turanian or Mediterranean distribution. Other endemic species of plants include the vulnerable chenopod, *Salsola tamamschjaniae* Iljin. In Armenia, Caucasian endemic mammals include the two vulnerable taxa, the Caucasian bear, *Ursus arctos* L. subsp. *syriacus* Hemprich & Ehrenberg, and the Armenian mouflon, *Ovis ammon* L. subsp. *gmelini* Blyth. The Caucasian or Persian leopard, *Panthera pardus* L. subsp. *ciscaucasica* Satunin, is the most endangered, with less than 1300 individuals across its range. Among insects, 20 beetle species are known as endemic to Armenia. Throughout the Caucasus, such animal taxa have become threatened with extinction because of human activities but remain in refuge populations in Armenia.

The vegetation of semi-deserts is usually called phrygana, phryganoid vegetation, or tomillares. It occurs in the lower mountain zones and is quite similar to Mediterranean ecosystems. Phrygana associations occur particularly in the Ararat plain, one of the largest on the regional Armenian plateau, in Vayots Dzor Province in the southeast, bordering Azerbaijan, and around Meghri in Syunik Province, the southernmost in Armenia. These areas are covered by sagebrush (*Artemisia fragrans* Willd., Asteraceae) and *Astragalus araxina* Takht. (Fabaceae), with various Mediterranean species such as *Capparis spinosa* L. (Capparidaceae) and *Zygophyllum fabago* L. (Zygophyllaceae) also common. Among the short-lived annuals of sagebrush communities are the drooping brome grass, *Anisantha tectorum* (L.) Nevski (Poaceae), the buttercup, *Ceratocephala falcata* (L.) Pers. (Ranunculaceae), and the spring draba, *Erophila verna* (L.) DC. (Brassicaceae).

A small area in the Ararat plain is characterized by gypsum and tufa-sand soil types, which produces plant communities similar to those seen in Spain, but with different representatives of the mint family Lamiaceae. These include *Stachys inflata* Benth., *Thymus kotschyanus* Boiss. & Hohen., *Eremostachys laciniata* (L.) Bunge, *Nepeta mussinii* Spreng. ex Henckel, *Teucrium polium* L., and *Phlomis orientalis* Mill. The hills in the southern part of the Ararat plain have plant communities dominated by distinctive cushion plants of *Gypsophila aretioides* Boiss. (Caryophyllaceae), *Acantholimon karelinii* (Stschegl.) Bunge, *Acantholimon armenum* Boiss. & A. Huet (Plumbaginaceae), *Onobrychis cornuta* (L.) Desv. (Fabaceae), and *Camphorosma lessingii* Litv. (Amaranthaceae). In the transitional areas between semi-desert and steppe, one can find the wild ancestors of our modern grain crops such as wheat (*Triticum* L.), rye (*Secale* L.), and barley (*Hordeum* L.), as well as goatgrass, *Aegilops* L., one species of which, *Aegilops tauschii* Coss., contributed its diploid D-genome to the parentage of the modern bread wheat, *Triticum aestivum* L. (all Poaceae). The crown of thorns, *Paliurus spina-christi* Mill. (Rhamnaceae), a low-growing, spiny tree, is one of the main components of shibliak vegetation, which consists of drought-tolerant, deciduous shrubs and small trees. In Armenia, shibliak is characterized by Pallas's buckthorn, *Rhamnus pallasii* Fisch. & C. A. Mey. (Rhamnaceae), the cherry plum, *Cerasus incana* (Pall.) Spach (Rosaceae), and the gymnosperm shrub, *Ephedra procera* Fisch. & C. A. Mey. (Ephedraceae). The herbaceous layer in shibliak communities here resembles steppe. Shibliak in both southern and northeastern Armenian sites contains many Mediterranean elements, including, sometimes, wild pomegranates, *Punica granatum* L. (Punicaceae), and figs, *Ficus carica* L. (Moraceae).

Thorny cushion formations are widespread throughout Armenia at elevations from 1400 to 2300 m. Principal components are goat's thorn species of *Astragalus* L., such as *Astragalus microcephalus* Willd. and *Astragalus lagurus* Willd., as well as the spiny legume *Onobrychis cornuta* (L.) Desv. (all Fabaceae). Prickly thrift species in *Acantholimon* Boiss. (Plumbaginaceae) also predominate.

Meadows in Armenia can be seen in high mountain areas, mainly above 2200 m. The vegetation of subalpine and alpine meadows is quite diverse; however, there are fewer endemic species here than seen at lower elevations. Abundant in high mountain meadows are grasses such as the orchard grass, *Dactylis glomerata* L., timothy grass, *Phleum pratense* L., and the violet-tinged barley, *Hordeum violaceum* Boiss. & Hohen. [= *Hordeum brevisubulatum* subsp. *violaceum* (Boiss. & Hohen.) Tzvelev]. Other meadow denizens include *Anemone fasciculata* L. (Ranunculaceae), *Doronicum oblongifolium* DC. and *Taraxacum stevenii* DC. (both Asteraceae), *Cephalaria gigantea* (Ledeb.) Bobrov and *Scabiosa caucasica* M. Bieb. (both Dipsacaceae), *Campanula tridentata* Schreb. (Campanulaceae), and the sedge *Carex tristis* M. Bieb. (Cyperaceae). The alpine meadow exhibits a number of ornamental plants where gentians, *Gentiana pyrenaica* L. (Gentianaceae), bellflowers, *Campanula tridentata* Schreb. (Campanulaceae), primulas, *Primula algida* Adams (Primulaceae), and gentian speedwells, *Veronica gentianoides* Vahl (Scrophulariaceae), beautify the landscape. Cliffs and talus slopes also host many interesting species such as the cherry plum, *Cerasus incana* (Pall.) Spach (Rosaceae), the succulent, *Sempervivum transcaucasicum* Muirhead (Crassulaceae), a dwarf catmint, *Nepeta mussinii* Spreng. ex Henckel (Lamiaceae), the legume, *Astragalus microcephalus* Willd. (Fabaceae), the bladder fern or fragile fern, *Cystopteris fragilis* (L.) Bernh. (Cystopteridaceae), the juniper, *Juniperus depressa* Steven, the rosid shrub, *Cotoneaster integerrimus* Medik., and the shrubby gymnosperm, *Ephedra procera* Fisch. & C. A. Mey.

Salt marshes and boggy areas occupy a very small territory in Armenia (less than 0.14% of the land area) and occur mainly in the middle and upper mountain belts. There are very interesting salt marshes in the Ararat valley, which are dominated by *Juncus acutus* L. (Juncaceae) and *Iris musulmanica* Fomin (Iridaceae). These marshes include endemics, such as *Linum barsegjanii* Gabrieljan & Dittrich (Linaceae) and *Sonchus araraticus* Nazarova & Barsegyan (Asteraceae), both critically endangered species.

Table 1. Distribution of endemic species within different types of vegetation.

HABITAT	SPECIES
Deserts	*Scorzonera gorovanica* Nazarova
Semi-deserts	*Acantholimon vedicum* Mirzoeva, *Allium schchianae* Ogan., *Allochrusa takhtajanii* Gabrieljan & Dittrich, *Astragalus holophyllus* Boriss., *Bufonia takhtajanii* Nersesian, *Carthamus tamamschjaniae* Gabrieljan, *Centaurea alexandri* Bordz., *Centaurea arpensis* (Czerep.) Wagenitz, *Cotoneaster armenus* Pojark., *Cousinia daralaghezica* Takht., *Gypsophila takhtadzhanii* Schischk. ex Ikonn., *Isatis buschiorum* Hovh., *Papaver roseolum* M. V. Agab. & Fragman, *Verbascum gabrielianae* (Hub.-Mor.) Hub.-Mor., *Verbascum horticolum* Hub.-Mor.
Open forests and shibliak	*Amygdalus nairica* Fed. & Takht., *Centaurea alexandri* Bordz., *Centaurea arpensis* (Czerep.) Wagenitz, *Cotoneaster armenus* Pojark., *Cousinia fedorovii* Takht., *Cousinia takhtajanii* Tamanian, *Crataegus armena* Pojark., *Crataegus zangezura* Pojark., *Dianthus grossheimii* Schischk., *Isatis sevangensis* N. Busch, *Linaria zangezura* Grossh., *Merendera mirzoevae* Gabrieljan, *Polygala urartu* Tamamsch., *Psephellus geghamensis* (Gabrieljan) Gabrieljan, *Psephellus pambakensis* Sosn., *Psephellus zangezuri* Sosn., *Pseudolysimachion spicatum* (L.) Opiz subsp. *transcaucasicum* (Bordz.) Gabrieljan, *Pyrus browiczii* Mulk., *Pyrus chosrovica* Gladkova, *Pyrus complexa* Rubtzov, *Pyrus daralaghezi* Mulk., *Pyrus elata* Rubtzov, *Pyrus gergerana* Gladkova, *Pyrus hajastana* Mulk., *Pyrus sosnovskyi* Fed., *Pyrus tamamschianiae* Fed., *Pyrus voronovii* Rubtzov, *Rosa hracziana* Tamamsch., *Rosa sjunikii* Jarosch., *Rosa sosnovskyana* Tamamsch., *Rosa zangezura* Jarosch., *Rubus takhtadjanii* Mulk., *Rubus zangezurus* Mulk., *Scorzonera safievii* Grossh., *Smyrniopsis armena* Schischk., *Tragopogon segetus* Kuth.
Steppes	*Alcea grossheimii* Iljin, *Alchemilla smirnovii* Juz., *Alchemilla venosa* Juz., *Allium schchianae* Ogan., *Bromopsis zangezura* Ogan., *Centaurea fajvuschii* Gabrieljan, *Centaurea takhtajanii* Gabrieljan & Tonjan, *Centaurea vavilovii* Takht. & Gabrieljan, *Cephalaria armeniaca* Bordz., *Colchicum gohariae* Gabrieljan, *Cousinia daralaghezica* Takht., *Cousinia fedorovii* Takht., *Cousinia takhtajanii* Tamanian, *Euphorbia vedica* Ter-Chatsch., *Isatis sevangensis* N. Busch, *Linaria zangezura* Grossh., *Merendera greuteri* Gabrieljan, *Merendera mirzoevae* Gabrieljan, *Minuartia armena* Schischk. ex Ikonn., *Minuartia daralagexica* Schischk. ex Ikonn., *Myosotis daralaghezica* T. N. Popova, *Onobrychis takhtajanii* Sytin, *Orobanche Orobanche connata* K. Koch, *Orobanche*

HABITAT	SPECIES
Steppes (continued)	*schelkovnikovii* Tzvelev, *Papaver roseolum* M. V. Agab. & Fragman, *Polygala urartu* Tamamsch., *Psephellus cronquistii* (Takht. & Gabrieljan) Gabrieljan, *Psephellus geghamensis* (Gabrieljan) Gabrieljan, *Psephellus manakyanii* (Gabrieljan) Gabrieljan, *Psephellus pambakensis* Sosn., *Pseudolysimachion spicatum* (L.) Opiz subsp. *transcaucasicum* (Bordz.) Gabrieljan, *Pyrus elata* Rubtzov, *Rhaponticoides hajastana* (Tzvelev) M. V. Agab. & Greuter, *Rhaponticoides tamanianiae* (M. V. Agab.) M. V. Agab. & Greuter, *Scorzonera aragatzi* Kuth., *Scorzonera safievii* Grossh., *Smyrniopsis armena* Schischk., *Stipa canescens* P. A. Smirn. ex Roshev., *Stipa gegarkunii* P. A. Smirn., *Stipa pennata* L. subsp. *lejophylla* (P. A. Smirn.) Tzvelev, *Tragopogon armeniacus* Kuth., *Tragopogon segetus* Kuth., *Verbascum horticolum* Hub.-Mor., *Verbascum patris* (Bordz.) Bordz.
Forests	*Cephalaria armeniaca* Bordz., *Colchicum gohariae* Gabrieljan, *Cotoneaster armenus* Pojark., *Linaria zangezura* Grossh., *Merendera mirzoevae* Gabrieljan, *Myosotis daralaghezica* T. N. Popova, *Polygala urartu* Tamamsch., *Psephellus debedicus* (Gabrieljan) Gabrieljan, *Psephellus pambakensis* Sosn., *Psephellus zangezuri* Sosn., *Pseudolysimachion spicatum* (L.) Opiz subsp. *transcaucasicum* (Bordz.) Gabrieljan, *Pyrus complexa* Rubtzov, *Pyrus daralaghezi* Mulk., *Pyrus elata* Rubtzov, *Pyrus hajastana* Mulk., *Pyrus megrica* Gladkova, *Pyrus sosnovskyi* Fed., *Pyrus tamamschianiae* Fed., *Pyrus voronovii* Rubtzov, *Ribes armenum* Pojark., *Rosa sosnovskyana* Tamamsch., *Rosa zangezura* Jarosch., *Rubus takhtadjanii* Mulk., *Rubus zangezurus* Mulk.
Meadows	*Alchemilla heteroschista* Juz., *Alchemilla sevangensis* Juz., *Alchemilla smirnovii* Juz., *Bromopsis zangezura* Ogan., *Cephalaria armeniaca* Bordz., *Colchicum gohariae* Gabrieljan, *Colchicum ninae* Sosn., *Erodium sosnowskianum* Fed., *Gladiolus hajastanicus* Gabrieljan, *Gladiolus kotschyanus* Boiss. subsp. *distichus* Gabrieljan, *Grossheimia caroli-henrici* (Gabrieljan & Dittrich) Gabrieljan, *Merendera mirzoevae* Gabrieljan, *Myosotis daralaghezica* T. N. Popova, *Nepeta alaghezi* Pojark., *Ornithogalum gabrielianiae* Agapova, *Orobanche schelkovnikovii* Tzvelev, *Papaver gabrielianae* M. V. Agab., *Papaver roseolum* M. V. Agab. & Fragman, *Poa greuteri* Gabrieljan, *Psephellus zuvandicus* Sosn. subsp. *gegharkunikensis* (Gabrieljan) Gabrieljan, *Psephellus zuvandicus* Sosn. subsp. *jelenewskyi* (Gabrieljan) Gabrieljan, *Scorzonera aragatzi* Kuth., *Symphytum hajastanum* Gvin., *Trisetum geghamense* Gabrieljan, *Verbascum sevanense* Hub.-Mor.
Wetlands (incl. salt marshes)	*Agrostis trichantha* (Schischk.) Tzvelev, *Linum barsegjanii* Gabrieljan & Dittrich, *Sonchus araraticus* Nazarova & Barsegyan, *Sonchus sosnowskyi* Schchian
Petrophilous	*Acantholimon gabrieljaniae* Mirzoeva, *Alchemilla smirnovii* Juz., *Allium struzlianum* Ogan., *Allium vasilevskajae* Ogan., *Astragalus agassii* Manden., *Astragalus bylowae* Elenevsky, *Astragalus coelestis* Boiss., *Astragalus sangesuricus* Boriss., *Bupleurum kosopoljanskyi* Grossh., *Campanula bayerniana* Rupr. subsp. *choziatovskii* (Fomin) Ogan., *Hypericum eleonorae*

HABITAT	SPECIES
Petrophilous (continued)	Elenevsky, *Isatis bungeana* Seidl., *Onosma gehardica* T. N. Popova, *Oxytropis armeniaca* Sosn. ex Mulk., *Papaver sjunicicum* M. V. Agab., *Potentilla seidlitziana* Bien., *Ribes achurjanii* Mulk., *Rubus takhtadjanii* Mulk., *Rubus zangezurus* Mulk., *Sameraria odontophora* Bordz., *Scrophularia olgae* Grossh., *Scrophularia takhtajanii* Gabrieljan, *Seseli leptocladum* Woronow, *Silene chustupica* Nersesian, *Thlaspi zangezuricum* Tzvelev, *Verbascum drymophiloides* P. P. Gritz.

MAIN THREATS

There are currently two major threats to plant species in Armenia.

1. Habitat loss has resulted from human activity (shifting agriculture, mining industry development, forest cutting, infrastructure development, overgrazing).

2. Natural disasters, such as drought and climate change, also impact plant distributions.

Botanists consider that some plant species are threatened because of their utility. Because these species are widespread beyond the borders of Armenia, they have been excluded from the List but are threatened for a variety of reasons, especially overexploitation for fuel or medicinal purposes.

LISTED IN THIS RED LIST 244

Extinct (EX) 0
Critically Endangered (CR) 29
Endangered (EN) 90
Vulnerable (VU) 43
Near Threatened (NT) 25
Least Concern (LC) 20
Data Deficient (DD) 37

CONSERVATION LEGISLATION

The Government of Armenia, recognizing the importance of biological diversity of the country, has established various policies and legislation, which include the following: Republic of Armenia Law "On plant world" (November 23, 1999); RA Law "On especially protected nature areas" (December 16, 2006); RA Law "On Lake Sevan" (June 14, 2001); "RA Land code" (May 04, 2001); "RA Water code" (July 01, 2002); "RA Forest code" (October 24, 2005); RA Governmental decision "National Forest program" (2005).

There are 34 protected areas of different categories and different jurisdictions in Armenia. The Ministry of Nature Protection manages the Sevan, Arpi-lich, Arevik, and Dilizhan National Parks; Khosrovsky Les (Forest), Shikakhokh, and Erebuni State Reserves; and some state sanctuaries (Khor Virap, Gorovanskie Peski [Gorovan Sands], Gilan, Platanovaja Roscha [Plane Grove], Tissovaja Roscha [Taxus Grove], Sev Lich, Vordan Karmir, Zangezur, etc.) where many rare and endemic plant species are growing. In addition, some protected areas belong to the Ministry of Agriculture, among them territories with special plants such as *Rhododendron caucasicum* Pall. (Ericaceae), *Corylus colurna* L. (Betulaceae), and other forest areas. The alpine meadow on the top of Mount Aragats belongs to the Institute of Physics, which is situated there.

TOP 50 NATIONAL CONSERVATION PRIORITIES FOR ARMENIA

APIACEAE

1. *Smyrniopsis armena* Schischk. **EN**

There are several populations only in the Darelegis Floristic Region. This species is known to grow usually in the vicinity of historical monuments, and the main threat would be recreational trampling.

ASTERACEAE

2. *Centaurea alexandri* Bordz. **EN**

There is only one population in the Meghri Floristic Region. The main threat would be loss and degradation of habitats caused by mining industry development.

3. *Centaurea arpensis* (Czerep.) Wagenitz **EN**

There are several populations in two floristic regions (Erevan and Darelegis). The main threat for this species would be land development for agricultural purposes.

4. *Centaurea takhtajanii* Gabrieljan & Tonjan **CR**

There is only one population noted for the Shirak Floristic Region. It grows in an area where agricultural activity is very high and land development would be the main threat for this species.

5. *Centaurea vavilovii* Takht. & Gabrieljan **CR**

There is only one population of this species in the Erevan Floristic Region. The main threats would be loss and degradation of habitats caused by the expansion of arable lands and by over-grazing.

6. *Cousinia fedorovii* Takht. **EN**

The entire population of this species is situated in the basin of Lake Sevan. The main threat would be loss and degradation of habitats caused by the expansion of arable lands.

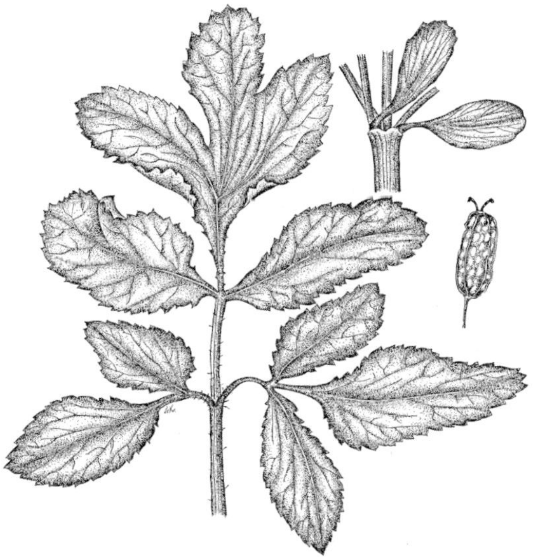

Smyrniopsis armena Schischk., endemic to the southern Caucasus.

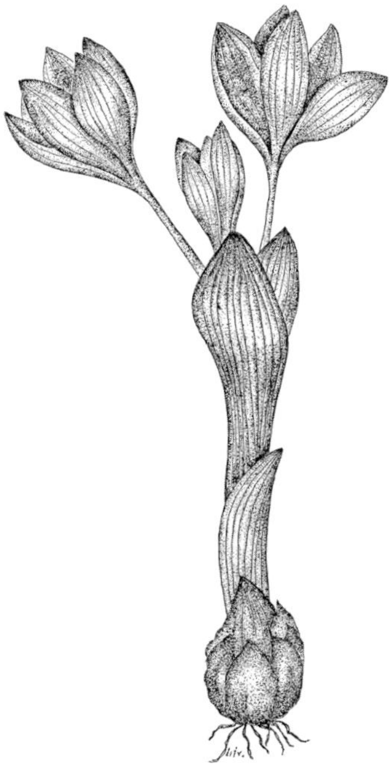

Colchicum ninae Sosn., endemic to Armenia.

7. *Grossheimia caroli-henrici* (Gabrieljan & Dittrich) Gabrieljan **CR**
This is a local narrow endemic and only one population is known in the Darelegis Floristic Region. It grows in the neighborhood of the Dzhermuk resort, and the main threats would be recreational trampling and the resort's infrastructure development.

8. *Rhaponticoides hajastana* (Tzvelev) M. V. Agab. & Greuter (*Centaurea hajastana* Tzvelev) **EN**
There are several populations only in the Shirak Floristic Region. This grows in mountain steppes, and the main threat would be habitat loss from agricultural activity.

9. *Rhaponticoides tamanianiae* (M. V. Agab.) M. V. Agab. & Greuter (*Centaurea tamanianiae* M. V. Agab.) **CR**
There are only two known populations in the two floristic regions of Shirak and Darelegis. The main threats would be loss and degradation of habitats caused by the expansion of arable lands and by overgrazing.

10. *Scorzonera safievii* Grossh. **EN**
This species is endemic to southern Transcaucasia. Four subpopulations are known in Armenia and Nakhichevan. The main threats would be loss and degradation of habitats caused by the expansion of arable lands and by overgrazing.

11. *Sonchus araraticus* Nazarova & Barsegyan **CR**
This species is a local narrow endemic of Armenia, with only one population known from the Erevan Floristic Region. It grows in salt marshes of the Ararat valley, and the main threats would be loss and degradation of habitats caused by land development for agricultural purposes, by the drainage of swamps and marshes, and by grazing and vegetation burning by local people.

12. *Tragopogon armeniacus* Kuth. **EN**
There is only one very fragmented population known in the Shirak Floristic Region. The main threat would be loss and degradation of habitat caused by the expansion of arable lands.

BRASSICACEAE

13. *Isatis sevangensis* N. Busch **EN**
This species is known only from the northeastern shore of Lake Sevan. The main threat would be loss and degradation of habitat caused by the expansion of arable lands.

14. *Rorippa spasskajae* V. I. Dorof. **CR**
This is known only from the northern portion of Zangezur Floristic Region. It grows in the subalpine belt, at altitudes ca. 2000 m above sea level, along riversides. Its main threat would be degradation of the subalpine habitats caused by global climate change.

15. *Sameraria odontophora* Bordz. **EN**
Only one fragmented population is known from the Darelegis Floristic Region. The main threat would be loss and degradation of habitat caused by the expansion of arable lands.

16. *Thlaspi zangezuricum* Tzvelev **CR**
Only one fragmented population is known from the Zangezur and Meghri floristic regions where it grows in alpine and subalpine belts, at altitudes from 2000 to 3100 m above sea level. The plants grow in soil accumulating between rock fissures. The main threat would be degradation of habitat caused by global climate change.

CARYOPHYLLACEAE

17. *Allochrusa takhtajanii* Gabrieljan & Dittrich **CR**
Only one population in the Erevan Floristic Region is known. The main threat would be loss and degradation of habitat caused by land development for agricultural purposes.

18. *Bufonia takhtajanii* Nersesian **CR**
Only one population in the Erevan Floristic Region is known. The main

threat would be loss and degradation of habitat caused by cattle trampling and road construction.

19. *Dianthus grossheimii* Schischk. EN

Two subpopulations in the Sevan and Darelegis floristic regions are known. The species grows at altitudes from 1800 to 2200 m above sea level, in mountain steppes, meadow-steppes, and meadows. The main threat would be loss and degradation of habitats caused by the expansion of arable lands and by grazing, haymaking, and recreational trampling.

20. *Silene chustupica* Nersesian CR

Only one small population is known on Khustup Mountain in Zangezur Floristic Region. The species grows in the alpine belt, at altitudes from 3000 to 3200 m above sea level, on scree. Its main threat would be degradation of habitats caused by global climate change.

COLCHICACEAE

21. *Colchicum gohariae* Gabrieljan EN

It is known only from the Idjevan Floristic Region. It grows in forests and meadows, and the main threats would be deforestation and land development for agricultural purposes.

22. *Colchicum ninae* Sosn. EN

The population is very fragmented and is known from altitudes of 2000–2100 m above sea level, in wetlands, and the main threat would be degradation of habitats caused by global climate change.

23. *Merendera greuteri* Gabrieljan CR

Only one population is known from the Shirak Floristic Region where it grows at altitudes from 1500 to 1900 m above sea level, on dry stony slopes, in tragacanth communities and mountain steppes. The main threat would be loss and degradation of habitats caused by land development for agricultural purposes and by overgrazing.

FABACEAE

24. *Astragalus agassii* Manden. EN

Only one fragmented population is known from the Zangezur Floristic Region. The species grows at altitudes from 2900 to 3000 m above sea level, on stony and rocky slopes in alpine meadows, and the main threat would be overgrazing.

25. *Astragalus bylowae* Elenevsky CR

Only one small population is known in the Zangezur Floristic Region where it grows in steppes. The main threat would be land development for agricultural purposes and overgrazing.

26. *Astragalus sangesuricus* Boriss. EN

Only one population is known on Zangezur Mountain Range where it grows at altitudes from 2100 to 3500 m above sea level, on dry stony slopes and screes, in meadows, and in oak forests. The main threat would be the degradation of habitat caused by global climate change.

27. *Onobrychis takhtajanii* Sytin EN

Only one fragmented population is known from the Erevan and Darelegis floristic regions. The species grows at altitudes from 1600 to 2200 m above sea level, in steppes, meadow-steppes, and on stony slopes. The main threat would be land development for agricultural purposes.

GERANIACEAE

28. *Erodium sosnowskianum* Fed. CR

Only two populations are known from the alpine belt of Aragats and Gegham mountains. The main threat would be degradation of habitat caused by global climate change.

GROSSULARIACEAE

29. *Ribes achurjanii* Mulk. EN

Only one fragmented population is known from the upper Akhuryan and Shirak floristic regions. The species grows at altitudes from 1500 to 2400 m above sea level, in rocky places and in river gorges. The main threat would be degradation of habitat caused by global climate change.

30. ***Ribes armenum*** Pojark. **EN**

Two subpopulations are known and are separated by a distance of 200 km. The species grows at altitudes from 1600 to 2200 m above sea level, in rocky and stony places, at forest edges. The main threat would be degradation of habitat caused by global climate change.

HYACINTHACEAE

31. ***Ornithogalum gabrielianiae*** Agapova **CR**

Only one population is known at altitudes from 2000 to 2800 m above sea level, in meadows and along the upper forest edge on Mount Aragats. The main threat would be the degradation of habitats caused by global climate change and intensive grazing.

HYPERICACEAE

32. ***Hypericum eleonorae*** Elenevsky **CR**

Two subpopulations are known, one in southern Armenia, with the other in northern Armenia, separated by 230 km. The species grows at altitudes from 2100 to 3100 m above sea level, on rocky and stony slopes, and the main threat would be degradation of habitats caused by global climate change.

LAMIACEAE

33. ***Nepeta alaghezi*** Pojark. **EN**

This is an endemic of southern Transcaucasia, with only one very fragmented population known. The expansion of arable land would be the main threat for this species.

LINACEAE

34. ***Linum barsegjanii*** Gabrieljan & Dittrich **CR**

Only one population of this species is known from the Erevan Floristic Region where it grows on salt marshes on the Ararat plain. The main threat would be loss and degradation of habitat caused by land development for agricultural purposes, by the drainage of swamps and marshes, by grazing, and by vegetation burning by local people.

POACEAE

35. ***Bromopsis gabrielianae*** Ogan. **CR**

There is only one small population of this species in the Zangezur Floristic Region where it grows at altitudes from 3300 to 3500 m above sea level, on scree and in rocky places. The main threat would be degradation of habitat caused by global climate change.

POLYGALACEAE

36. ***Polygala urartu*** Tamamsch. **EN**

Only one fragmented population is known. The species grows at altitudes from 1000 to 1700 m above sea level, in steppes, light forests, felled oak forests, and the main threat would be loss and degradation of habitat caused by overgrazing.

ROSACEAE

37. ***Amygdalus nairica*** Fed. & Takht. **EN**

Endemic to the southern Caucasus, this species grows in lower and middle mountain belts, at altitudes from 600 to 1800 m above sea level, on dry stony slopes, on scree, among phryganoid vegetation, in arid open forests. The main threat would be loss and degradation of habitat caused by intensified efforts of the mining industry and by the development of lands for agricultural purposes.

38. ***Crataegus armena*** Pojark. **EN**

Endemic to southern Transcaucasia, the population of this species is very fragmented. It grows in oak forests, along forest edges, and in shibliak, and the main threat would be habitat loss and degradation caused by expansion of arable lands.

39. *Crataegus zangezura* Pojark. EN

Endemic to the southern Caucasus, three subpopulations are known for this species. It grows in oak forests, along forest edges, and in shibliak, as well as the rocky slopes of river gorges. The main threat would be habitat loss and degradation caused by expansion of arable lands.

40. *Potentilla seidlitziana* Bien. CR

Two subpopulations are known in the alpine belt of the Aragats and Gegham mountains. It grows at altitudes from 3500 to 4000 m above sea level, on rocks and screes, and the main threat would be the degradation of habitat caused by global climate change.

41. *Pyrus browiczii* Mulk. CR

Only one population is known in the Darelegis Floristic Region. It grows at altitudes from 1700 to 2100 m above sea level, in broad-leaved forests, at the edges of fields and forests, and in meadows. The main threat would be habitat loss and degradation caused by expansion of arable lands.

42. *Pyrus daralaghezi* Mulk. EN

The population is very fragmented, growing at altitudes from 1700 to 2200 m above sea level. It is found in broad-leaved forests, at the edges of fields and forests, and in meadows. The main threat would be habitat loss and degradation caused by expansion of arable lands.

43. *Pyrus elata* Rubtzov EN

The population is very fragmented, growing at altitudes from 1400 to 2300 m above sea level, in broad-leaved forests. The main threat would be habitat loss and degradation caused by expansion of arable lands and deforestation.

44. *Pyrus gergerana* Gladkova CR

The population is very fragmented, and the species grows at altitudes from 1600 to 2200 m above sea level, in broad-leaved forests. The main threat would be habitat loss and degradation caused by expansion of arable lands and deforestation.

45. *Pyrus hajastana* Mulk. EN

The population is very fragmented, and the species grows at altitudes from 1700 to 2200 m above sea level, in broad-leaved forests. The main threat would be habitat loss and degradation caused by expansion of arable lands and deforestation.

46. *Pyrus sosnovskyi* Fed. EN

The population is very fragmented, and the species grows at altitudes from 1700 to 2200 m above sea level, in broad-leaved forests. The main threat would be habitat loss and degradation caused by expansion of arable lands and deforestation.

47. *Pyrus tamamschianiae* Fed. EN

The population is very fragmented. The species grows at altitudes from 1300 to 2200 m above sea level, in broad-leaved forests, and at the edges of fields and forests. The main threat would be habitat loss and degradation caused by expansion of arable lands.

48. *Pyrus voronovii* Rubtzov CR

Two subpopulations in the Zangezur and Meghri floristic regions are known. The species grows at altitudes from 1400 to 2000 m above sea level, at the edges of broad-leaved forests. The main threat would be habitat loss and degradation caused by expansion of arable lands.

49. *Rubus takhtadjanii* Mulk. CR

Only one population from the Zangezur Floristic Region is recorded. The species grows at altitudes from 800 to 1400 m above sea level, on the slopes of mountain river gorges, at forest edges, on stony slopes and rocks, and along roadsides. The main threat would be the restricted extent of occurrence and the reduced area of occupancy.

SCROPHULARIACEAE

50. *Scrophularia takhtajanii* Gabrieljan **CR**

Only one small population from the Meghri Floristic Region is known. The species grows at altitudes from 2300 to 2700 m, in the cracks and fissures between rocks. The main threat would be the degradation of habitat caused by global climate change.

ARMENIA MAPS

1. *Smyrniopsis armena* Schischk.

2. *Centaurea alexandri* Bordz.

3. *Centaurea arpensis* (Czerep.) Wagenitz

4. *Centaurea takhtajanii* Gabrieljan & Tonjan

5. *Centaurea vavilovii* Takht. & Gabrieljan

6. *Cousinia fedorovii* Takht.

7. *Grossheimia caroli-henrici* (Gabrieljan & Dittrich) Gabrieljan

8. *Rhaponticoides hajastana* (Tzvelev) M. V. Agab. & Greuter

9. *Rhaponticoides tamanianiae* (M. V. Agab.) M. V. Agab. & Greuter

10. *Scorzonera safievii* Grossh.

11. *Sonchus araraticus* Nazarova & Barsegyan

12. *Tragopogon armeniacus* Kuth.

13. *Isatis sevangensis* N. Busch

14. *Rorippa spasskajae* V. I. Dorof.

15. *Sameraria odontophora* Bordz.

16. *Thlaspi zangezuricum* Tzvelev

17. *Allochrusa takhtajanii* Gabrieljan & Dittrich

18. *Bufonia takhtajanii* Nersesian

19. *Dianthus grossheimii* Schischk.

20. *Silene chustupica* Nersesian

21. *Colchicum gohariae* Gabrieljan

22. *Colchicum ninae* Sosn.

23. *Merendera greuteri* Gabrieljan

24. *Astragalus agassii* Manden.

25. *Astragalus bylowae* Elenevsky

26. *Astragalus sangesuricus* Boriss.

27. *Onobrychis takhtajanii* Sytin

28. *Erodium sosnowskianum* Fed.

29. *Ribes achurjanii* Mulk.

30. *Ribes armenum* Pojark.

31. *Ornithogalum gabrielianiae* Agapova

32. *Hypericum eleonorae* Elenevsky

33. *Nepeta alaghezi* Pojark.

34. *Linum barsegjanii* Gabrieljan & Dittrich

35. *Bromopsis gabrielianae* Ogan.

36. *Polygala urartu* Tamamsch.

37. *Amygdalus nairica* Fed. & Takht.

38. *Crataegus armena* Pojark.

39. *Crataegus zangezura* Pojark.

40. *Potentilla seidlitziana* Bien.

41. *Pyrus browiczii* Mulk.

42. *Pyrus daralaghezi* Mulk.

43. *Pyrus elata* Rubtzov

44. *Pyrus gergerana* Gladkova

45. *Pyrus hajastana* Mulk.

46. *Pyrus sosnovskyi* Fed.

47. *Pyrus tamamschianiae* Fed.

48. *Pyrus voronovii* Rubtzov

49. *Rubus takhtadjanii* Mulk.

50. *Scrophularia takhtajanii* Gabrieljan

BIBLIOGRAPHY

Agaronyan, Z. A., K. G. Tamanian & G. M. Fayvush. 1990. Dilizanskij gosudarstvennyi Zapovednik. [Dilidjan State Reserve.] Pp. 340–347 *in* V. E. Sokolov (editor), Zapovedniki Kavkaza. [Reserves of the Caucasus.] Mysl', Moscow [in Russian].

Bagdasaryan, A. B. 1958. Klimat Armjanskoi SSR. [Climate of the Armenian SSR.] National Academy of Sciences of the Republic of Armenia, Yerevan [in Russian].

Fayvush, G. M. 2007. Endemichnye rastenia flory Armenii. [Endemic plants of Armenian flora.] Pp. 185–207 *in* Flora, Rastitel'nost' i Rastitel'nye Resursy Armenii, Tom. 16. [Flora, Vegetation and Plant Resources of Armenia, Vol. 16.] Academy of Sciences of the Republic of Armenia, Yerevan [in Russian].

Gabrielyan, E. T. & Z. A. Agaronyan. 1990. Erebuniysky gosudarstvennyi zapovednik. [Erebuni State Reserve.] Pp. 348–350 *in* V. E. Sokolov (editor), Zapovedniki Kavkaza. [Reserves of the Caucasus.] Mysl', Moscow [in Russian].

Gabrielyan, E. T. & G. M. Fayvush. 1989. Floristic links and endemism in the Armenian Highlands. Pp. 191–206 *in* K. Tan (editor), The Davis and Hedge Festschrift. Edinburgh University Press, Edinburgh.

Gabrielyan, E. T. & G. M. Fayvush. 1990. Shikahohskyi gosudarstvennyi zapovednik. [Shikahoh State Reserve.] Pp. 351–357 *in* V. E. Sokolov (editor), Zapovedniki Kavkaza. [Reserves of the Caucasus.] Mysl', Moscow [in Russian].

Gabrielyan, E. T., V. O. Gelikman & A. K. Unanjan. 1990. Khosrovskyi gosudarstvennyi zapovednik. [Khosrov State Reserve.] Pp. 323–329 *in* V. E. Sokolov (editor), Zapovedniki Kavkaza. [Reserves of the Caucasus.] Mysl', Moscow [in Russian].

Gevorgyan, A. & A. Ahasyan. 2008. Modern Situation with Specially Protected Areas in Armenia. Transboundary Joint Secretariat, Yerevan.

Magakyan, A. K. 1941. Rastitel'nost' Armjanskoi SSR. [Vegetation of Armenian SSR.] USSR Academy of Sciences, Moscow–Leningrad [in Russian].

Takhtajan, A. L. 1941. Botaniko-Geographicheskii ocherk Armenii. [Phyto-geographical survey of Armenia.] Trudy Bot. Inst. Akad. Nauk. Armyansk. S.S.R., Vol. 2 [in Russian].

Takhtajan, A. L. (editor). 1954–2011. Flora Armenii, Tt. 1–11. [Flora of Armenia, Vols. 1–11.] Nauka, Moscow–Leningrad [in Russian, distributed by Koeltz Scientific Books, Germany].

Takhtajan, A. L. 1978. Floristicheskie Oblasti Zemli. [Floristic Regions of the World.] Nauka, Moscow–Leningrad [in Russian].

Takhtajan, A. L. 1986. Floristic Regions of the World. University of California Press, Los Angeles.

Tamanian, K. G. & G. M. Fayvush. 2010. K voprosu o floristicheskom raionirovanii Armenii. [On the problem of floristic regions of Armenia.] Pp. 73–78 *in* Flora, rastitel'nost' i rastitel'nye resursy Armenii, Tom. 17. [Flora, Vegetation and Plant Resources of Armenia, Vol. 17.] Academy of Sciences of the Republic of Armenia, Yerevan [in Russian].

Tamanian, K. G., G. M. Fayvush, S. G. Nanagulyan & T. S. Danielyan (editors). 2010. Krasnaja Kniga Armenii. [The Plant Red Data Book of Armenia.] Academy of Sciences of the Republic of Armenia, Yerevan [in Russian].

Vardanyan, Z. H. 2008. Razvitie botanicheskoi nauki v Armenii. [Development of the botanical science in Armenia.] Pp. 10–16 *in* Aktual'nye problemy botaniki v Armenii. Mat. Int. Conf., Akad Nauk, Yerevan [in Russian].

Vavilov, N. I. 1926. Zentry proisxozhdenia kul'turnyx rastenii. [Centers of origin of cultivated plants.] Trudy Prikl. Bot., Ser. A., Socialist. Rasteniev, Vol. 16 [in Russian].

Vavilov, N. I. 1987. Proiskhozhdenie I geografiía kul'turnykh rastenii. [Origin and Geography of Cultivated Plants.] Nauka, Leningrad. [English translation by D. Löve, 1992. Cambridge University Press, Cambridge.]

Vavilov, N. I. 2009. IV Nazional'nyi doklad po bioraznoobraziu. [Fourth National Report on Biodiversity] Ministry of Nature Protection of the Republic of Armenia, Yerevan [in Russian].

AZERBAIJAN

Azərbaycan Respublikası
V. C. Hacıyev, S. H. Musayev, V. M. Əli-zadə
V. Kərimov, R. Abdıyeva, V. S. Fərzəliyev

Dünya florasının qaynar nöqtələrindən biri olan Qafqazın bitkilər aləmi regionun coğrafi və geoloji tarixi xüsusiyyətlərinə görə olduqca zəngindir. Qafqaz florasında 6300-ə qədər ali bitki növü vardır ki, bunlardan 1600-ə qədəri bu region üçün endem hesab olunur.

Respublikamızda bitki örtüyünə dair elmi məlumatlar XVII əsrdən meydana gəlməyə başlamış, sonralar daha sistemli öyrənilməyə başlanmışdır. Azərbaycanda bitki örtüyünün öyrənilməsində A.A.Qrossheym, L.İ.Prilipko, İ.S.Səfərov, V.C.Hacıyev və b. alimlərin xüsusi rolu olmuşdur. Hal-hazırda Azərbaycanda botanika elmi üzrə aparıcı tədqiqat mərkəzi rolunu zəngin herbari kolleksiyasına (600 000 nüsxə) malik olan Azərbaycan Milli Elmlər Akademiyasının Botanika İnstitutu həyata keçirir. Respublika florasının planlı və daha əsaslı surətdə tədqiqinə məhz Azərbaycanda Botanika İnstitutu təşkil olunduqdan sonra başlanmışdır. Bununla yanaşı Azərbaycanda bitki örtüyünün öyrənilməsində bir sıra institut və təşkilatların da (Mərkəzi Nəbatat Bağı, Mərdəkan Dendrarisi, Naxcıvan Bioresurslar İnstitutu, Genetik Ehtiyatlar İnstitutu, Bakı, Gəncə və Naxçıvan Dövlət Universitetlərinin Botanika kafedraları və Azərbaycan Dövlət Aqrar Universiteti) böyük rolu olmuşdur. Respublikamızda bu sahədə mühüm işlər görülməsinə baxmayaraq bir çox nadir bitki növlərinin təbii bitmə yeri, biologiyası, ekologiyası, müxtəlif antropogen təsirlərə davamlılığı və s. hələ də öyrənilməmişdir.

140 nadir və nəsli kəsilməkdə olan bitki növünü əhatə edən "Azərbaycanın Qırmızı Kitabı" hələ 1989-cu ildə çap olunmuşdur. Respublika alimləri tərəfindən nadir, nəsli kəsilməkdə olan və endem bitki növlərinin tədqiqi üzrə son illər monitorinqlər keçirilmiş, onların beynəlxalq səviyyədə müasir qorunma statusu müəyyənləşdirilmişdir. Hal-hazirda akademik C.Əliyev rəhbərliyi altında bitki və heyvan müxtəlifliyinin qorunması üçün mühüm olan "Qırmızı Kitab"ın II nəşri hazirlanir.

2006-2009-cu illər ərzində Qafqaz regionu ölkələrinin botanikləri Missuri Botanika Bağından (ABŞ) olan həmkarları ilə birlikdə Qafqazda

Bitkilərin Qırmızı Siyahısının Qiymətləndirilməsi Koordinasiyası və İnkişafı üzrə layihəni həyata keçirmişlər. Aparılmış iş çox böyük əhəmiyyət daşımışdır. "Qafqazın Endem Bitkilərinin Qırmızı Kitabı"nın nəşri ƏMMBİ -nin (İUCN) kateqotiya və kriteriyalarına uyğun olaraq müasir məlumatlarla, milli və Qafqaz endemlərinin qiymətləndirilməsi əsasında reallaşdırılmışdır.

Təqdim olunan kitabın tərtibində əsas məqsəd Qafqaz regionuna aid endem bitki növlərinin məhv olma təhlükəsi dərəcəsini qiymətləndirmək məqsədilə onların müasir vəziyyətinin müəyyənləşdirməkdən ibarət olmuşdur. Kitabda Qafqazda yayılmış bəzi endem bitki növləri haqqında ətraflı məlumatlar verilmiş, həmçinin təhlükə dərəcəsinin qiymətləndirilməsi Ətraf Mühitin Mühafizəsi üzrə Beynəlxalq İttifaq (ƏMMBİ) tərəfindən qəbul edilmiş kateqoriya və kriteriyalar sistemi (nəsli kəsilmişlər (EX), yabanı təbiətdə nəsli kəsilmişlər (EW), kritik təhlükə həddində olanlar (CR), nəsli kəsilmək təhlükəsində olanlar (EN), nəsli kəsilməyə həssas olanlar (VU), təhlükəli həddə yaxın olanlar (NT), daha az təhlükəli vəziyyətdə olanlar (LC), az öyrənilənlər (DD), qiymətləndirilməmişlər (NE) məlumatların kifayət etməməsi, az öyrənilənlər (DD) əsasında həyata keçirilmişdir.

Azərbaycan Respublikası Qafqazın cənub-şərqində və İran yaylasının şimal-qərbində Avropa və Asiya qitələrinin təması zonasında yerləşən, 86.6 min km^2 ərazini əhatə edən qədim bir ölkədir. Azərbaycan ən qədim insan məskənlərindən biri olmaqla bəşəriyyətin tərəqqisi və mədəniyyət dialektikasına öz töhvəsini vermişdir. Doğma diyarımızın təbiəti öz zənginliyi, rəngarəngliyi, təravəti ilə bütün dünyada tanınmışdır. Təbii-coğrafi baxımdan Respublikanın əlverişli mövqeyə malik olmasını hər şeydən əvvəl, məhsuldar torpaqları olan böyük düzənliklərin, geniş dağətəyi və zəngin təbii şəraiti və ehtiyatları ilə seçilən dağ sistemlərinin, məhsuldar Xəzər dənizinin mövcudluğu və d. müəyyən edir. Respublikanın əhalisinin sayı 9 milyondan çoxdur ki, bunun da 51,5% şəhər və 48,5% kənd əhalisidir. Ermənistanın hərbi təcavüzü Azərbaycan ərazisinin 20%-nin işğalı ilə nəticələnmişdir və bir milyondan çox qaçqın və məcburi köçkünlər öz torpaqlarını və evlərini itirmiş və yaşama vəsaitlərindən məhrum olmuşlar. İşğal olunmuş ərazilərdə yayılmış qiymətli bitki növləri məhv edilmiş, meşələrin əksəriyyəti kəsilmiş və talan edilmişdir. Burada yerləşən meşələrimizə, xüsusilə də dağ meşələrimizə ciddi ziyan vurulmuşdur. Hazırda

261 min hektar meşə fondu sahəsi Ermənistan tərəfindən işğal edilmişdir.

Azərbaycan Qafqazın ən zəngin botaniki-coğrafi regionlarından biridir. Azərbaycan meşələrinin ümumi sahəsi 1213,7 min hektardır. Bundan meşə ilə örtülü sahə 989,5 min hektar təşkil etməklə, ümumi ərazinin 11,4% qədəridir. Tariximizin səhifələrini vərəqlədikdə görürük ki, hələ yaxın keçmişdə respublikamızın bitki və heyvanat aləmi olduqca zəngin olmuşdur. Bir sıra bitki və heyvan növlərinin nəsli kəsilərək adları flora və faunamızdan silinmişdir. Bir çox qiymətli ağac cinslərindən ibarət olan meşələrin sahəsi xeyli azalmışdır. Respublikamızın flora və faunasının mühafizə olunub saxlanması məqsədilə 8 Milli park, 11 dövlət təbiət qoruğu, 24 dövlət təbiət yasaqlığı yaradılmışdır. Mühafizəsinə ciddi ehtiyac olan bitki növlərinin yalnız 50%-ə qədəri Xüsusi Mühafizə olunan Təbiət Ərazilərində mühafizə olunur. Həmin bitkiləri qoruyub saxlamaq və artırmaq üçün ictimaiyyəti təbiətin mühafizəsinə geniş cəlb etmək və həmin bitkiləri insanlara tanıtmaq mühüm məsələdir. Təbiətin və onun bitki aləminin mühafizəsi yer üzündə canlı aləmin, həyatın, insan varlığının qorunması deməkdir.

Azərbaycanın təbii landşaftlarının və bitki örtüyünün inkişafı nəticəsində bir neçə əsas bitki qrupları təşəkkül tapmışdır. Bunlardan ən mühümləri düzənlik və dağətəyi zonaların yarımsəhraları və dağlıq meşə-çəmən bitkiləri sistemidir. Relyef-iqlim vəhdətinin yaratdığı şərait nəticəsində ölkə ərazisində yarımsəhra, bozqır, dağ-kserofit, friqana, şiblyək, qariqa, çala-çəmən, su-bataqlıq, düzənlik və dağ-meşə bitki tipləri, tuqay meşəlikləri, iynəyarpaqlı və enliyarpaqlı meşəliklər, hündür otluqlar, subalp, alp çəmənlikləri və xalıları formalaşmışdır. Azərbaycan florasının tərkibində bütün areal tiplərinə (qədim meşə, boreal, bozqır, kserofit, səhra, Qafqaz) və adventiv (gəlmə) bitki tipləri nümunələrinə rast gəlinir.

Azərbaycan florasının növ müxtəlifliyi və yüksək dərəcəli endemizmi hər şeydən əvvəl respublika ərazisində bitkilərin ekoloji şəraitinin çox müxtəlif olması və müxtəlif dövrlərdə landşaftın biotik komponentlərinin inkişafı tarixi ilə əlaqədardır. Ağac və kol bitkilərinin ən böyük endemizmi Naxçıvan dağlıq sahəsində, Kiçik Qafqazın cənub hissəsində və Lənkəran dağlarında müşahidə edilir. Azərbaycanın florası tərkibində mövcud olan relikt bitkilər daha çox Talış vilayətində, Qismən Orta Kür çökəkliyində,

Böyük Qafqazın cənub ətəklərində yayılmışdır. Talış dağlarında bitən üçüncü dövrün relikt və nadir ağaclarından dəmirağac (*Parrotia persica*), Lənkəran akasiyası (*Albizia julibrisssin*), şabalıdyarpaq palıd (*Quercus castaneifolia*), azat (*Zelkowa carpinifolia*), Hirkan ağcaqayını (*Acer hyrcanum*) və s. təbiətin nadir incilərindəndir.

Azərbaycanın 45%-dən çox ərazisini dağlar və yaylalar, 55%-ni isə dağətəyi və düzənliklər tutur. Yüksək dağlarda sabit iqlim zonaları var və bitki aləmi torpağa uyğun olaraq və dağ ətəyi istiqamətində dəyişir. Həmçinin dəniz səviyyəsindən 26 metr aşağıda yerləşən geniş çökəkliklər, düzənlik ərazilər və Xəzər dənizinin sahil zonaları vardir.

İnsanların əməli fəaliyyəti nəticəsində son 60-70 il ərzində (*Quercus macranthera, Acer trautvetterii, Betula pendula, B.litvinovii, Paeonia mlokosewitschii, Laurocerasus officinalis* və s.) bir çox bitkilərin arealları azalmış, bəzilərinin isə çox az sahədə rast gəlməsi və ya tamamilə məhv olması müşahidə edilmişdir. Bozdağ, Dibrardağ, Beşbarmaq dağ silsiləsinin buzlaşma dövrünün sonunda bitki senozlarında *Erysimum, Xeranthemum, Stipa, Dianthus* cinslərinə aid olan növlər geniş yayıldıqları halda, hazırda onların çoxu sıradan çıxmışdır. Sənayenin, tikintinin, kənd təsərrüfatının inkişafı, əhalinin istirahəti üçün yeni ərazilərin istifadəsi nəticəsində müxtəlif bitkilərin həyat şəraiti dəyişilir, bəzən də bir çox bitkilər sıradan çıxır.

Hazırda respublikamızın əksər təbii zonalarında meşə örtüyünün antropogen dəyişilməsi və bitki qurşaqlarının əvəz olunması müşahidə olunur. Dağ yamaclarında eroziya prosesi güclənir, dağ ərazilərinin və çaylarının su rejimi kəskin pisləşir. Uzun illərin təcrübəsi bir daha göstərmişdir ki, Azərbaycan florasında yayılmış nadir, relikt, endem, nəsli kəsilmək təhlükəsində olan bitki növlərinin mühafizəsi üçün aidiyyatlı təşkilatlar, bu sahədə çalışan mütəxəssislər əlaqəli fəaliyyət göstərməli və öz əməli təkliflərini verməlidirlər.

Son illərdə bioloji müxtəlifliyin problemləri önə çəkilərək bu sahədə xeyli işlər görülmüşdür. Azərbaycan hökuməti ölkədə bioloji müxtəlifliyinin mühümlüyünü nəzərə alaraq müxtəlif qanunvericiliklər təsis etmişdir.

Azərbaycan Respublikasında təbiətin mühafizəsi bir sıra qanunverici aktlarla tənzim olunur. Ətraf mühitin mühafizəsi üzrə Milli Fəaliyyət Planı, Ekoloji cəhətdən dayanıqlı sosial-iqtisadi inkişafa dair Milli Proqram, Azərbaycan Respublikasında bioloji müxtəlifliyin qorunması və davamlı istifadəsinə dair Milli Strategiya və Fəaliyyət Planı və digər dövlət proqramları qəbul edilmişdir.

Bioloji müxtəlifliyin mühafizəsi və davamlı istifadəsi Azərbaycan da daxil olmaqla, dünya birliyi ölkələrinin birgə fəal səyi nəticəsində mümkündür.

AZERBAIJAN

Valida Ali-zade, Vahid Hajiev†, Vugar Kerimov, Salim Musayev†, Rena Abdiyeva, and Vahid Farzaliyev

CAPITAL
Baku, the largest city and main port.

OTHER CITIES
Ganja, Sumgayit, Nakhchivan, Mingechevir, Ali-Bayramli.

POPULATION
9,590,159 [July 2013 est.], according to CIA World Factbook, with 90.6% Azerbaijanis, 2.2% Lezgians, 1.8% Russians, 1.5% Armenian, 1% Talysh, and others.

LANGUAGE
Mostly Azerbaijani, also with other national languages; about 50% of population know Russian, and 30% speak English.

FLORA
4500 (total species of vascular plants); 200 species (total national endemics); 950 species (Caucasian region endemics).

BOTANICAL INSTITUTIONS
The Institute of Botany of the Azerbaijan National Academy of Sciences has herbarium collections that number ca. 600,000 specimens, and the herbarium (BAK) dates back to 1827. Major collectors that are represented at BAK include A. Grossheim, L. Prilipko, I. Koryagin, R. Rza-zade, and Y. Isayev. The Botanical Institute has two field stations in the regions of Zakatala and Kyurdamir. Address: 40 Batamdar Highway, Baku AZ1073, Azerbaijan; Tel.: (994 12) 5024330, FAX/Tel.: (994 12) 4970994; e-mail: director@botany. az; vm_alizade@yahoo.com; www.botany. az.

The Central Botanical Garden, Mardakyan Dendrary (Arboretum), and the Nakhchivan Botanical Garden (Department of the Institute of Bioresources of the Azerbaijan National Academy of Sciences) support research and conservation in the plant sciences.

PROTECTED AREAS
There are eight national parks, 11 state reserves, and 24 state sanctuaries in the Republic of Azerbaijan. The major state reserves include the Ag-Gel, Eldar Pine, Gey-Gel, Girkan, Gobustan, Shirvan, and Zagatala natural reserves.

In the territory of Azerbaijan, humans have been present at every stage of their historical development. Some of the most ancient wall pictures in the world are found in Azykh Cave, dating back to the Stone Age. The Gobustan National Park is famous for its petroglyphs and rock engravings and has been designated as a UNESCO World Heritage Site. The populations that lived here in prehistoric times were mostly fire-worshippers, and the word *Azer* means "fire." Azerbaijan has been populated, with evidence of intensive land use, as far back as 5000–40,000 years ago and has passed through the basic historical stages throughout the development of civilization. In the third millennium BC, the tribal system was initiated, and at the beginning of the first millennium BC, more political structures began. In the ninth century BC, the Manna Empire appeared, with high levels of economy and culture. In the ancient period, the states of Atropatena and Albania played central roles in strengthening the independent statehood of the territory of Azerbaijan. The initial name of the country was Atropatena, and the country was politically aligned with the Roman Empire, as represented by the diversity of ancient archeological monuments in Azerbaijan. The rich history and culture of the country is directly derived and essentially based on the diversity of its natural resources.

Azerbaijan is situated in the southeastern portion of the Caucasus region. It is bounded in the north by Georgia and Russia (Dagestan), in the east by the Caspian Sea, in the south by Iran, and in the west by Armenia. The Araz River flows along the southern border and separates Azerbaijan from Iran and Turkey. The Autonomous Republic of Nakhchivan, which lies within the borders of Armenia, and Daglig Garabakh are also part of Azerbaijan territory. The country includes the southeastern portion of the Greater Caucasus and part of the Lesser Caucasus, the Talysh Mountains, and the Kyur-Araz lowlands situated between these ranges. Azerbaijan also includes some islands located along the western coast of the Caspian Sea. About 55% of the country is covered by foothills and plains while 45% is occupied by mountains above 1000 m, with 3.5% of these considered to be high mountains above 2500 m. The average elevation in the country is 384 m above sea level; elevation ranges from a maximum of 4466 m (the peak of Mt. Bazar-Duzi) to a minimum of 27 m below sea level along the Caspian Sea (about 19% of the country). The region is seismically active, with frequent earthquakes, and is subject to floods.

CLIMATE

The climate of Azerbaijan is determined by its geographical position, its varied topography, and its proximity to the Caspian Sea. There are several climatic zones, varying by altitude and distance from the Caspian Sea, and these include dry subtropical, wet subtropical, temperate, and montane climates. In general, great differences exist between the western and eastern Transcaucasus. Moving eastward, the climate changes from humid and warm to one that is dry and hot. In some parts of southeastern Azerbaijan, the summers are hot and may last four to five months of the year. The annual average precipitation is ca. 200 mm, which is regarded as semi-desert, e.g., in the Kyur-Araz lowland and on the Absheron Peninsula. However, in other places in southeastern Azerbaijan, the precipitation is much higher and during one month in the autumn can reach 300 mm. A wet subtropical climate is found only in the south Talysh Mountains and is typical there, in the foothills and Lankaran lowlands. A temperate climate dominates the mountain slopes of the Greater and Lesser Caucasus, in alpine and subalpine zones. The average annual temperature can vary from 14°C in the lowlands to 0°C and below in the highlands. The average temperature for July in the lowlands is 25°–27°C and in the highlands is 5°C. The absolute maximum temperature reaches 43°C, and the absolute minimum temperature falls to –30°C, with these extremes found in the Nakhchivan depression and in the highlands, respectively. The prevailing winds are from the north on the Absheron Peninsula, from the northwest/southwest in the Kyur-Araz lowlands, and from the west in the Lankaran depression. Rainfall varies from less than 200 mm per year along the south coast of the Absheron Peninsula to 300–900 mm in the foothills and mountainous zones, 1000–1300 mm on the southern slopes of the Greater Caucasus, and as much as 1200–1400 mm in the Lankaran region.

FLORA AND VEGETATION

Azerbaijan demonstrates an extremely diverse native flora, consisting of about 4500 species of vascular plants that represent 64% of the total species known for the Caucasus region. No less than 10% of the species are regarded as rare or endangered and are subject to protection. Representatives of 125 plant families

Campanula trautvetteri Grossh. ex Fed., endemic to the northern and eastern Caucasus.

Pseudovesicaria digitata (C. A. Mey.) Rupr., generic endemic to the Greater Caucasus region.

Sorbus subfusca (Ledeb.) Boiss., endemic to the Caucasus region.

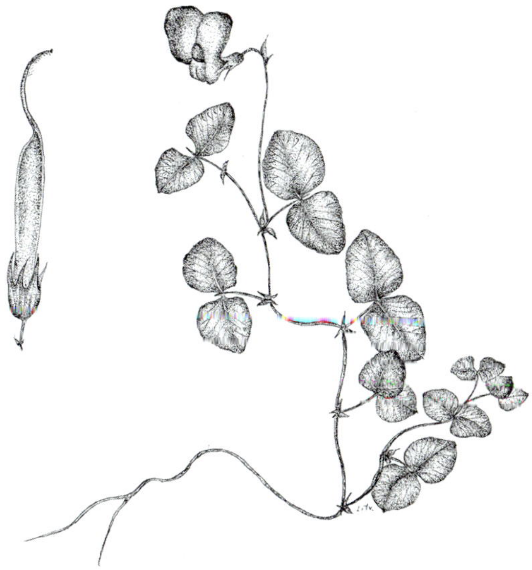

Vavilovia formosa (Steven) Fed., found in the southern Caucasus region.

grow in Azerbaijan, or 80% of the 156 family total for the Caucasus region; 930 genera are represented in Azerbaijan, or 72.3% of the 1286 total genera in the Caucasus. By species count, the major plant families are represented by the Asteraceae (566 species), Fabaceae (443), Poaceae (380), Lamiaceae (337), Brassicaceae (231), Rosaceae (229), Caryophyllaceae (194), Apiaceae (176), and Chenopodiaceae (102) (Grossheim, 1939–1967; Grossheim, 1949; Karyagin, 1950–1961).

The diversity of the Azerbaijan flora and its high endemism are due to the variable physical geography and the past geological complexity, as well as the past natural history of the territory, which was influenced by neighboring and even quite remote floristic regions. The Azerbaijan flora is rich in relictual species, which originated in the Tertiary period. These relictual taxa are found in all floristic provinces, although most are concentrated in the Talysh region (Akhundov, 1972; Gadzhiev et al., 1979). Examples of Tertiary relicts are *Albizia julibrissin* Durazz. (Fabaceae); *Diospyros lotus* L. (Ebenaceae); *Parrotia persica* (DC.) C. A. Mey. (Hamamelidaceae); *Quercus castaneifolia* (DC.) C. A. Mey. (Fagaceae); and also species of *Buxus* L. (Buxaceae); *Danae* Medik. and *Ruscus* L. (both Asparagaceae); *Ilex* L. (Aquifoliaceae); and *Zelkova* Spach. (Ulmaceae). According to the *Flora of Azerbaijan* (Karyagin, 1950–1961), there were ca. 370 endemic species, although these data have been revised. Now the flora of the Republic of Azerbaijan includes 200 national and 950 Caucasian endemics. The flora is also a source of useful and valuable products that includes about 800 medicinal plants, 100 of which are already of commercial importance (Gadzhieva, 1969; Iskenderov, 1969; Gadzhieva & Iskenderov, 1971; Asadov, 1981; Garakhani et al., 1996; Asadov & Asadov, 2001; Kasumov, 2004).

Forests occupy 11.4% of the total territory of Azerbaijan, predominantly in the Greater and Lesser Caucasus and also in the Talysh region. Plains, steppes, and deserts occupy 31.8%. More than half the land area is occupied by mountains and highlands (Gadzhiev & Aliev et al., 1990).

In the mountains of the Greater and the Lesser Caucasus as well as in Talysh (although rarely in the mountains of Nakhchivan) at 800–1800 m above sea level, broad-leaved forests occur where the major component is *Fagus orientalis* Lipsky (Fagaceae) (Prilipko, 1961–1970). These beech forests occupy 32% of all the forest area of the Republic and are found in various associations, particularly with *Acer platanoides* L. (Aceraceae), *Carpinus caucasica* Grossh. (*Carpinus betulus* L. p.p., Betulaceae), *Danae racemosa* (L.) Moench (Asparagaceae), *Euonymus latifolius* (L.) Mill. (Celastraceae), *Tilia begoniifolia* Steven (Tiliaceae), and *Ulmus elliptica* K. Koch (Ulmaceae). Holly, ferns, blackberry, Pontic azalea, and numerous herbaceous perennials, including *Festuca drymeja* Mert. & W. D. J. Koch (Poaceae), *Galium odoratum* (L.) Scop. (Rubiaceae), and *Sanicula europaea* L. (Apiaceae), among many others, are frequently found in the beech forests. Oak trees, just as beech trees, are widespread. *Quercus iberica* Steven ex M. Bieb. (Fagaceae) together with *Carpinus caucasica* Grossh. (Betulaceae) form a wide band in the middle mountain belt and occur on the southern and southeastern hillsides in all the mountain areas. In the alpine zone (1800–2200 m) *Quercus iberica* Steven ex M. Bieb. is replaced by *Quercus macranthera* Fisch. & C. A. Mey. ex Hohen., forming park-like forests. Birches (*Betula litwinowii* Doluch., *Betula pendula* Roth, Betulaceae) also occur in this zone.

Among the conifers, *Pinus eldarica* Medw. and *Pinus kochiana* Klotzsch ex K. Koch (Pinaceae) are widespread in the Republic of Azerbaijan. The first species, *Pinus eldarica* Medw., occurs on the steppe plateau at 600 m above sea level in the western part of the Republic, while *Pinus kochiana* Klotzsch ex K. Koch is found at higher elevations, at 1600 m above sea level in the region of Gyok Gyol (the Lesser Caucasus) and in the middle mountain zone in the Belokany region (the Greater Caucasus). This species is also widespread over the southern part of the Lesser Caucasus. Shrubs rarely form such autonomous formations. However, among the evergreen shrubs in the subalpine zone of the Greater Caucasus in the Zakataly and Belokany regions of northwestern Azerbaijan, *Rhododendron caucasicum* Pall. (Ericaceae) forms small thickets, sometimes with an admixture of blueberries (also Ericaceae), e.g., *Vaccinium myrtillus* L. and *Vaccinium vitis-idaea* L. (Gadzhiev, 1954).

In the foothills of the Greater and the Lesser Caucasus, arid open forests with pistachio, juniper, and pomegranate are observed (Asadov, 1981). Within these open forests also occur *Celtis caucasica* Willd. (Ulmaceae), *Cotinus coggygria* Scop. (Anacardiaceae), *Cerasus incana* (Pall.) Spach (Rosaceae), and *Jasminum fruticans* L. (Oleaceae). In the Eldar Oyugu Range there is an isolated spot of typical open forest formed by *Pinus eldarica* Medw. (Pinaceae). Along the major rivers (Kyur, Araz, Tori, Alazan) in Azerbaijan, riparian (tugai) forests made up of southern willow, mulberry trees, elms, hybrid poplars, English elm, and others form discontinuous narrow bands. Riparian forests also contain a considerable number of shrubs, among which are *Berberis vulgaris* L. (Berberidaceae); *Elaeagnus angustifolia* L. and *Hippophae rhamnoides* L. (both Elaeagnaceae); *Pyracantha coccinea* M. Roem. (Rosaceae); and *Tamarix ramosissima* Ledeb. (Tamaricaceae). Trees and shrubs are interwoven with vines, especially *Cynanchum acutum* L. and *Periploca graeca* L. (both Apocynaceae); *Solanum persicum* Willd. ex Roem. & Schult. (Solanaceae); and *Vitis sylvestris* C. C. Gmel. (Vitaceae). It should be mentioned that *Alnus subcordata* C. A. Mey. (Betulaceae) sometimes forms pure stands. Another species of alder, *Alnus barbata* C. A. Mey. (Betulaceae), is more characteristic of wetlands in Talysh. Typically found in the coastal forests of Talysh are species such as *Humulus lupulus* L. (Cannabaceae), *Sambucus ebulus* L. (Caprifoliaceae), and *Smilax excelsa* L. (Smilaceae), and the local endemic, *Ficus hyrcana* Grossh. (Moraceae). Herbaceous cover consists of numerous species of *Carex* L. (Cyperaceae), *Cardamine* L. (Brassicaceae), and *Poa* L. (Poaceae). In the marshes the plant communities are dominated by *Juncus* L. (Juncaceae).

In the southeastern Caucasus, oak species play a significant role in the formation of forest cover from the lower to the upper and even subalpine zones. *Quercus pedunculiflora* K. Koch (Fagaceae) occurs in the riparian forests of the Kyur River basin, whereas *Quercus boissieri* Reut. and *Quercus macranthera* Fisch. & C. A. Mey. ex Hohen. (Fagaceae) are common along the Armenian border, close to the Zangezur Range, in the eastern part of the Ordubad region (Nakhchivan) and in the Kubatlinsk region. In the plains, in the Kuba-Khachmas and Karabakh forest areas, and in the Alazan-Avtaraa valley, forests dominated by *Quercus pedunculiflora* K. Koch (Fagaceae), with other trees such as *Acer velutinum* Boiss. (Aceraceae), *Pyrus caucasica* Fed. (Rosaceae), *Tilia begoniifolia* (Tiliaceae), and species of *Fraxinus* L. (Oleaceae) are found. Shrubs here include species of *Crataegus* L. and *Mespilus germanica* L. (both Rosaceae); vines include *Clematis vitalba* L. (Ranunculaceae), *Hedera helix* L. (Araliaceae), *Smilax excelsa* L. (Smilaceae), and *Vitis sylvestris* C. C. Gmel. (Vitaceae) (Akhundov et al., 1978). This composition somewhat resembles that found in the forests on the plains of Talysh. In the Kuba Khachmas region, besides *Quercus pedunculiflora* K. Koch, *Quercus pubescens* Willd. (both Fagaceae) also occurs. The most common component of the lowland forests is *Ulmus minor* Mill. (Ulmaceae), sometimes in association with *Acer campestre* L. (Aceraceae); *Carpinus betulus* L. and *Corylus avellana* L. (both Betulaceae); species of *Crataegus* L., *Mespilus germanica*, and *Pyrus caucasica* (all three, Rosaceae); and the vine *Hedera pastuchovii* Woronow ex Grossh. (Araliaceae).

Oak forests in the Karabakh region are park-like with arboreal components of species of *Crataegus* L. and *Prunus divaricata* Ledeb. (both Rosaceae); *Elaeagnus angustifolia* L. (Elaeagnaceae); *Morus alba* L. (Moraceae); *Paliurus spina-christi* Mill. (Rhamnaceae); and *Punica granatum* L. (Lythraceae). Vines include *Periploca graeca* L. (Apocynaceae), *Smilax excelsa* L. (Smilaceae), and *Vitis sylvestris* C. C. Gmel. (Vitaceae).

Of special note are the relictual Hyrcanian forests of the Talysh plain with *Parrotia persica* (DC.) C. A. Mey. (Hamamelidaceae) and *Quercus castaneifolia* C. A. Mey. (Fagaceae) (Prilipko & Gogina, 1978). Paleobotanical evidence shows that *Parrotia persica* (DC.) C. A. Mey., Persian ironwood, did occur in many regions in Europe but is now rare, occurring only in the southeastern Caucasus in Azerbaijan and Iran. When Persian ironwood forms a nearly pure stand the overall impression is of an enchanted forest, with trunks and branches fusing and growing together and lichens hanging from the branches. Heavy shade dominates the forest floor under these tall trees and undergrowth is almost absent. In autumn, its leaves first become yellow and later faintly tinged with lilac. Also found in the Hyrcanian forests are *Carpinus betulus* L. (Betulaceae); *Populus hyrcana* Grossh. (Salicaceae); *Prunus divaricata* subsp. *caspica* Browicz. (Rosaceae); and *Ulmus glabra* Huds. and *Zelkova carpinifolia* (Pall.) K. Koch (both Ulmaceae). In the lowermost understory of the woods are found the evergreen shrubs *Ruscus hyrcanus* Woronow and *Danae racemosa* (L.) Moench (both Asparagaceae). Species of *Crataegus* L., *Cydonia* Mill., and *Mespilus* L. (all three, Rosaceae); *Smilax* L. (Smilaceae); *Periploca* L. (Apocynaceae); *Vitis* L. (Vitaceae); *Hedera* L. (Araliaceae); and many other genera are also found in the undergrowth. The date plum, *Diospyrus lotus* L. (Ebenaceae), is a tree found in groves usually on more shaded and humid hillsides, frequently admixed with Caspian honey locust (*Gleditsia caspia* Desf., Fabaceae). The latter dominates a special type of forest in the Talysh foothills on the hillsides facing the sea. *Albizia julibrissin* Durazz. (Fabaceae) along with *Tilia platyphyllos* Scop. and *Tilia begoniifolia* Steven (both Tiliaceae) can also be found there. At somewhat higher elevations, *Acer velutinum* Boiss. (Aceraceae) occurs on northern slopes. *Fagus orientalis* Lipsky (Fagaceae) and *Taxus baccata* L. (Taxaceae) are widely represented on low and medium sea terraces in the foothills and in the lower mountain zone of Talysh, along with groves of *Buxus hyrcana* Pojark. (Buxaceae). Vines such as the silk-vine *Periploca graeca* L. (Apocynaceae) as well as *Tamus communis* L. (Dioscoreaceae) are abundant in the alder woods of Talysh.

In the southern Caucasus, especially in the Nakhchivan and Zuvand regions with very limited precipitation, other vegetation types such as steppe, semi-steppe, phrygana, and shibliak occur. These areas are rich in herbaceous endemic plants (Akhundov, 1972).

Vast areas of the Kyur-Araz, the Caspian coastal plain, and other plains in Azerbaijan are covered with semi-deserts of wormwood (*Artemisia* L., Asteraceae) and Russian thistle (*Salsola* L., Chenopodiaceae). On the Kyur-Araz plain, a mixed variety of plant communities develop. These have been characterized by Movsumova (1991) as the Artemisieto-Salsoletum dendroides type, which is a widely distributed community, along with the Artemisieto-Salsoletum ericoides and the Artemisieto-Salsoletum gemmascens communities. Here and there in the foothills of Gobustan, in the Jeyrankechmaz basin, and in the region of Nakhchivan, Artemisieto-Stipetum lessinglana communities and Artemisieto-Bothriochloa ischaemum-type steppes occur. Ephemeral plants (annual and perennial) predominate throughout the wormwood community types. These plants form a natural semi-desert landscape and include *Anisantha rubens* (L.) Nevski, *Bromus japonicus* Thunb., *Eremopyrum orientale* (L.) Jaub. & Spach, *Er-*

emopyrum triticeum (Gaertn.) Nevski, and *Poa bulbosa* L. (all Poaceae); *Erodium cicutarium* (L.) L'Hér. (Geraniaceae); *Neotorularia contortuplicata* (Stephan) Hedge & J. Léonard (Brassicaceae); *Medicago minima* (L.) Bartal. and *Medicago caerulea* Less. ex Ledeb. (both Fabaceae); and many others. The above formations develop on the brown, light chestnut, and sierozem soils and in the sands along the Caspian shore. In hummocky soils are found species of *Artemisia* L. (Asteraceae); *Leymus racemosus* (Trin.) Tzvelev (Poaceae); *Convolvulus persicus* L. (Convolvulaceae); *Argusia sibirica* (L.) Dandy (Boraginaceae); *Astragalus hyrcanus* Pall., *Astragalus igniarius* Popov, and the endemic *Astragalus bakuensis* Bunge, as well as other species of *Astragalus* (Fabaceae). In the sands may be found such interesting and rare species as *Convolvulus erinaceus* Ledeb. (Convolvulaceae); *Stipa pellita* (Trin. & Rupr.) Tzvelev and *Ammochloa palestina* Boiss. (both Poaceae); *Calligonum bakuense* Litv. (Polygonaceae); and strictly local species such as the nitre bush, *Nitraria schoberi* L. (Zygophyllaceae) and Caspian sweet clover, *Melilotus polonicus* (L.) Pall. (Fabaceae), among others (Gadzhiev & Musaev, 1996).

In the semi-steppes, in addition to grasses, also present are a considerable number of perennial plants, such as *Medicago caucasica* Vassilcz. (Fabaceae), and *Centaurea reflexa* Lam. (Asteraceae) and annual plants such as *Sideritis montana* L. (Lamiaceae) and *Minuartia wiesneri* (Stapf) Schischk. (Caryophyllaceae), among others. At somewhat higher levels above the Caspian sea there are admixtures of grasses, such as *Festuca sulcata* (Hack.) Nyman and *Stipa szovitsiana* (Trin. & Rupr.) Griseb., as well as the annual *Trachynia distachya* (L.) Link (all Poaceae). In the region of the Sheki highland, there are large thickets with the perennial bunchgrass *Chrysopogon gryllus* (L.) Trin. (Poaceae). The floristic composition of these thickets varies with the hillside aspect, with perennial grasses found on slopes of southern exposure (*Stipa caspia* K. Koch, *Agropyrum cristatum* (L.) Beauv., Poaceae) forming golden bovodnik steppes. On slopes with northern exposure, forbs are more abundant, forming semi-steppes with *Potentilla recta* L. (Rosaceae), *Galium vernum* L. (Rubiaceae), *Medicago caucasica* Vassilcz. (Fabaceae), and *Salvia nemorosa* L. (Lamiaceae), among others. Steppe vegetation also occurs in the middle montane zones with species of feather grass, *Stipa* L. (Poaceae); however, in the alpine zone, feather grass species mix with species of sheep's fescue, *Festuca* L. (Poaceae). In the alpine zones, there are also steppe-like meadows where the principal components are species of *Festuca* L. and *Stipa capillata* L. Also represented in all the variations of the steppe just mentioned are the grass *Koeleria cristata* (L.) Pers. (Poaceae); species of *Thymus* L. (Lamiaceae), *Pyrethrum* Zinn (Asteraceae), and *Dianthus* L. (Caryophyllaceae); as well as *Scabiosa caucasica* M. Bieb. (Dipsacaceae) and *Pimpinella rhodantha* Boiss. (Apiaceae), among others.

In the warm regions of the Republic of Azerbaijan, as in Nakhchivan and Zuvand, various montane xerophytic vegetation types develop that include the characteristic phrygana, shibliak, and tragacanth formations. Small areas of montane xerophytic vegetation of the Nakhchivan region are occupied by phrygana that is diverse and present in a number of variants. In Nakhchivan, phrygana types are found on different substrates and sites. On coarse limestone may be seen *Salsola glauca* M. Bieb. (Chenopodiaceae), *Salvia dracocephaloides* Boiss. (Lamiaceae), and *Astragalus szovitsii* Fisch. & C. A. Mey. (Fabaceae). On clay substrates and rock waste hillsides can be found *Hedysarum formosum* Fisch. & C. A. Mey. ex Basiner and *Hedysarum atropatanum* Bunge ex Boiss. (Fabaceae), as well as *Stachys fruticulosa* M. Bieb. (Lamiaceae). On stony and stony-rocky habitats, *Stachys inflata* Benth. (Lamiaceae) imparts a special grayish green aspect to the vegetation. Finally, phrygana may be dominated by the scented *Thymus kotschyanus* Boiss. & Hohen. and *Satureja macrantha* C. A. Mey. (Lamiaceae). In all of these phrygana variants, other species of *Achillea* L., *Cousinia* Cass., *Lactuca* L., and *Pyrethrum* Zinn (these four, Asteraceae); *Marrubium* L. and *Phlomis* L. (both Lamiaceae); as well as other genera are widely represented.

Shibliak vegetation occupies a smaller area compared to that covered by phrygana and often results from the conversion of these landscapes to agriculture, such as vineyards. In some parts of the Nakhchivan region, examples of shibliak are well preserved, with *Amygdalus fenzliana* (Fritsch) Lipsky (Rosaceae) present. In the low mountain belt of the Greater and the Lesser Caucasus, scrub formations include *Rhamnus pallasii* Fisch. & C. A. Mey (Rhamnaceae), *Atraphaxis spinosa* L. (Polygonaceae), *Cotoneaster suavis* Pojark. (Rosaceae), as well as *Colutea komarovii* Takht. (Fabaceae).

There are many lakes, bogs, and marshes in the Kyur-Araz and Caspian lowlands, as well as other lowlands. This intrazonal type of vegetation includes some endemic taxa in the Red List. Sometimes small, isolated wet areas may be situated within semi-arid vegetation, within the coastal woods and along the major rivers, channels, and estuaries. These particular habitats contain an unusual set of plants. In the backwaters and estuaries, such as the Divichi estuarine lowland (along the Caspian shore in northern Azerbaijan), an abundance of *Typha angustifolia* L. (Typhaceae) and other species of *Typha* L. occur. Rather typical for the lakes in the lower Kura river basin are lotus, *Nelumbo caspica* (DC.) Fisch. (Nelumbonaceae); water lily, *Nymphaea alba* L. (Nymphaeaceae); *Nymphoides peltata* L. (Menyanthaceae); and bladderwort, *Utricularia vulgaris* L. (Lentibulariaceae); along with the floating fern *Salvinia natans* (L.) All. (Salviniaceae) and other aquatic plants. Lotus forms dense stands with a small admixture of other wetland plants and when blooming in the early autumn beautifies the landscape. Lotus is a relictual species in Azerbaijan that has survived there since the Tertiary period. *Arundo donax* L. and *Phragmites australis* (Cav.) Trin. ex Steud. (both Poaceae) grow in boggy places and along the coastline. In certain places, these reed grasses form dense, almost impenetrable thickets, up to 3 m in height. Some taxa in the reed fields include meadow foxtail grass, *Alopecurus myosuroides* Huds., and pampas grass, *Erianthus ravennae* (L.) Beauv. (both Poaceae); alkaliweed, *Cressa cretica* L. (Convolvulaceae); *Cynanchum acutum* L. (Apocynaceae); and *Salsola soda* L. (Chenopodiaceae); as well as species of *Typha* L. and *Sparganium* L. (both Typhaceae); and the minute, aquatic duckweed, *Wolffia* Horkel ex Schleid. (Araceae).

Various vegetation types are more or less widespread throughout the ditches across the Kura-Araks plain. These include the *Cynodonetum*-type characterized by the grass species *Cynodon dactylon* (L.) Pers. (Poaceae); the *Glycyrrhizetum*-type with *Glycyrrhiza glabra* L. (Fabaceae), licorice; and *Bolboschoenus maritimus* (L.) Palla (Cyperaceae), bulrush. The latter sedge or bulrush forms a vegetation that is most similar to the grassy reed fields in its structure. In wet habitats in the Karabakh plain in southwestern Azerbaijan, *Limonium scoparium* (Pall. ex Willd.) Stankov (Plumbaginaceae), *Polygonum patulum* M. Bieb. (Polygonaceae), *Stachys palustris* L. (Lamiaceae), and *Lythrum salicaria* L. (Lythraceae) often occur. On the moist saline sands and wet solonchaks, characterized by salty, poorly drained soils of the eastern Transcaucasian plain, a vegetation of small shrubs develops that is distinguished by *Halostachys belangeriana* (Moq.) Botsch., *Halocnemum strobilaceum* (Pall.) M. Bieb., *Salicornia europaea* L., *Suaeda altissima* (L.) Pall., and *Kalidium capsicum* (L.) Ung.-Sternb. (all Amaranthaceae).

Marshy areas are widely spread in the region of the Talysh plain, notable for the aquatic plants *Potamogeton pectinatus* L. (Potamogetonaceae), *Myriophyllum spicatum* L. (Haloragaceae), *Trapa hyrcana* Woronow (Trapaceae), and *Ceratophyllum demersum* L.

(Ceratophyllaceae). Another widespread type of boggy vegetation in the Talysh plain is represented by thickets of *Iris pseudacorus* L. (Iridaceae), usually with an admixture of the bur-weed *Sparganium erectum* L. (Typhaceae), and sedges such as *Scirpus lacustris* L. and *Eleocharis palustris* (L.) Roem & Schult. (both Cyperaceae). Seasonally drying marshes scattered across the Talysh plain are typically characterized by short plants that include *Alisma plantago-aquatica* L. (Alismataceae); *Buschia lateriflora* (DC.) Ovcz. and *Ranunculus ophioglossifolius* Vill. (both Ranunculaceae); *Lippia nodiflora* (L.) Michx. (Verbenaceae); *Mentha aquatica* L. (Lamiaceae); and *Persicaria minor* (Huds.) Opiz (Polygonaceae). Wetland plants also occur in the foothills and mountainous zones at different elevations and are especially numerous in the alpine zone, where more than 100 marshes and marshy habitats are represented. Most typical are those from the Alagollar area in southwestern Azerbaijan and Zalkhy Gyol, Gara Gyol, Maral Gyol, Zali Gyol in the Lesser Caucasus, as well as Khala Gyol and Gabolan Gyol in the Greater Caucasus. Others include Bata Bat, Ganly Gyol, Gyoy Gyol, Solvavty Gyol, and Gazanyk.

Subalpine and alpine meadows and meadow steppes occur at elevations from 1800 to 3200 m above sea level. Places that were formerly woodlands are quickly populated in the higher forest belt by grassy vegetation, thus forming post-woodland meadows with tall grasses. This subalpine meadow vegetation is quite diverse, due to the variety of topographic and microclimatic conditions (Gadzhiev, 1954). These post-woodland meadows contain representatives from both forest and meadow elements, including the grasses *Agrostis tenuis* Sibth., *Brachypodium sylvaticum* (Huds.) Beauv., and *Calamagrostis arundinacea* (L.) Roth. (all, Poaceae) (Musaev, 1991). Among these grass communities are representatives of primary and secondary ecological succession. Secondary tall grasses form weedy plots of exclusively anthropogenic landscapes, associated with human mismanagement. However, in primary tall grass communities dicots are also found that include all the species of *Heracleum* L. (eight species are known in the Republic, Apiaceae); *Aconitum orientale* Mill. and *Delphinium flexuosum* M. Bieb. (both Ranunculaceae); *Cephalaria gigantea* (Ledeb.) Bobrov (Caprifoliaceae); and *Knautia heterotricha* K. Koch (Dipsacaceae). Meanwhile, secondary tall grass communities differ with dicot species of *Urtica* L. (Urticaceae), *Rumex* L. (Polygonaceae), and *Cirsium* Mill. (Asteraceae), as well as the fern *Dryopteris* Adans. (Dryopteridaceae), among others.

Diverse alpine meadows form a distinct zone on many mountain ridges. In the subalpine zone, these meadows may be variably distinguished as moderately humid grass-dominated, mixed-grass, humid-motley grass, mesophytic legumes and sedges, excessively humid subalpine cryophytic, and steppe-like communities. Characteristic elements of meadows are grasses such as *Agrostis tenuis* Sibth., *Bromopsis variegata* (M. Bieb.) Holub, and *Festuca karabaghensis* Mussajev (all Poaceae); the sedge *Carex tristis* M. Bieb. (Cyperaceae); *Stachys macrantha* (K. Koch) Stearn (Lamiaceae); and the legume *Amoria ambigua* (M. Bieb.) Soják (Fabaceae).

The alpine zone vegetation is widespread at elevations from 2400 to 3200 m above sea level and is represented by two community types: alpine meadow and carpet grass. Vegetation is formed at these higher altitudes primarily on the sloping hillsides, cupola-like peaks, mountain passes, and ridge saddles. Short-grass, heavy-sod alpine meadows are dominated by Cyperaceae, especially the sedge *Carex tristis* M. Bieb., and mixed grasses such as *Festuca ovina* L. (Poaceae) (Musaev, 1991). Among the *Kobresia* Willd. sedge vegetation of alpine meadows occur short plants such as *Amoria ambigua* (M. Bieb.) Soják (Fabaceae); *Myosotis alpestris* F. W. Schmidt (Boraginaceae); *Omalotheca supina* (L.) DC. and *Taraxacum stevenii* DC. (both Asteraceae); and *Veronica gentia-*

noides Vahl (Scrophulariaceae). Carpet grasses are represented by two community types: typical alpine carpet grasses on a fine earth substrate with *Carum* L. (Apiaceae), *Alchemilla* L. (Rosaceae), and *Plantago* L. (Plantaginaceae) and by stony carpet grasses on a stony substrate with *Sibbaldia* L. (Rosaceae), *Campanula* L. (Campanulaceae), and *Macrotomia* DC. ex Meisn. (Boraginaceae). Both carpet grass formations contain *Carum caucasicum* (M. Bieb.) Boiss. and *Campanula tridentata* Schreb.

At more than 3200 m above sea level in the domain of high cliffs and mounds, a sparse primitive vegetation develops. Here the representatives of higher plants are rare, and mosses and lichens are prevalent instead. Mosses, lichens, and fungi are also well represented in lower-lying belts.

Table 1. List of assessed national endemics by their ecological groups.

HABITAT	SPECIES
Meadows	*Centaurea hyrcanica* Bornm., *Centaurea kobstanica* Tzvelev, *Euphorbia grossheimii* Prokh., *Euphrasia karjaginii* Kem.-Nath., *Papaver talyshense* Grossh., *Tragopogon nachitschevanicus* Kuth., *Verbascum erivanicum* E. Wulff, *Verbascum paniculatum* E. Wulff, *Verbascum stachydiforme* Boiss. & Buhse
Alpine and subalpine meadows and rocks	*Alchemilla jaroschenkoi* Grossh., *Amoria bobrovii* (Khalilov) Roskov, *Anthemis sachokiana* Sosn. ex Grossh., *Astragalus insidiosus* Boriss., *Astragalus kubensis* Grossh., *Campanula fominii* Grossh., *Dianthus schemachensis* Schischk., *Erodium schemachense* Grossh., *Marrubium nanum* Knorr., *Nepeta grossheimii* Pojark., *Pyrethrum ordubadense* Manden., *Rosa abutalybovii* Gadzh., *Scorzonera pulchra* Lomak., *Stachys talyschensis* Kapeller, *Thymus fedschenkoi* Ronniger, *Vicia rafigae* Tamamsch.
Steppes	*Achnatherum roshevitzii* Mussajev, *Ammania pubiflora* (Koehne) Sosn., *Bellevalia fominii* Woronow, *Dactylis hyrcana* (Tzvelev) Mussajev, *Fritillaria grandiflora* Grossh., *Gagea improvisa* Grossh., *Iris camillae* Grossh., *Merendera candidissima* Miscz. ex Grossh., *Muscari elegantulum* Schchian, *Sternbergia colchiciflora* Waldst. & Kit. var. *alexandrae* (Sosn.) Artjush., *Stipa karjaginii* Mussajev & Sadychov, *Veronica amoena* M. Bieb.
Deserts and semi-deserts	*Allium mariae* Bordz., *Anabasis eugeniae* Iljin, *Papaver schelkovnikovii* N. Busch, *Polygonum caspicum* Kom., *Salsola flavovirens* Iljin, *Salsola futilis* Iljin, *Salsola tamamschjaniae* Iljin, *Taraxacum desertorum* Schischk.
Forests	*Amoria talyschensis* (Khalilov) Roskov, *Crataegus eriantha* Pojark., *Dryopteris raddeana* (Fomin) Fomin, *Ficus hyrcana* Grossh., *Muscari grossheimii* Schchian, *Onobrychis schuschajensis* O. D. Agajeva, *Ornithogalum hyrcanum* Grossh., *Rosa azerbajdzhanica* Novopokr. & Rzazade, *Rosa gadzhievii* Chrshan. & Iskend., *Rosa isaevii* Gadzh. & Iskend., *Rosa komarovii* Sosn., *Rosa nisami* Sosn., *Rosa sosnovskyi* Chrshan., *Rosa zakatalensis* Gadzh., *Silene praestans* Schischk.
Mountain xerophytes	*Acantholimon schemachense* Grossh., *Acantholimon tenuiflorum* Boiss., *Aethionema levandowskyi* N. Busch, *Alcea lenkoranica* Iljin, *Alcea sachsachanica* Iljin, *Allium leonidis* Grossh., *Allium materculae* Bordz., *Anemone kusnetzowii* Woronow ex Grossh., *Astragalus albanicus* Grossh., *Astragalus maraziensis* Rzazade, *Astragalus*

HABITAT	SPECIES
Mountain xerophytes (continued)	*nachitschevanicus* Rzazade, *Astragalus regelii* Trautv., *Astragalus zangelanus* Grossh., *Erysimum caspicum* N. Busch, *Erysimum strictisiliquum* N. Busch, *Euphorbia hyrcana* Grossh., *Galium czerepanovii* Pobed., *Iris schelkownikowii* (Fomin) Fomin, *Linaria schirvanica* Fomin, *Malva grossheimii* Iljin, *Nepeta noraschenica* Grossh., *Oxytropis lupinoides* Grossh. & Fed., *Peltariopsis grossheimii* N. Busch, *Phlomis lenkoranica* Knorr., *Salvia andreji* Pobed., *Scorzonera grossheimii* Lipsch. & Vassilcz., *Scrophularia zuvandica* Grossh., *Scutellaria darriensis* Grossh., *Scutellaria grossheimiana* Juz., *Scutellaria rhomboidalis* Grossh., *Silene talyschensis* Schischk., *Stachys fominii* Sosn. ex Grossh., *Thlaspi rostratum* N. Busch, *Thymus karamarjanicus* Klokov & Des.-Shost.
Hill xerophytes	*Astragalus barnassari* Grossh., *Astragalus biebersteinii* Bunge, *Astragalus chalilovii* Grossh. & Fed., *Astragalus husseinovii* Rzazade, *Astragalus rostratus* C. A. Mey., *Astragalus schemachensis* Karjagin, *Hypericum theodorii* Woronow, *Iris pseudocaucasica* Grossh., *Seseli cuneifolium* M. Bieb.
Psamophytes (sand-loving), littoral	*Astragalus bakuensis* Bunge, *Astragalus igniarius* Popov, *Calligonum bakuense* Litv., *Raphanus rostratus* DC., *Thesium maritimum* C. A. Mey.

THREATS

There are four major threats to plant species in Azerbaijan.

1. Habitat loss through human activity (agriculture, livestock, construction, overcollection, etc.)
2. Overgrazing
3. Haying
4. Soil erosion

The natural vegetation of Azerbaijan has long been influenced by constant human pressure. During this period, forests of piedmont, highlands, and mountains have been destroyed, as well as the natural landscape of the Kyur Araz valley altered. In recent times, there have been major development projects in steppe, desert, semi-desert, and montane xerophytic vegetation types. Among cultivated areas, only small fragments of virgin coenotypes persist. As the pressure from intensive grazing leads to vegetation changes, numerous alien plant taxa invade the coenotype with the result that only an insignificant percentage of the original vegetation survives (Gadzhiev & Musaev, 1996).

It is always a challenge to maintain the balance in the relationship between humans and nature. The establishment of national parks and other protected areas supports this concern and makes it possible to benefit from nature through its conservation. Azerbaijan has been taking its first steps in this direction, and this is a new trend for our country and citizens. Shirvan was the first national park to open in 2003, and now the total area of protected sites in Azerbaijan, which includes national parks, is 8.7% of the country's land area. The future of national parks in Azerbaijan lies in the further development of ecotourism, as part of sustainable development and the protection of natural resources. A key priority is to involve the public in this process by increasing environmental awareness on issues of nature conservation.

The *Red Book of Azerbaijan* was was published in 1989 (Adighesal et al., 1989) and was based on the national resolution #11-21-21,

04/11/1989 that was approved by the State Commission on Environmental Protection. The Red Book includes the rare and threatened species of both animals and plants, with 140 rare and endangered plant species considered.

The preparation of the second edition of the *Red Book of Azerbaijan* has been completed. Under the support of the Science Development Foundation, established by the President of the Azerbaijan Republic, the majority of rare and disappearing species of plants were monitored in recent years, with their conservation status assessed following the IUCN (2001) categories and criteria. This updated Red List will include 315 plants and fungi species of Azerbaijan and was planned to be published by the Ministry of Ecology and Natural Resources of Azerbaijan in 2011.

LISTED IN THIS RED LIST 129

Extinct (EX) 0
Critically Endangered (CR) 20
Endangered (EN) 18
Vulnerable (VU) 22
Near Threatened (NT) 25
Least Concern (LC) 18
Data Deficient (DD) 26

Botanists also consider other species to be threatened through their usage for fuel, medicinal purposes, and horticultural or ornamental qualities. These species are widespread beyond the borders of Azerbaijan and have been excluded from this national list. However, a few important plants should be mentioned and include *Inula helenium* L. and *Helichrysum rubicundum* (K. Koch) Borm. (both Asteraceae); *Origanum vulgare* L. (Lamiaceae); and *Valeriana officinalis* L. (Caprifoliaceae).

Land degradation, which has long-term consequences, is very apparent in Azerbaijan, especially from oil pollution. In addition to contamination of groundwater, severe oil pollution has rendered vast areas of land useless that might otherwise have agricultural, commercial, or residential value. Much of this land is in or near Baku and could now or in the future be valuable residential land. Close to 5000 hectares of productive crop lands and more than 15,000 hectares of pastures have also been flooded. This, in combination with increased salinity and water logging from poor irrigation practices, as well as soil pollution and soil erosion, is causing serious productivity losses on agricultural lands.

Soils are polluted by atmospheric depositions of heavy metals from industries, power plants, and motor vehicles. In addition, large areas of land are contaminated by oil. Discarded oil industry wastes have resulted in some (drilling-associated) radiation contamination in localized areas. Mineral deposit exploitation, particularly oil extraction and pipeline construction, has contributed to the pollution of about 30,000 hectares of land. The most acute soil degradation problems are on the Absheron Peninsula where a century of oil production, combined with environmental neglect, has left more than 10,000 hectares of heavily contaminated land. The national government of the Republic of Azerbaijan needs to explore the clean-up of metal- and oil-contaminated soils and to develop regulatory recommendations and suggestions for priority actions.

It is important to note that the destruction of forest resources from over-cutting and erosion is the most urgent problem for natural resources. An estimated 11.4% of the total land area is reportedly under some type of forest or shrub cover. The actual figure is probably lower, as remaining forests are declining rapidly. Azerbaijan has relatively few old-growth forests, which are concentrated mainly in the Greater and the Lesser Caucasus. Forest resources suffer from rapidly increasing and uncontrolled exploitation that leads to large-scale deforestation, soil erosion with subsequent risks of floods, and loss of ecosystem resources and services. In addition, 10% of the strictly protected state forest preserves are located in areas that border Armenia. Azeri management in these forests has ceased, and there is a major refugee problem in these and adjacent areas. Shortages of cooking gas and fuel have led to an increase in the use of firewood by refugee and local rural populations.

CONSERVATION LEGISLATION

The Constitution of the Republic of Azerbaijan outlines the main principles of environmental protection, and the legislative basis for this has been put in place, along with an improved environmental management structure. Approximately 20 laws have been adopted to bring the country in line with international standards on environmental protection. The Government of Azerbaijan, recognizing the importance of the biological diversity of the country, has established various policies and legislation. The Ministry of Ecology and Natural Resources of Azerbaijan provides management and implements priorities for environmental protection.

LEGISLATION CONCERNING THE FLORA OF AZERBAIJAN

Policies concerning environmental issues in the Republic of Azerbaijan are assured under the Constitution of Azerbaijan (1995). This constitution stipulates that the government is to take necessary measures for the protection and rational use of natural resources. Furthermore, the policy on environmental protection is also assured through international conventions ratified by Azerbaijan. These include the Convention on Biological Diversity (CBD, 2002), ratified by Azerbaijan on 14 March 2000; the United Nations Framework Convention on Climate Change (UNFCCC; United Nations, 1992), ratified in 1 October 1995; the Convention on Wetlands of International Importance Especially as a Waterfowl Habitat or the Ramsar Convention on Wetlands (RCW, 1971), ratified on 18 July 2001; the Convention on International Trade in Endangered Species of Wild Flora and Fauna (CITES, 1979), ratified on 23 June 1998; and the United Nations Convention to Combat Desertification (UNCCD), ratified on 29 April 1998.

There are several laws regulating protection and use of the biodiversity of the country. The main law regulating the issues of plant protection and use of plant species in Azerbaijan is the Law on Plant Protection (316-IQD; 10 June 1997). The law defines the prevention of distribution of plant pests and diseases, mass propagation of weeds, and loss of productivity and availability of ecologically pure products. Its aim is the protection of the environment, population health, and useful fauna and flora from the deleterious impact of pesticides and the realization of purposeful state programs for the isolation, quarantine, and destruction of deleterious objects.

The Law on Protection of Environment (678-IQ; 8 June 1999) is the main legal act in the field of nature protection for Azerbaijan. It identifies the legal, economic, and social basis of environmental protection. The law aims to ensure environmental safety in the sphere of protection of environmental balance; prevent the hazardous impact of different activities on natural ecological systems; and ensure preservation of biological diversity and proper use of natural resources. The law defines the main principles of environmental protection; rights and duties (of authorities and citizens) in environmental protection; use of nature (user issues, limits and quotas, special permits, regulations, responsibilities); standardization and certification of environmental protection and natural resources, mapping, and monitoring; and balancing the economy in terms of environmental protection. The law regulates issues of ecological balance of environment and sets the requirements for industrial entities and for other activities.

In addition, the law defines the practices and procedure of environmental expertise, issues of environmental protection control and state ecological audits, and issues on education and research.

The Law on Environmental Protection and Use of Natural Resources (92; 25 February 1992) aims to ensure relations between nature and society while preserving features of nature in Azerbaijan, as well as plants and animals and their habitats. The law sets the objectives, as follows: restoration and proper use of natural resources; preservation of specific landscapes, certain territories, ecosystems, and natural objects; and protection of, and prevention of anthropogenic impacts on, the environment and natural resources. The law regulates cooperation in the field of environmental protection and use of natural resources. The law defines the rights of citizens (e.g., to live in a healthy and favorable environment); state management of nature protection and use of natural resources; economic issues about nature protection and regulation of natural resource use; rights of nature users; and control and monitoring over the state of the environment and use of nature. The law also defines issues concerning state ecological expertise; ecological requirements for different activities; and public participation, awareness, and research. It regulates issues of specially protected natural areas and objects and legal liability for infringement of legislation and international cooperation.

The Law on Specially Protected Natural Areas and Objects (840-IQ; 24 March 2000) sets the legal basis for establishment and protection of specially protected natural territories and protection of specially protected natural objects in the Republic of Azerbaijan. The law gives general provisions identifying main principles for organization, protection, and use of specially protected natural areas and ownership of these areas and objects. The law defines the forms and categories of specially protected natural areas; the forms of their use and rent; schemes for their development, location, and establishment; and issues of their protection, cadastre, and monitoring. The law defines authorities at different levels in the control system of the specially protected natural areas and objects and their power in the management of those areas and objects. The law also defines issues concerning financing, control over protection, and use and issues of international cooperation.

The Law for the Forest Code (424; 30 December 1997) establishes legal basis for protection, restoration, and use of forests; increase of their ecological and resource potential; and regulation of forest interactions, executed with the view of the forest as a complex of plants, soils, animals, and environmental components of ecological, economic, and social importance.

The *Red Book of Azerbaijan* (resolution approved by the State Commission on Environmental Protection 11-21-21; 4 November 1989) contains information about the biology, areal distribution, and number—as well as accepted and necessary actions for protection—of mammals, birds, amphibians, reptiles, invertebrate animals, and higher plants.

The legal act for the Environmental Impact Assessment (EIA) in Azerbaijan (1996) checks the conformity of a project, already in its initial phase, to the requirements of the nature protection legislation, including the requirements of flora and fauna protection, or other elements of biodiversity.

The Administrative Offence Code regulates the administrative measures available for violation of legislation on the use of fauna and flora and their habitats in the Republic of Azerbaijan. The legislation of the Criminal Code (14 July 2000) defines the measures of criminal responsibility for similar offences.

The National Strategy and Action Plan on Conservation and Sustainable Use of Biodiversity in the Republic of Azerbaijan (1368; 24 March 2006 approval by the President of Azerbaijan) aims to ensure regulation of environmental balance and conservation and sustainable use of biodiversity. Among several directions given by this strategy are conservation of genetic resources of plants and animals, development and implementation of programs on the sustainable use of biodiversity, and conservation of rare and endangered plant and animal species as a priority, considering the importance of an ecosystem approach. The action plan specifies the activities for conservation and sustainable use of biodiversity in the country, among them to review and revise the list of rare and endangered plant and animal species (including publication of the second edition of the *Red Book of Azerbaijan* on plants and animals), identify and assess the ecosystems with high biodiversity, undertake activities to conserve rare and threatened species, and expand protected areas.

The Law on Ecological Safety (677-IQ; 8 June 1999) regulates relations in the ecological safety field and establishes legal basis for protection of humans, society, and natural resources from possible ecological hazards.

PRIORITY ENDEMIC TAXA OF AZERBAIJAN OCCURRING IN PROTECTED AREAS

Hyrkan National Park

Dryopteris raddeana (Fomin) Fomin, *Ficus hyrcana* Grossh., *Iris hyrcana* Woronow & Grossh., *Muscari elegantulum* Schchian, *Ornithogalum hyrcanum* Grossh., *Podospermum grossheimii* (Lipsch. & Vassilcz.) Kuth., *Silene talyschensis* Schischk., and *Verbascum stachydiforme* Boiss. & Buhse

Shahdag National Park

Anthemis sachokiana Sosn. ex Grossh., *Astragalus kubensis* Grossh., *Campanula fominii* Grossh., *Hypericum theodorii* Woronow, and *Rosa abutalybovii* Gadh.

Absheron National Park

Astragalus bakuensis Bunge, *Astragalus igniarius* Popov, *Astragalus maraziensis* Rzazade, and *Calligonum bakuense* Litv.

Zangezur National Park

Anabasis eugeniae Iljin, *Rosa abutalybovii* Gadzh., *Rosa karjaginii* Sosn., *Rosa nisami* Sosn., *Salvia andreji* Pobed., and *Scutellaria rhomboidalis* Grossh.

Gei-Gel National Park

Acantholimon tenuiflorum Boiss., *Crataegus eriantha* Pojark., and *Rosa gadzhievii* Chrshan. & Iskend.

Shirvan National Park

Bellevalia albana Woronow and *Papaver schelkovnikovii* N. Busch

Altiagaj National Park

Astragalus biebersteinii Bunge

TOP 50 NATIONAL CONSERVATION PRIORITIES FOR AZERBAIJAN

ALLIACEAE

1. *Allium mariae* Bordz. **EN**

There are three known localities in Nakhchivan, in the region of Julfa. The distributional area of the species is continuing to decline, and the number of mature individuals is declining because of poor reproduction. This is a fodder species that is potentially ornamental, and its major threat is from overgrazing.

AMARYLLIDACEAE

2. *Sternbergia alexandrae* Sosn. **CR**

Only a single location is known for this taxon in the vicinity of the village of Shamakhi, in the Greater Caucasus. A continuing decline in the number of mature individuals has been observed. No measures are in place to protect the species in the wild. This is an ornamental plant that is under threat, through the over-collection of flowers and bulbs by the local population and tourists.

APIACEAE

3. *Ferula caucasica* Korovin **VU**

This is known from a single location in the Agdash area, in the Kyur-Araz lowland, and is an essential oil-bearing plant species with important medicinal properties. The small population is decreasing, with weak reproduction. The species assessment is likely to move to CR within a very short time period, as the management of the protected area needs improvement to address the need for species- and site-based conservation actions. The major threat is over-collection by the local population.

ASTERACEAE

4. *Centaurea kobstanica* Tzvelev **NT**

Only two locations for this taxon are known from Guba and Gobustan, in the Greater Caucasus. It is possible that this plant species will be re-assessed as CR or even EX within a very short time period. Although the species is common in its distributional areas and its populations are currently increasing, there are no conservation measures in place. The monitoring of population trends as well as controlled grazing are recommended. This is a potential ornamental plant. Its main threats are from human-induced habitat loss and degradation, from over-collection, and from trampling.

5. *Scorzonera grossheimii* Lipsch. & Vassilcz. **EN**

Only a single location is known in Hyrkan National Park for this species. The number of mature individuals continues to decline because of weak reproduction. Although this plant species is protected in a national park, no conservation measures are in place to promote its increased reproduction.

BRASSICACEAE

6. *Aethionema levandowskyi* N. Busch **NT**

This is a rare, narrowly local endemic species that occurs in Talysh, close to Pir Shakhverdi. This is a potentially ornamental species that is under threat from over-collection by the local population and religious tourists.

7. *Thlaspi rostratum* N. Busch **CR**

There is a single location known for this plant species in the mountains of the Nakhchivan region. A continuing decline in the number of mature individuals is caused by human-induced habitat degradation related to the clear-cutting of forests. No conservation measures are in place at the current time.

CAMPANULACEAE

8. *Campanula fominii* Grossh. **VU**

There is a single known location in the Guba region, in the Greater Caucasus. It is possible that in the near future this species will be re-assessed as CR or even EX category. Population monitoring and maintenance are necessary. This is a potential ornamental species, having significance as food. Its threat is from over-collection by the local population.

CARYOPHYLLACEAE

9. *Silene talyschensis* Schischk. **VU**

Only a single location for this plant species is known from Talysh. It may be susceptible to the effects of human activities (haymaking is identified as the major possible threat) or stochastic events within a very short time period in an uncertain future. Its assessment is thus capable of changing to CR or even EX. The monitoring of population trends as well as maintenance of the known site for the species and control over haymaking are necessary. Threats include both haymaking and overgrazing.

CHENOPODIACEAE

10. *Anabasis eugeniae* Iljin **VU**

There is only one location with a small population in Nakhchivan, in the regions of Julfa and Ordubad. Although currently the population is increasing, the monitoring of population trends as well as maintenance of the known site are necessary. Major threats include both overgrazing and trampling.

DRYOPTERIDACEAE

11. *Dryopteris raddeana* (Fomin) Fomin **EN**

This fern species is known from two locations in Hyrkan National Park. The number of mature individuals continues to decline. Its main threat is the loss of habitat because of the expansion of agricultural lands and deforestation.

12. *Polystichum kadyrovii* Askerov & A. E. Bobrov **VU**

This is a rare, local endemic species of fern that occurs in Hyrkan National Park and in other parts of Talysh, from Masalli-Yardimli, Astara, and Lenkeran. Its main threat is from deforestation.

FABACEAE

13. *Amoria talyschensis* (Khalilov) Roskov **VU**

This legume species is known from only one locality in Talysh. Although the population is currently stable, the monitoring of population trends as well as maintenance of the sites and controlled grazing are necessary. This is a valuable fodder species, with its major threat from over-grazing.

14. *Astragalus bakuensis* Bunge **CR**

This critically endangered species occurs at a single location in the Absheron region and shows a continuing decline. The number of subpopulations and mature individuals is decreasing because of construction. No conservation measures are in place at the current time. This legume species has importance as a fodder plant. Its main threat is from the active construction of coastal developments.

15. *Astragalus igniarius* Popov **EN**

The species occurs at only a single location in the Absheron region. Within a very short time period in an uncertain future, this legume species may become CR or even EX. Although currently the population is increasing, the monitoring of population trends is necessary. Its major threat is from the active construction of coastal developments.

16. *Astragalus kubensis* Grossh. **VU**

There is only a single location known for this legume species in the Guba region, which is in the Greater Caucasus. This species grows in alpine pastures and has importance as fodder. It may be sensitive to the effects of human activities (hay-making) or stochastic events (effects of global warming), so that within a very short time period in an uncertain future the plant species may become CR or even EX. Although the population is currently increasing, the monitoring of population trends is necessary. Threats include overgrazing and soil erosion.

17. *Astragalus maraziensis* Rzazade **EN**

A rare species growing in the Absheron and Gobustan areas, this legume occurs at only five known locations. No conservation measures are in place. This species has importance as fodder. Threats include infrastructure development, construction, and overgrazing.

18. *Astragalus schachbuzensis* Rzazade **CR**

This species is found at only a single location in the montane part of Nakhchivan. The distributional area of this legume is shrinking, and the number of mature plants is reduced every year. No conservation actions are in place, although this is a valuable fodder species. Its main threat is from overgrazing that may lead to soil erosion.

GERANIACEAE

19. *Erodium schemachense* Grossh. **NT**

A single location is known in the Guba region, in the Greater Caucasus. The plant species, which grows in an area trampled by cattle, may be subject to the effects of human activities as well as the effects of stochastic events within a very short time period in an uncertain future. Its assessment is thus capable of changing to CR or even EX. Although the population is currently stable, the monitoring of population trends as well as maintenance of the site are necessary. As a fodder species, its main threat is from human-induced degradation because the species grows in pastures and is subject to overgrazing and trampling.

HYACINTHACEAE

20. *Bellevalia albana* Woronow **VU**

There is a single location known in the Shirvan region for this potential ornamental species. The number of mature plants is continuing to decline because of poor reproduction after the annual collection of flowering stems by local people. Its principal threat is from over-collection.

21. *Muscari elegantulum* Schchian **NT**

This is known from only a single location in Hyrkan National Park and from a very restricted area. It may be subject to the effects of human activities (collection as an ornamental plant is a plausible threat) or stochastic events within a very short time period in an uncertain future, despite the current stability of the population and the fact that the species is protected in Hyrkan National Park. As a potential ornamental species, its major threat is from over-collection.

22. *Muscari grossheimii* Schchian **NT**

This is known from a single location in the Lerik region. The species may be subject to the effects of human activities (collection as an ornamental plant is a plausible threat) or stochastic events within a very short time period in an uncertain future, despite the current stability of the population and the fact that the species is protected in Hyrkan National Park. As a potential ornamental plant, its principal threats are from overgrazing and haymaking.

23. *Ornithogalum hyrcanum* Grossh. **VU**

There are no more than five locations known for this species in the Lerik region and Hyrkan National Park. This plant species may be subject to the effects of human activities (haymaking is identified as a plausible threat) or stochastic events within a very short time period in an uncertain future, despite the current stability of the population and the fact that the species is protected in Hyrkan National Park. It is a potential ornamental plant, and its principal threats are from over-collection and haymaking.

HYPERICACEAE

24. *Hypericum theodorii* Woronow **EN**

There are two or three locations in Gobustan and in other parts of the Greater Caucasus. The number of plants continue to decline or at least the number of mature individuals as a result of collection by the local population for medicinal use. There are still no conservation measures in place to protect the species, which is an important medicinal and ornamental plant. Its principal threat is from over-collection by the local population.

IRIDACEAE

25. *Crocus polyanthus* Grossh. **CR**

This species is known from a single location in Talysh, where the number of mature individuals is declining because of collection by the local population as an ornamental plant. An important medicinal plant, this species also has significance as a dye. Its major threat is from over-collection by the local population.

26. *Iris camillae* Grossh. **VU**

Known from five locations in three districts of Azerbaijan in the Lesser Caucasus, the total area of occupancy of this species is estimated to be approximately 50 km^2. The population is currently stable but is threatened by over-collection. This is a potential ornamental plant, with its principal threat from over-collection.

27. *Iris hyrcana* Woronow ex Grossh. **VU**

A single location is known from Hyrkan National Park (Lenkeran region) for this species. It may be subject to the effects of overgrazing or stochastic events within a very short time period in an uncertain future and is thus capable of becoming CR or even EX, despite the current stability of the population. This is a potential ornamental species, and its principal threats are over-collection by the local population and overgrazing.

28. *Iris schelkownikowii* (Fomin) Fomin **EN**

This is a rare species distributed in the Gazakh region of the Lesser Caucasus and in the Goychay region, in the foothills of the Greater Caucasus. As a potential ornamental species, it is threatened by over-collection by the local population and by tourists.

LAMIACEAE

29. *Scutellaria rhomboidalis* Grossh. **CR**

A single location is known in the mountainous part of Nakhchivan for this species. It has been observed and projected that there will be a continuing decline in the number of mature individuals caused by its harvesting by the local population for medicinal purposes. Its principal threat is from over-collection by the local population.

30. *Thymus karamarjanicus* Klokov & Des.-Shost. **NT**

There are five localities in the Greater and the Lesser Caucasus and in the Kyur-Araz lowland that are known for this species. The plants are intensively collected by the local population for medicinal purposes, which, along with haymaking, contribute to its continuing decline. Therefore, the monitoring of the populations is necessary. As a valuable medicinal plant, this species is threatened by over-collection by the local population and by haymaking.

LILIACEAE

31. *Fritillaria grandiflora* Grossh. **CR**

This species occurs at a single location in the Lerik district of Talysh. There is a continuing decline in the number of mature individuals as well as the number of subpopulations caused by over-collection of flowers and bulbs by the local population. No conservation measures are in place. Because of its wonderful ornamental value, it is threatened by over-collection.

MORACEAE

32. *Ficus hyrcana* Grossh. **NT**

This species grows in forests of Talysh, in Hyrkan National Park, where it is a common component of the understory. Currently the population is stable. However, plants outside the protected area may be subject to the effects of human activities such as habitat degradation from agricultural land development on the area still covered by Hyrcanian forests. Fruits of this tree species are edible. This species is threatened by habitat degradation caused by agricultural land development and deforestation.

ORCHIDACEAE

33. *Orchis schelkownikowii* Woronow **EN**

This species is known from three locations in Talysh. The number of mature individuals is decreasing because of over-grazing. An ornamental plant species with a small population and weak reproduction, this species is also threatened by over-collecting.

PAPAVERACEAE

34. *Papaver schelkovnikovii* N. Busch **EN**

A rare, ornamental species distributed in Kyur-Araz lowland, Gobustan, in the central part of the Lesser Caucasus, it is known from two locations. Despite the current stability of some populations, continuing decline is reported for others. The cause is overgrazing. No measures are in place to protect the species in the wild. Its principal threat is habitat degradation (soil erosion and desertification).

35. ***Papaver talyshense*** **Grossh. VU**

There is only a single location in Talysh. It may be subject to effects of human activities (overgrazing, over-collection, and possibly haymaking) or stochastic events within a very short time period in an uncertain future and is thus capable of becoming CR or even EX. Although currently the population is increasing, the monitoring of population trends as well as maintenance of the site are necessary. A potential ornamental plant, this species is threatened by over-collecting, overgrazing, and possibly haymaking.

PLUMBAGINACEAE

36. ***Acantholimon tenuiflorum*** **Boiss. EN**

A species distributed in the Azerbaijani part of the Lesser Caucasus, it is known from two or three localities. The distribution area is shrinking, and the number of mature individuals is diminishing as well. This species has potential ornamental value and is threatened by human-induced habitat loss and degradation, agricultural activities, and livestock.

POACEAE

37. ***Bromus tzvelevii*** **Mussajev CR**

A species known from only a single population in Nakhchivan (Ordubad region). This is a forage species whose major threat is overgrazing.

38. ***Dactylis hyrcana*** **(Tzvelev) Mussajev VU**

There is a single location in Talysh, and the number of plants continues to decline (at least the number of mature individuals) because of poor regeneration. This species is significant as fodder. Weak reproduction, small population size, and overgrazing are the main reasons for the declining population.

39. ***Stipa karjaginii*** **Mussajev & Sadychov VU**

There is a single location in the middle and high montane part of the Nakhchivan region. Despite current stability of the population, it is susceptible to the effects of human activities (grazing) or stochastic events within a very short time period and is thus capable of becoming CR or even EX. No conservation measures are in place. Its small population size and weak reproduction are major concerns. The main threat is overgrazing.

POLYGONACEAE

40. ***Calligonum bakuense*** **Litv. EN**

This species occurs at five known locations in Azerbaijan, and the number of individuals is continuing to decline. It is under threat from active coastal construction and infrastructure development.

41. ***Polygonum caspicum*** **Kom. CR**

This plant species is known from only a single location in Absheron. The number of mature individuals is continuing to decline because of human impact. The species grows in ruderal places, gardens, and along roads; and no measures are in place to protect it in the wild. Threats include human-induced habitat degradation (development and construction).

RANUNCULACEAE

42. ***Anemone kusnetzowii*** **Woronow ex Grossh. CR**

There are just a few populations at only one location in the Lesser Caucasus, in the Dashkesan area. The number of mature individuals is continuing to decline because of over-collection of this highly ornamental plant. Urgent measures should be undertaken to protect the species in the wild. Threats include human-induced habitat loss and degradation, agriculture, and over-collection.

ROSACEAE

43. ***Rosa abutalybovii*** **Gadzh. NT**

This is a rare species known from just a few locations in Nakhchivan, from the Shaxbuz area and in Guba from the Shakhdag area. The species is not protected in the wild, which makes it subject to the effects of human activities, such as cutting for fuel, or stochastic events, and its conservation assessment is capable of becoming CR or even EX in a very short time period. Current population trends need to be monitored. This is a valuable medicinal plant species. Its principal threats are cutting for fuel and over-collection by the local population.

44. ***Rosa alexeenkoi*** **Crép. ex Juz. EN**

This is a rare species, distributed in the Guba region of the Greater Caucasus. This is a species with valuable medicinal and ornamental significance, and its principal threat is from over-collection by the local population.

45. ***Rosa gadzhievii*** **Chrshan. & Iskend. NT**

This is a rare species known from only a single location close to Ganja city in open forests. Currently, the species is protected in the Gek-gel Reserve and used in horticulture. It may be subject to the effects of human activities in the wild, such as habitat degradation caused by clear-cutting, and needs stricter habitat management. This is a species of valuable medicinal and ornamental importance, with its principal threat from over-collection by the local population.

46. ***Rosa karjaginii*** **Sosn. NT**

The species is already included in the *Red Book of Azerbaijan* (Adighesal et al., 1989) and is a very rare endemic species. Only a single location is known from the Shamakhi region, Greater Caucasus. Although the population is currently stable and the species is common in the area of its distribution, it may be subject to the effects of human activities in the wild such as habitat degradation or over collecting of fruits. The species needs more strict habitat management. An important medicinal and ornamental species, it is under threat from over-collection for medicinal purposes.

47. ***Rosa nisami*** **Sosn. NT**

This species is included in the *Red Book of Azerbaijan* (Adighesal et al., 1989). It is known from no more than five locations, some in the Lesser Caucasus, in the Gey-Gol region, and Nakhchivan. Despite the fact that the population is currently stable, the population density is relatively low, and the species may be susceptible to the effects of human activities such as use for fuel or stochastic events. A valuable medicinal and ornamental plant, it is under threat from over-collection and by its use by the local population for fuel.

48. ***Rosa zakatalensis*** **Gadzh. NT**

This rose species is distributed in the Zakatala region, in the Greater Caucasus, where it is known from two locations. Despite the current stability of the populations, the population density is relatively low, and it may be susceptible to the effects of human activities or stochastic events. As a valuable medicinal and ornamental species, this is under threat from the over-collection of fruits for medicinal purposes by the local population.

SANTALACEAE

49. ***Thesium maritimum*** **C. A. Mey. VU**

There are from six to eight localities in Absheron and the Kyur-Araz lowland on sandy and clay soils. There is a continuing decline of habitat caused by habitat degradation related to construction. The main threat to this species is from active construction and infrastructure development along the coast.

SCROPHULARIACEAE

50. ***Verbascum stachydiforme*** **Boiss. & Buhse CR**

There is a single location known in the Talysh region for this species. There are a number of subpopulations, and the number of mature individuals is declining because of over-collecting by local people. No conservation measures are in place to protect the species, which is a valuable medicinal plant with small populations and weak reproduction. Its principal threat is from over-collection by the local population.

AZERBAIJAN MAPS

1. *Allium mariae* Bordz.

2. *Sternbergia alexandrae* Sosn.

3. *Ferula caucasica* Korovin

4. *Centaurea kobstanica* Tzvelev

5. *Scorzonera grossheimii* Lipsch. & Vassilcz.

6. *Aethionema levandowskyi* N. Busch

7. *Thlaspi rostratum* N. Busch

8. *Campanula fominii* Grossh.

9. *Silene talyschensis* Schischk.

10. *Anabasis eugeniae* Iljin

11. *Dryopteris raddeana* (Fomin) Fomin

12. *Polystichum kadyrovii* Askerov & A. E. Bobrov

13. *Amoria talyschensis* (Khalilov) Roskov

14. *Astragalus bakuensis* Bunge

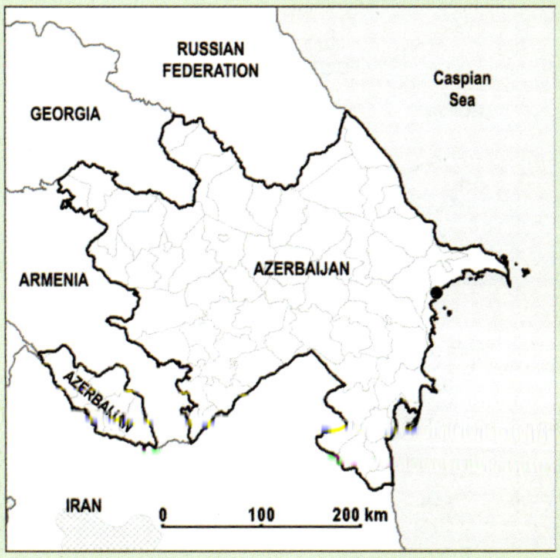

15. *Astragalus igniarius* Popov

16. *Astragalus kubensis* Grossh.

17. *Astragalus maraziensis* Rzazade

18. *Astragalus schachbuzensis* Rzazade

19. *Erodium schemachense* Grossh.

20. *Bellevalia albana* Woronow

21. *Muscari elegantulum* Schchian

22. *Muscari grossheimii* Schchian

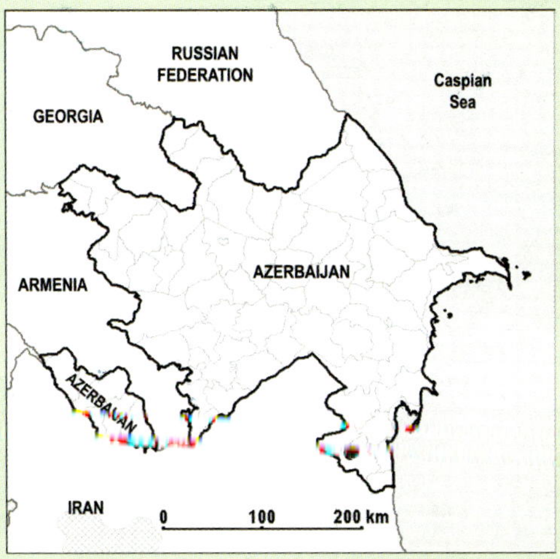

23. *Ornithogalum hyrcanum* Grossh.

24. *Hypericum theodorii* Woronow

25. *Crocus polyanthus* Grossh.

26. *Iris camillae* Grossh.

27. *Iris hyrcana* Woronow ex Grossh.

28. *Iris schelkownikowii* (Fomin) Fomin

29. *Scutellaria rhomboidalis* Grossh.

30. *Thymus karamarjanicus* Klokov & Des.-Shost.

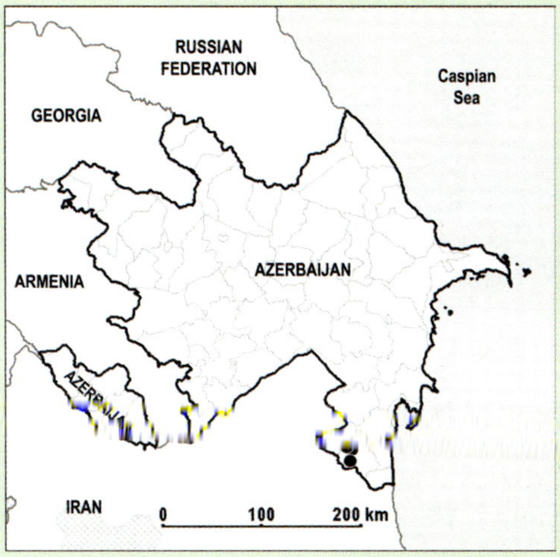

31. *Fritillaria grandiflora* Grossh.

32. *Ficus hyrcana* Grossh.

33. *Orchis schelkownikowii* Woronow

34. *Papaver schelkovnikovii* N. Busch

35. *Papaver talyshense* Grossh.

36. *Acantholimon tenuiflorum* Boiss.

37. *Bromus tzvelevii* Mussajev

38. *Dactylis hyrcana* (Tzvelev) Mussajev

39. *Stipa karjaginii* Mussajev & Sadychov

40. *Calligonum bakuense* Litv.

41. *Polygonum caspicum* Kom.

42. *Anemone kusnetzowii* Woronow ex Grossh.

43. *Rosa abutalybovii* Gadzh.

44. *Rosa alexeenkoi* Crép. ex Juz.

45. *Rosa gadzhievii* Chrshan. & Iskend.

46. *Rosa karjaginii* Sosn.

47. *Rosa nisami* Sosn.

48. *Rosa zakatalensis* Gadzh.

49. *Thesium maritimum* C. A. Mey.

50. *Verbascum stachydiforme* Boiss. & Buhse

BIBLIOGRAPHY

Adighesal, B. M., U. K. Alakparov, G. A. Aliev, V. D. Hajiyev, S. A. Isirafilov, M. A. Musaev, I. D. Mustafaev & I. S. Safarov (editors). 1989. Red Book of Azerbaijan. State Committee of Nature Conservation of Azerbaijan and Azerbaijan Academy of Sciences. Elm Publishing House, Baku.

Akhundov, G. F. 1972. Endemy Azerbaijana. [Endemics in Azerbaijan.] Institut Botaniki, Baku [in Russian].

Akhundov, G. F., E. E. Gogina & L. I. Prilipko. 1978. Uzkoendemichnye i redkie vidy prirodnoi flory Nakhichevanskoi SSR. [Local and rare endemics native to Nakhchivan SSR flora.] Trudy Bot. Sada Akad. Nauk SSSR 107: 54–62 [in Russian].

Asadov, K. S. 1981. Dikorastuschie plodovye rastenia Bol'shogo Kavkaza. [Wild Fruit Trees in the Greater Caucasus.] Elm Publishing House, Baku [in Russian].

Asadov, K. S. & A. K. Asadov. 2001. Dikorastuschie plodovye rastenia Azerbaijana. [Wild Fruit Plants of Azerbaijan.] Azerbaijan Milli Encylopedia, Baku [in Russian].

Convention on Biological Diversity (CBD). 2002. Conference of the Parties, COP 6 Decision VI/9. Global Strategy for Plant Conservation. 2002–2010. Sixth Ordinary Meeting of the Conference of the Parties to the Convention on Biological Diversity (COP 6), The Hague, Netherlands. <http://www.cbd.int/decision/cop/?id=7183>. Azerbaijan as approving country, <http://www.cbd.int/countries/?country=az>, accessed 9 September 2013.

Convention on International Trade in Endangered Species of Wild Fauna and Flora (CITES). 1979 [22 June]. [Signed at Washington, D.C., on 3 March 1973; amended at Bonn, 22 June 1979.] <http://www.cites.org/eng/disc/E-Text.pdf>, accessed 9 September 2013.

Gadzhiev, V. D. 1954. Rastitel'nost' Bolshogo Kavkaza. [The Greater Caucasus Vegetation.] Elm Publishing House, Baku [in Russian].

Gadzhiev, V. D. 1990. Vysokogornaja rastitel'nost' Malogo Kavkaza (v predelakh Azerbaijana). [High Mountain Vegetation in the Lesser Caucasus (within Azerbaijan Borders).] Elm Publishing House, Baku [in Russian].

Gadzhiev, V. D. & S. Musaev. 1996. Rastenia i rastitel'nye formatsii rekomenduemye dlj vnesenia v "Krasnyu" i "Zelenyu" knigi Azerbaijana. [Plants and Plant Communities Recommended to Include in "Red" and "Green" Books of Azerbaijan.] Elm Publishing House, Baku [in Russian].

Gadzhieva, G. G. 1969. Shipovniki Uzhnogo sklona Bol'shogo Kavkaza (v predelakh Azerbaijana). [Roses on the Southern Slope of the Greater Caucasus (within Azerbaijan Borders).] Dissertatsia Kand. Nauk. In-t Botaniki, Baku [in Russian].

Gadzhieva, G. G. & A. T. Iskenderov. 1971. Materialy k izucheniju shipovnikov Malogo Kavkaza (v predelakh Azerbaijana). [Essay on Rosa studies in the Lesser Caucasus (within Azerbaijan borders).] Izv. Akad. Nauk Azerbaidzhansk. SSR 7: 18–25 [in Russian].

Garakhani, P., P. Askerova & O. Ibadov. 1996. Issledovanie roda Ornithogalum L. v Azerbaijane. [Studies on the genus Ornithogalum L. in Azerbaijan.] Trudy pervoi Vserossiskoi konf. po botanicheskomu resursov. Sankt-Peterburg, Russia [in Russian].

Grossheim, A. A. 1939–1967. Flora Kavkaza, 2-izdanie. Tt. 1–7. [Flora of the Caucasus, 2nd ed. Vol. 1–7.] Trudy Bot. Inst. Akad. Nauk, SSSR Azerbaidzhansk. Fil., Baku. [in Russian].

Grossheim, A. A. 1949. Opredelitel' rastenii Kavkaza. [Manual of the Caucasian Plants.] Sovetskaia Nauka, Moscow [in Russian].

International Union for Conservation of Nature and Natural Resources (IUCN). 2001. IUCN Red List Categories and Criteria, Version 3.1. Prepared by the IUCN Species Survival Commission. IUCN, Gland, Switzerland, and Cambridge, United Kingdom.

Iskenderov, A. T. 1969. Bioecological osobennosti nekotoryx vidov shipovnikov Azerbaijana, ix introdukzia i ispol'zovanie. [Biological characters of some species of the genus Rosa occurring in Azerbaijan, their introduction and usage.] Bot. Inst., Baku [in Russian].

Karyagin, I. I. 1950–1961. Flora Azerbaijana. [Flora of Azerbaijan.] Acad. Sci. Azerbaijan SSR, Baku [in Russian].

Kasumov, F. Y. 2004. Issledovaniae khimicheskogo sostava efirnykh masel v rode Thymus L. proizrastaushikx v Azerbaijane. [Chemical studies on essential oil in Thymus L. occurring in Azerbaijan.] Pp. 274–277 *in* I ix poleznye svoistva. V: Geneticheskie resursy lekarstvennykx i aromaticheskikx rastenii. [Genetical Resources of Medicinal and Aromatic Plants.] VILAR, All-Russian Institute of Medicinal and Aromatic Plants, Moscow [in Russian].

Movsumova, F. G. 1991. Rastitel'nost' soljankovykh pustyn'. [Vegetation in saltwort desert.] Pp. 79–80 *in* Flora i rastitel'nost' soljankovyx pustyn' Nakhchivanskoi AR. [Flora and Vegetation of Saltwort Desert in Nakhchivan AR.] Shams, Baku [in Russian].

Musayev, S. G. 1991. Zlaki Azerbaijana. [Grasses in Azerbaijan.] Elm Publishing House, Baku [in Russian].

Prilipko, L. I. 1961–1970. Derev'ja i kustarniki Azerbaijana. Tt. 1–3. [Trees and Shrubs of Azerbaijan, Vols. 1–3.] Acad. Sci. Azerbaidzhansk. SSR, Baku [in Russian].

Prilipko, L. I. & E. Gogina. 1978. Redkie vidy prirodnoi flory Talysha zasluzhivaushie okhrany. [Rare species in Talysh flora that deserve protection.] Bull. Main Bot. Gard. 107: 62–68 [in Russian].

Ramsar Convention on Wetlands (RCW). 1971 [2 Feb.]. Convention on Wetlands of International Importance especially as Waterfowl Habitat. Ramsar (Iran). UN Treaty Series No. 14583. As amended by the Paris Protocol, 3 December 1982, and Regina Amendments, 28 May 1987. <http://www.ramsar.org/cda/en/ramsar-about-about-ramsar/main/ramsar/1-36%5E7687_4000_0__>, accessed 9 September 2013.

United Nations. n.d. United Nations Convention to Combat Desertification (UNCCD). <http://www.unccd.int/en/Pages/default.aspx>, accessed 4 September 2013.

United Nations. 1992. United Nations Framework Convention on Climate Change. FCCC/INFORMAL/84. GE.05-6222 (E) 200705. <http://unfccc.int/resource/docs/convkp/ conveng.pdf>, accessed 9 September 2103. [Azerbaijan as Party to the Convention, 1 October 1995.]

GEORGIA

საქართველო

გ. ნახუცრიშვილი[1], †რ. გაგნიძე[1], შ. შეთეკაური[2], ზ. მანველიძე[3], ნ. მემიაძე[4], დ. ხარაზიშვილი[4], ქ. ბაცაცაშვილი[1]

[1]ილიას სახელმწიფო უნივერსიტეტი, [2]ივანე ჯავახიშვილის სახ. თბილისის სახელმწიფო უნივერსიტეტი, [3]ბათუმის შოთა რუსთაველის სახელმწიფო უნივერსიტეტი, [4] ბათუმის ბოტანიკური ბაღი

კავკასიის ეკორეგიონის (საქართველო, სომხეთი, აზერბაიჯანი, ჩრდ. კავკასია (რუსეთის ფედერაცია), ჩრდ.-აღმ. თურქეთი, ჩრდ.-დას. კასპიისპირა ირანი) ფლორა დიდი მრავალფეროვნებით გამოირჩევა და ჭურ-ჭლოვან მცენარეთა 7,000-მდე სახეობას ითვლის.

ეს წიგნი ეძღვნება კავკასიის ენდემებს – მცენარეებს, რომლებიც მსოფლიოში მხო-ლოდ ამ რეგიონში გვხვდება. ამ სამუშაოს მიზანი, რომლის შედეგებიც ამ წიგნშია მოცემული, იყო დაგვეზუსტებინა ამ მცენარეთა გავრცელება კავკასიის ქვეყნებში, დაგვედგინა, თუ რა საფრთხეები ემუქრება ამ მცენარეებს და შეგვეფასებინა მათი გაქრობის საშიშროება.

შეფასება მოხდა ბუნების დაცვის საერთაშორისო კავშირის (IUCN) მიერ შემუშავებული კატეგორიებისა და კრიტერი-უმების სისტემით: გადაშენებული (EX), ველურ ბუნებაში გადაშენებული (EW), კრიტიკულ საფრთხეში მყოფი (CR), საფრთხეში მყოფი (EN), მოწყვლადი (VU), საფრთხეში მყოფთან მიახლოვებული (NT), დაცვას არ საჭიროებს (LC), არასაკმარისი მონაცემების მქონე (DD).

საქართველოს მცენარეული სამყარო მდიდარი და მრავალფეროვანია. ეს განპირობებულია ქვეყნის ტერიტორიის ფიზიკურ-გეოგრაფიული პირობების ნაირგვა-რობითა და კონტრასტულობით, მისი მდებარეობით განსხვავებული ბოტანიკურ-გეოგრაფიული პროვინციების შესაყარზე, გეოგრაფიული წარსულის სირთულით.

საქართველოს ტერიტორიაზე აღრიც-ხულია ჭურჭლოვანი მცენარეების (გვიმრა-ნაირები, შიშველ- და ფარულთესლოვანები) 4,100-ზე მეტი, ხავსების 800-მდე და წყალმცენარეების 2,000-მდე, სოკოების 6,500-ზე მეტი და ლიქენების 700-მდე სახეობა. მაღალია საქართველოს ფლორის ენდემიზმის დონე: 1,300-მდე სახეობა კავკასიის ენდემია, მათგან 275 – საკუთრივ საქართველოსი.

საქართველოში გავრცელებულია მცენარეულობის მრავალნაირი ტიპი შავიზღვისპირა ქვიშნარების მცენარეულობით დაწყებული, კავკასიონის თოვლისა და ყინვარისპირა მცენარეულობით დამთავრებული, აღმოსავლეთ საქართველოს უდაბნოს მცენარეული დაჯგუფებებით დაწყებული კოლხეთის უღრანი ტყეებით დამთავრებული.

საქართველოსთვის, როგორც მთაგორიანი ქვეყნისთვის, დამახასიათებელია მცენარეულობის ვერტიკალური სარტყლიანობა. აღმოსავლეთ საქართველოში ძირითადი სარტყლებია: ნახევრად-უდაბნოს, მშრალი სტეპისა და არიდული მეჩხერი (ნათელი) ტყის, მთის ტყის, სუბალპური, ალპური, სუბნივალური და ნივალური; დასავლეთ საქართველოში: ტყის, სუბალპური, ალპური, სუბნივალური და ნივალური.

წარსულში საქართველოს ტერიტორია თითქმის მთლიანად ტყით იყო დაფარული. ამჟამად, ბევრგან ტყე გაჩეხილია და მიწა სასოფლო-სამეურნეო საქმიანობისთვის გამოიყენება. საქართველოს ტყის ძირითადი ფორმაციებია: წიფლნარები, მუხნარები, რცხილნარები, წაბლნარები, სოჭნარები, ნაძვნარები, ფიჭვნარები (განსაკუთებით აღსანიშნავია კავკასიის ენდემის – ბიჭვინთის ფიჭვის კორომები, რომლებიც შემორჩენილია მხოლოდ შავი ზღვის სანაპიროზე – ბიჭვინთა, მიუსერა), არყნარები, უთხოვრიანები (ერთი ასეთი ტყე დაცულია ბაწარის ნაკრძალში), ძელქვიანები (დაცულია კოლხეთის, აჯამეთის, ბაბანეურის ნაკრძალებში) და სხვ. კოლხეთის ტყეებისათვის განსაკუთრებით დამახასიათებელია ხშირი ქვეტყე, რომელსაც ქმნის მესამეული პერიოდის თბილი და ტენიანი კლიმატის დენდროფლორის მარადმწვანე რელიქტები.

აღმოსავლეთ საქართველოში, დაბლობსა და მთისწინეთში დიდ ტერიტორიაზე გავრცელებულია მშრალი სტეპები.

სუბალპური სარტყლის მცენარეულობის ძირითადი ტიპებია: სუბალპური ტანბრეცილი ტყეები, სუბალპური მაღალბალახეულობა, სუბალპური მდელოები, სუბალპური ბუჩქნარები (ძირითადად დეკიანი). ალპურ სარტყელში უფრო ხშირია მარცვლოვანთა და ისლისებრთაგან შექმნილი მდელოები.

სუბნივალურ სარტყელში მეჩხერი მცენარეულობაა, რომელსაც გართხმული ძლიერ შებუსვილი, სქელფოთლიანი, ბალიშისმაგვარი ფორმის ბალახოვანი მცენარეები ქმნის. მცენარე ნივალურ სარტყლეში იშვიათია. მუდმივი თოვლისა და ყინულისაგან თავისუფალ ადგილებში იზრდება ხავსები, ლიქენები, წყალმცენარეები და თითოოროლა ყვავილოვანი მცენარე.

ტბის ნაპირებსა და თხელ წყლებში, მოდებულია ლერწამი, ლაქაში და სხვ. მდორე წყლებში, იზრდება დუმფარა, წყლის პერი, წყლის ვაზი, წყლის კაკალი, წყლის ბაია; ჭაობებში – ისლისა და ჭილის სახეობები, წყლის ზამბახი და სხვ.

უკანასკნელ ათწლეულებში საქართველოს მცენარეთა მრავალფეროვნებაზე ადამიანის საქმიანობის გავლენა გაძლიერდა – იჩეხება ტყეები, გადატვირთულია საძოვრები, დაუნდობლად მოიპოვება ეკონომიკური მნიშვნელობის მცენარეები. აუცილებელია მათ გადასარჩენად პრევენციული ღონისძიებების განხორციელება, რისთვისაც ერთ-ერთი ყველაზე ეფექტური გზა ახალი დაცული ტერიტორიების შექმნაა.

საქართველოს ენდემურ მცენარეთა წითელი ნუსხის შედგენა ამ ღონისძიებათაგან ერთ-ერთი ყველაზე მნიშვნელოვანია. საქართველოს ენდემები გადაშენების საფრთხის შემდეგ კატეგორიებში გადანაწილდა: EW – 3; CR – 6; EN – 21; VU – 80; NT – 31; LC – 17; DD – 117.

ამჟამად მოქმედ საქართველოს "წითელ ნუსხაში" შეტანილია მცენარეთა 56 სახეობა (მერქნიანი მცენარეები). საქართველოს ბევრი ენდემური მცენარე, რომლებსაც გადაშენების საფრთხე ემუქრება (CR, EN, VU), კანონით დაცული არ არის. რეკომენდაცია ამ სახეობათა "წითელ ნუსხაში" შეტანის შესახებ წარედგინება საქართველოს გარემოს დაცვის სამინისტროს.

GEORGIA

Georgi Nakhutsrishvili, Revaz Gagnidze†, Shamil Shetekauri, Zurab Manvelidze, Nino Memiadze, David Kharazishvili, and Ketevan Batsatsashvili

CAPITAL

Tbilisi.

OTHER CITIES

Telavi (Kakheti region), Kutaisi (Imereti region), Batumi (Adjara region), Sokhumi (Abkhazeti region).

POPULATION

According to CIA World Factbook, 4,555,911 (July 2013 estimate).

LANGUAGE

Georgian.

Georgians call themselves *kartveli* (pl. *kartvelebi*), and the name of the country in the Georgian language is *Sakartvelo*. The language, which belongs to the group of Kartvelian languages within the family of Iberian-Caucasian languages, possibly is related to West Asian languages and the Basque language (of the Basque region, located in Spain and France on the northern Iberian Peninsula, adjacent to the Pyrenees), while connections to the Indo-European languages are less certain. The written language in Georgia dates back to the third century BC, and its development was supported by the adoption of Christianity as the official religion in 337 AD. Ancient Georgian manuscripts, as well as other artifacts, indicate an early economic, social, and cultural development of the Georgian people (Lortkipanidze, 2002), who were especially skilled in the fields of farming, cattle breeding, and metallurgy.

FLORA

4130 (total species of vascular plants); 276 (total national endemics); ca. 1330 (Caucasian region endemics).

BOTANICAL INSTITUTIONS

The Tbilisi Institute of Botany was established in 1933 by combining several departments of the Tbilisi Botanical Garden, in existence since the 18th century. The herbarium collections were initiated in the middle of the 19th century. The central mission of the scientific activities conducted at the institute is to study Georgia's plant diversity and to develop scientific principles and practical recommendations for its protection and stable utilization.

From its foundation, the Tbilisi Institute of Botany has published comprehensive works on the floristic diversity of Georgia. *The Flora of Georgia*, consisting of eight volumes (Makashvili et al., 1941–1952), was the most significant of these published works. Since the 1970s, the Institute has been working on the second edition of *The Flora of Georgia*, the 17th volume of which is now in press (Ketskhoveli et al., 1971–2011). Monographs on the systematic structure of separate taxa (genera, tribes, families) within Georgia's flora, as well as checklists of certain groups of cryptogams, gymnosperms, and angiosperms, have also appeared. *Vascular Plants of Georgia: A Nomenclatural Checklist* represents the most recent work (Gagnidze, 2005). *The Vegetation of Georgia* was the first English-language book on Georgia's plant assemblages (Nakhutsrishvili, 1999). Comparative analyses of the vegetation diversity of the Caucasus with other high-mountain regions of Europe have been conducted and are reflected in a number of scholarly publications. The Institute staff collaborated in the production of the *Map of the Reconstructed Vegetation of Central and Eastern Europe* (Komarov Botanical Institute, 1996) and *Map of the Natural Vegetation of Europe* (Dolukhanov & Nakhutsrishvili,

2003; Nakhutsrishvili, 2005). Scientists of the Institute have investigated the reproduction of rare and threatened plant species and have carried out ex situ conservation work for many decades. In 2004, The Caucasus Regional Seed Bank was established at the Institute. Its staff is involved in seed stock activities within the framework of the Millennium Seed Bank Project, Royal Botanic Gardens, Kew. The Tbilisi Institute of Botany cooperates with botanical institutions in the United States, Great Britain, Austria, Germany, Japan, Switzerland, and other countries. In recent years, the Institute has received ca. 50 grants from international organizations, foundations, and foreign botanical institutions. The Institute staff provides ecological expertise for activities associated with large development projects within Georgian territory.

The Bakuriani Alpine Botanical Garden was established in 1910 and has long been affiliated with the Tbilisi Institute of Botany. Situated at 1650 m, the Garden specializes in plants of the Caucasus region. Its living collections include ca. 500 species collected in the Caucasus, an arboretum of ca. 125 species of trees and shrubs (including 75 woody taxa from the Caucasus), and ca. 250 species of herbaceous plants from elsewhere in the world grown from seed. The Centre for Environmental Education was built at the Garden in 1997 with support from the World Wildlife Fund (Nakhutsrishvili et al., 2006). Address: 1, Botanikuri St., 0105 Tbilisi, Georgia; Tel.: +995 32 272 21 14; FAX: +993 32 272 34 09.

The botanical gardens in Georgia include the National Botanical Garden of Georgia in Tbilisi, the Batumi Botanical Garden in Adjara, the Sokhumi Botanical Garden in Abkhazeti, and the Kutaisi Botanical Garden in Kutaisi.

PROTECTED AREAS

Protected areas cover a total of 495,892 hectares, representing ca. 7% of the country's territory, ca. 75% of which is forested. There are 14 strict nature reserves (Batsara-Babaneuri, Bichvinta-Miusera, Kintrishi, Kobuleti, Lagodekhi, Liakhvi, Mariamjvari, Ritsa, Sataplia, Tusheti, and Vashlovani Strict Nature Reserves, and Pskhu-Gumista Nature Reserve, including Gumista, Pskhu, and Skrucha Strict Nature Reserves), nine national parks (Algeti, Borjomi-Kharagauli, Djavakheti, Kazbegi, Kolkheti, Mtirala, Tbilisi, Tusheti, and Vashlovani), and 18 managed nature reserves (such as Ajameti, Chachuna, Gardabani, Ilto, Katsoburi, Kobuleti, Ktsia-Tabatskuri, Korughi, Lagodekhi, and Tetrobi). There are 24 natural monuments (NM) (such as Vashlovani NM, Artsivis-Khevi NM, Takhti-Tepa mud volcanoes NM, Juma Bay NM, and Alazani Floodplains NM) and a further two protected landscapes (PL) (Kintrishi PL and Tusheti PL) in Georgia (http:// www.apa.gov.ge).

Georgia occupies an interesting geobotanical position as a central part of the Caucasus region. The country is characterized by quite diverse natural conditions that account for the very high degree of divergence of plant communities in a relatively restricted geographic area. The landscape of the country ranges from different desert and semi-desert vegetation types, which occur mainly in eastern Georgia, to the lush, Colchic forests supported by the moist, almost subtropical climate in western Georgia, and then to the high-mountain plant communities, which occur in the north and the south.

Georgia is an ancient Christian country, which occupies the central and southwestern portions of the Caucasus. It lies on an isthmus between the Black and the Caspian seas and extends over a land area of 69,700 km^2. In the west, it is bounded by the Black Sea, and in the east, by the intermountain depression of the Mtkvari (Kura) River in the Caspian lowland, also known as the South Caucasian Depression. The northern border is shaped by the peaks of the Greater Caucasus; the southern border transects the Lesser Caucasus in its eastern extent and separates the Lesser Caucasus from the adjacent mountain ranges of northeastern Asia Minor. Politically, Georgia borders the Russian Federation to the north, with Azerbaijan to the east, Turkey to the southwest, and Armenia to the southeast. The national borders have shifted several times during the history of Georgia.

Geographically the intermontane lowlands of Georgia, consisting of the Colchic lowland, adjacent to the Black Sea in western Georgia, and the foothills between the Greater Caucasus in the north and the Lesser Caucasus in the southwest, including the Mtkvari (Kura) valley in eastern Georgia, have environmental conditions that have long been favorable for settlement and farming.

Moving eastward from the Black Sea, the Colchic lowland merges with the Imeretian upland, which connects the Greater Caucasus to the Lesser Caucasus. The highest part of the upland, the Likhi or Surami range, divides the country into the two natural geographic regions of western and eastern Georgia, the former having been historically identified as Colchis and the latter as Iberia by ancient Greek and Roman writers.

The diversity of climate and relief in Georgia (as well as rich mineral resources) has been important for its economic development since early human history. Numerous rivers and streams (more than 25,000) supported intensive agriculture, without the need for irrigation systems. In addition, these rivers served as early trade routes and, due to their rapid current, were used for transportation of timber, particularly in Colchis.

OROGRAPHY

Geologically the country of Georgia belongs to the alpine system of Eurasia. Structurally the area may be divided into the following major landforms: (1) the Greater Caucasus known as the Kavkasioni; (2) Georgia's intermontane area, between the Greater and the Lesser Caucasus; and (3) the mountain system of the Lesser Caucasus known as the Meskheti-Trialeti Range, which includes the South Georgian volcanic upland. The Likhi or Surami range is a sub-meridian mountain range connecting the Greater and the Lesser Caucasus mountain systems. This range divides Georgia into eastern and western parts, and the Likhi Range at the same time forms the watershed between the Black and the Caspian seas (Abashidze, 1981).

CLIMATE

Various climatic zones can be distinguished, ranging from almost subtropical and humid conditions in western Georgia to semi-arid climes in eastern Georgia, yet with permanent snow and glaciers on the high mountains. Such a broad range of climatic conditions is caused by the influence of the Black Sea and also by the complex orographic structure of the country. The highest mean annual temperature is ca. 15°C on the Black Sea coast in Abkhazeti, western Georgia, and the lowest ca. –6°C at about 3700 m in the Greater Caucasus. Annual precipitation ranges from 200 to 300 mm in the Eldari lowland, East Georgia, to 4500 mm on Mt. Mtirala, Adjara, western Georgia (Abashidze, 1981).

SOILS

Soil types are quite variable in Georgia and the principal soil types are distinguished by corresponding altitudes. In western Georgia bog and podzolic soil zones are found in the lowlands, the red and yellow soil zone is known from the foothills, and a zone of montane forest and montane meadow soils appears at higher altitudes. Chestnut, black,

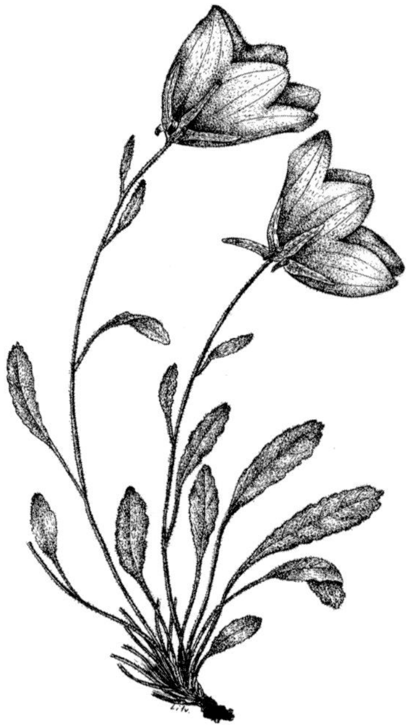

Campanula argunensis Rup., endemic to the Greater Caucasus region.

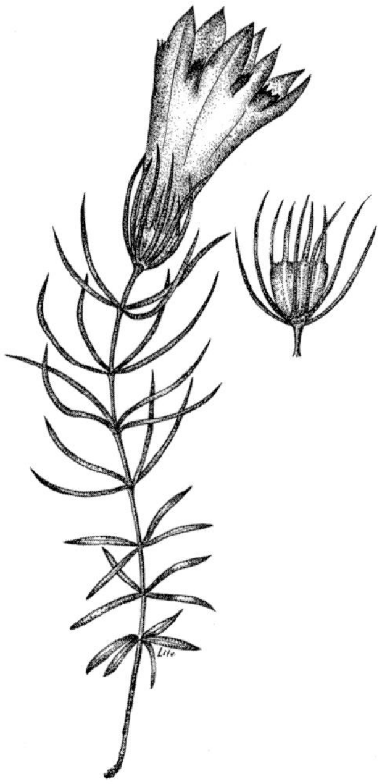

Gentiana paradoxa Albov, endemic to the western Caucasus region.

Leptopus colchicus (Fisch. & C. A. Mey. ex Boiss.) Pojark., endemic to the central Caucasus region.

Lonicera steveniana Fisch. ex Pojark., endemic to the western Caucasus region.

brown humic-sulphate, and saline soils of steppe and semi-desert, as well as interme-di-ate forest-steppe and montane meadow soils are found in eastern Georgia. Much of the Lesser Caucasus system, in southern Georgia, is distinguished by montane black soils and meadow black-like soils. In the highlands these are replaced by montane meadow soils. In addition, alluvial soils, redzinas, and brown soils, as well as meadow brown soils, occur here, with the predominance of brown forest soil type in the montane forest zone (Urushadze, 1997).

FLORA AND VEGETATION

The differences between the climates of eastern and western Georgia determine the diversity of their vegetation, which is also impacted by altitudinal zonation. Arid and semi-arid vegetation zones are absent from western Georgia. This accounts for a less complex profile of altitudinal zonation, represented by five main zones: forest (0–1900 m), subalpine (1900–2500 m), alpine (2500–3100 m), subnival (3100–3600 m), and nival (above 3600 m). In eastern Georgia, altitudinal zonation is more complicated. Here, six main zones can be recognized: desert, steppe, and arid light forest (150–600 m), forest (600–1900 m), subalpine (1900–2500 m), alpine (2500–3000 m), subnival (3000–3600 m), and nival (above 3600 m). The border between the semi-arid and forest zones varies, determined by climatic conditions and slope exposure. For example, within the forest and subalpine zones of the south Georgian upland, there are small areas occupied by semi-arid ecosystems, but the montane steppe vegetation is prevalent. Overall, almost all the altitudinal zones typical for the nemoral zone of the Northern Hemisphere are represented in Georgia.

The differentiation of vegetation along vertical gradients can be illustrated in the two markedly different parts of eastern and western Georgia. In eastern Georgia, the vegetation of deserts, semi-deserts, and hemi-xerophytic light forests is characterized by *Pistacia mutica* Fisch. & C. A. Mey. (pistachio tree, Anacardiaceae), *Celtis caucasica* Willd. (hackberry, Cannabaceae), and *Juniperus communis* L. (juniper, Cupressaceae). This is replaced by *Quercus iberica* Steven ex M. Bieb. (Georgian oak, Fagaceae) forests with an admixture of *Carpinus orientalis* Mill. (Caucasian hornbeam, Betulaceae) on southern slopes. At altitudes from 1000 to 1200 m, there is a narrow zone of Caucasian hornbeam–beech forest (*Carpinus caucasica* Grossh.–*Fagus orientalis* Lipsky), which is replaced higher up by *Quercus macranthera* Fisch. & C. A. Mey. ex Hohen. (high mountain oak, Fagaceae), which constitutes the treeline dominant. Above this, relatively dry meadows can be found at altitudes of 2250–2300 m. However, on northern slopes, above the hemi-xerophytic vegetation zone, there is a subzone of Caucasian hornbeam–oak forest (*Carpinus caucasica* Grossh.–species of *Quercus* L.), replaced higher up by beech forest, which occupies large areas from 600 to 1800 m. Here, *Acer trautvetteri* Medw. ex Trautv. (mountain maple, Aceraceae) is a common associate of the dominant beech, *Fagus orientalis* Lipsky (Fagaceae). At even higher elevations, these communities are replaced by birch forest made up of *Betula litwinowii* Doluch. (Betulaceae) with *Sorbus caucasigena* Kom. ex Gatsch. (mountain ash, Rosaceae) and tall herbaceous vegetation confined to glades. Above the treeline, all slopes are covered with *Rhododendron caucasicum* Pall. (Ericaceae) scrub or dekiani and carpet-like alpine meadows. In more continental climates (e.g., in Tusheti, eastern Georgia) beech forest is displaced by pine forest, composed of *Pinus kochiana* Klotzsch ex K. Koch (Pinaceae) (Nakhutsrishvili, 1999).

The climate of the Colchis region (western Georgia), which is located in the eastern portion of the Black Sea catchment basin, is characterized as having moderately warm summers (24–25°C) and cool winters (4–6°C), with abundant annual precipitation (1800–2200 mm, up to 4500 mm). Mountainous ranges along the coastline trap a large portion of the moisture arising from oceanic air masses on their windward side. These physical barriers are formed by a topographical triangle created by the intersection of the western part of the Greater Caucasus, the western part of the Lesser Caucasus, and the Likhi Range. The warm and humid climate of the Colchis region has been present since the late Tertiary period. This is the primary reason the Caucasus acted as a shelter for hygro-thermophilous relictual plant taxa during the previous ice age. Therefore, Colchic forests along with Hyrcanian forests are the most significant refugia for the Arcto-Tertiary forests in western Eurasia, with many relictual and endemic plants (Nakhutsrishvili et al., 2010).

Unlike eastern Georgia, in South Colchis, lowland forests begin at sea level. The most characteristic species of these forests are *Pterocarya pterocarpa* (Michx.) Kunth ex Iljinsk. (Caucasian wingnut, Juglandaceae), *Alnus barbata* C. A. Mey. (alder, Betulaceae), and a relict pine from the Tertiary *Pinus pityusa* Steven (Pinaceae) in northwestern Colchis. Southern slopes are covered by forests of *Quercus hartwissiana* Steven (Colchic oak, Fagaceae) and *Carpinus caucasica* Grossh. (Caucasian hornbeam, Betulaceae). At 600–700 m, there are beech forests with *Abies nordmanniana* (Steven) Spach (Nordmann's fir, Pinaceae), while farther up, spruce–fir forests of *Picea orientalis* (L.) Peterm. (spruce, Pinaceae) and *Abies nordmanniana* (Steven) Spach appear from 1200 to 1800(to 1900) m. Near the treeline, spruce gives way to birch forests (*Betula medwediewii* Regel, *Betula megrelica* Sosn., Betulaceae) and *Quercus pontica* K. Koch (Pontic oak, Fagaceae). On northern slopes, oak forests are replaced by those of *Carpinus caucasica* Grossh. (Caucasian hornbeam, Betulaceae) and *Castanea sativa* Mill. (sweet chestnut, Fagaceae). The distinguishing feature of these forests is the half-prostrate shrubs characterized by vegetative reproduction, forming a dense and high understory, to as much as 4 m. Forests are marked by broad-leaved evergreens, including several rhododendrons (*Rhododendron ponticum* L., *Rhododendron ungernii* Trautv. ex Regel, *Rhododendron smirnowii* Trautv. ex Regel, Ericaceae), with the last two local endemics to South Colchis, *Lauro-cerasus officinalis* M. Roem. (cherry laurel, Rosaceae) and *Ilex colchica* Pojark. (Colchic holly, Aquifoliaceae), as well as *Vaccinium arctostaphylos* L. (deciduous mountain cranberry, Ericaceae) and *Viburnum orientale* Pall. (Oriental viburnum, Caprifoliaceae) (Dolukhanov, 1980). The alpine meadows occur higher up.

Another variant of altitudinal zonation in western Georgia can be seen in Svaneti. There, *Alnus barbata* C. A. Mey. (alder, Betulaceae) forests grow in swampy lowlands. This dominant species is often accompanied in these communities by *Carpinus caucasica* Grossh. (Caucasian hornbeam, Betulaceae), with rich fern undergrowth of *Matteuccia struthiopteris* (L.) Tod. (ostrich fern, Onocleaceae). At 600 m, beech forests start to predominate with *Rhododendron ponticum* L. (Ericaceae) represented in the understory. At ca. 1300 m, beech usually mixes with *Abies nordmanniana* (Steven) Spach (Nordmann's fir, Pinaceae). The subalpine zone is dominated by *Acer trautvetteri* Medw. (mountain maple, Aceraceae), growing together with *Fagus orientalis* Lipsky (Oriental beech, Fagaceae), *Betula litwinowii* Doluch. (Litwinow's birch, Betulaceae), and *Sorbus caucasigena* Kom. ex Gatsch. (mountain ash, Rosaceae). The development of the herbaceous vegetation is also characteristic of this zone. In several gorges the lower portion of the subalpine zone is covered by crook-stem beech forest, and the upper portion is occupied by decumbent beech forest and rhododendron scrub. On southern slopes, lowland forests are replaced by oak forests, which are in turn replaced by fir–spruce forests at about 1000 m in elevation, and rocky habitats are covered by pine forests. Near the treeline, crook-stem beech forests are found, while subalpine and alpine meadows appear from 1850 to 2400 m in elevation (Ketskhoveli, 1960; Nakhutsrishvili, 1999; Dolukhanov, 2010).

Unlike many other mountain systems, tall herbaceous communities are well represented in the subalpine zone of Georgia as well as in the Caucasus region, in general. Here these tall herbaceous communities are particularly confined to the lower part of the subalpine zone. This vegetation is formed from tall herbs (3–4 m), composed mainly of dicots that do not form rosettes and have short taproots and rhizomes (Gagnidze, 1974, 1977). The subnival zone is represented by nanocoenoses from 2900 to 3000 m and 3500 to 3700 m. A number of endemic species and genera occur in the subnival zone including monotypic genera represented by *Symphyoloma graveolens* C. A. Mey. (Apiaceae), *Pseudovesicaria digitata* (C. A. Mey.) Rupr. (Brassicaceae), and *Pseudobetckea caucasica* (Boiss.) Lincz. (Valerianaceae) (Nakhutsrishvili, 1999).

It is noteworthy that wheat (*Triticum* L., Poaceae) shows high taxonomic diversity in Georgia. Nineteen of the 26 known species in the genus *Triticum* L. have been historically distributed in Georgia. Some of these species are country endemics, including *Triticum timopheevii* (Zhuk.) Zhuk., *Triticum zhukovskyi* Menabde & Eritzjan, *Triticum macha* Dekapr. & Menabde, *Triticum carthlicum* Nevski, and *Triticum palaeo-colchicum* Menabde. In all, 16 species, 144 varieties, and 150 forms of wheat were registered in Georgia in the 1940s, according to Menabde (1948; as cited by Akhalkatsi et al., 2012). Three species, among them *Triticum timopheevii* (Zhuk.) Zhuk. and *Triticum macha* Dekapr. & Menabde, were considered by Menabde as primary species (close to the first domesticated wheat species); while another nine native species, among them *Triticum palaeo-colchicum* Menabde, *Triticum carthlicum* Nevski, and *Triticum zhukovskyi* Menabde & Eritzjan, were considered by Menabde as secondary species that originated by hybridization with wild primary species of *Triticum* L., *Aegilops* L. species, and *Thinopyrum* A. Löve species (Akhalkatsi et al., 2012). Although the above endemic wheat species are crop plants, they are included in the Caucasus Red List as EW or Extinct in the Wild because all five (*Triticum carthlicum* Nevski, *Triticum macha* Dekapr. & Menabde, *Triticum palaeo-colchicum* Menabde, *Triticum timopheevii* (Zhuk.) Zhuk., and *Triticum zhukovskyi* Menabde & Eritzjan) are known only from ex situ collections.

Table 1. Georgia's endemic plant species in different ecotopes. Asterisks (*) indicate taxa confined to calcareous substrates.

HABITAT	SPECIES
Subnival rocky areas	DICOTS: *Arabis kazbegi* Mtskhvet., *Erysimum subnivale* Prima (Brassicaceae); *Cerastium svanicum* Charadze (Caryophyllaceae); *Corydalis vittae* Kolak. (Fumariaceae); *Euphrasia svanica* Kem.-Nath., *Veronica tumadzhanovii* Mardal. (Scrophulariaceae).
Montane, subalpine to alpine, rocky and stony areas	FERNS: *Asplenium hermanni-christii* Fomin* (lower montane, Aspleniaceae). DICOTS: *Heracleum egrissicum* Gagnidze*, *Heracleum ossethicum* Manden. (subalpine), *Pimpinella schatilensis* Otsch. (middle montane), *Polylophium*

HABITAT	SPECIES
Montane, subalpine to alpine, rocky and stony areas (continued)	*panjutinii* Manden. & Schischk.* (subalpine, Apiaceae); *Anthemis emiliae* Sosn. (montane), *Centaurea bagadensis* Woronow* (montane), *Cirsium sosnowskyi* Charadze*, *Jurinea exuberans* (Trautv.) Sosn. (alpine), *Kemulariella tugana* (Albov) Tamamsch. (middle montane to alpine), *Psephellus kacheticus* Rehmann ex Boiss. (montane), *Psephellus kolakovskii* (Sosn.) Greuter* (alpine), *Scorzonera ketzkhowelii* Sosn. ex Grossh.* (subalpine), *Scorzonera kozlowskyi* Sosn. ex Grossh.* (subalpine, Asteraceae); *Corylus colchica* Albov*, *Corylus kachetica* Kem.-Nath. (montane, Betulaceae); *Omphalodes kusnetzovii* Kolak.* (montane, Boraginaceae); *Arabis colchica* Kolak.* (alpine), *Barbarea ketzkhovelii* Mardal. (alpine), *Callothlaspi abchasicum* F. K. Mey.* (montane), *Draba meskhetica* Chinth.* (montane), *Draba mingrelica* Schischk. ex Grossh.* (alpine), *Erysimum caucasicum* Trautv. (middle montane), *Erysimum contractum* Sommier & Levier (montane, Brassicaceae); *Campanula armazica* Charadze (montane), *Campanula bzybica* Jabr.-Kolak.*, *Campanula calcarea* (Albov) Charadze*, *Campanula dzaaku* Albov*, *Campanula engurensis* Charadze (lower montane), *Campanula fondervisii* Albov * (subalpine), *Campanula hieracioides* Kolak.* (lower montane), *Campanula imeretina* Rupr.* (montane), *Campanula irinae* A. I. Kuth. * (montane), *Campanula jadvigae* Kolak.* (alpine), *Campanula kachethica* Kantsch.* (lower montane), *Campanula kantschavelii* Zagareli* (lower montane), *Campanula kemulariae* Fomin* (montane), *Campanula kolakovskyi* Charadze* (lower montane), *Campanula letschchumensis* Kem.-Nath.* (lower montane), *Campanula mirabilis* Albov* (lower to middle montane), *Campanula paradoxa* Kolak.* (lower montane), *Campanula radchensis* Charadze* (subalpine), *Campanula raddeana* Trautv. (middle montane), *Campanula schistosa* Kolak. (alpine), *Symphyandra antiqua* Kolak.* (lower montane, Campanulaceae); *Cerastium argenteum* M. Bieb. (montane), *Dianthus abchasicus* Gvinian.* (montane), *Dianthus azkurensis* Sosn. (montane), *Minuartia subuniflora* (Albov) Woronow ex Grossh.* (alpine), *Silene alexeji* Kolak.* (montane), *Silene boissieri* Panjut.* (subalpine, Caryophyllaceae); *Sempervivum charadzeae* Gurgen. (montane), *Sempervivum ermanicum* Gurgen. (alpine, Crassulaceae); *Scabiosa adzharica* Schchian (upper montane), *Scabiosa colchica* Steven (middle montane), *Scabiosa imeretica* (Sommier & Levier) Sulak. (montane, Dipsacaceae); *Astragalus argillosus* Manden. (montane), *Astragalus aspindzicus* Manden. & Chinth. (montane), *Astragalus atenicus* Ivan. (montane), *Astragalus doluchanovii* Manden. (alpine), *Astragalus hirtulus* Ledeb. (montane), *Astragalus kemulariae* Grossh. (montane), *Astragalus kozlovskyi* Grossh. (montane), *Astragalus leonidae* Manden. (middle montane), *Astragalus magnificus* Kolak.* (lower montane), *Astragalus meskheticus* Manden. (montane), *Astragalus raddeanus* Regel (montane), *Astragalus schischkinii* Grossh., *Astragalus vardziae* Charadze & Chinth., *Cicer caucasicum* Bornm. (montane), *Genista adzharica* Popov (lower montane), *Genista sachokiana* A. I. Kuth.* (montane), *Onobrychis kemulariae* Chinth. (alpine, Fabaceae);

HABITAT	SPECIES
Montane, subalpine to alpine, rocky and stony areas (continued)	*Gentiana kolakovskyi* Doluch.*, *Gentiana rhodocalyx* Kolak.* (alpine, Gentianaceae); *Galeopsis nana* Otsch. (alpine), *Nepeta iberica* Pojark. (montane), *Salvia compar* Trautv. ex Grossh. (montane), *Satureja bzybica* Woronow* (lower montane), *Scutellaria helenae* Albov* (alpine), *Thymus ladjanuricus* Kem.-Nath.* (lower montane), *Thymus sosnowskyi* Grossh., *Ziziphora borzhomica* Juz. ex Grossh. (montane, Lamiaceae); *Polygala albowii* Kem.-Nath. (montane, Polygalaceae); *Aquilegia gegica* Jabr.-Kolak.* (montane), *Ranunculus migaricus* Kem.-Nath. & Chinth.* (alpine, Ranunculaceae); *Rhamnus cordata* Medw.* (Rhamnaceae); *Potentilla kemulariae* Kapeller & A. I. Kuth.* (montane), *Potentilla sosnowskyi* Kapeller (montane, Rosaceae); *Galium praemontanum* Mardal.* (montane, Rubiaceae); *Saxifraga abchasica* Oett. (alpine), *Saxifraga kusnezowiana* Oett. (alpine), *Saxifraga trautvetteri* Manden. (alpine, Saxifragaceae); *Scrophularia imerethica* Kem.-Nath. (montane, Scrophulariaceae); *Parietaria kemulariae* Schchian* (montane, Urticaceae). MONOCOTS: *Allium chevsuricum* Tscholok., *Allium otschiauriae* Tscholok. (montane, Alliaceae); *Muscari alpanicum* Schchian* (Hyacinthaceae); *Crocus autranii* Albov* (alpine, Iridaceae); *Alopecurus longifolius* Kolak.* (montane, Poaceae).
Upper montane, subalpine to alpine meadows	DICOTS: *Astrantia colchica* Albov, *Carum grossheimii* Schischk., *Chymsydia colchica* (Albov) Woronow ex Grossh.* (Apiaceae); *Anthemis schischkiniana* Fed., *Cirsium fominii* Petr., *Podospermum grigoraschvilii* Sosn., *Psephellus adjaricus* (Albov) Grossh., *Pyrethrum marionii* Albov, *Tragopogon colchicus* Albov, *Tragopogon ketzkhovelii* Kuth., *Tragopogon makaschwilii* Kuth., *Tragopogon meskheticus* Kuth., *Tragopogon otschiaurii* Kuth., *Tripleurospermum* szowitzii (DC.) Pobed. (Asteraceae); *Campanula megrelica* Manden. & Kuth., *Campanula panjutinii* Kolak.*, *Campanula symphytifolia* (Albov) Kolak.* (Campanulaceae); *Cephalaria sosnowskyi* Kolak. (Dipsacaceae); *Genista mingrelica* Albov, *Onobrychis grossheimii* Kolak. ex Fed. (Fabaceae); *Hypericum nordmannii* Boiss. (Hypericaceae); *Potentilla sommieri* Siegfr. & R. Keller (Rosaceae); *Euphrasia kemulariae* Juz., *Melampyrum alboffianum* Beauverd, *Pedicularis elisabethae* T. N. Popova, *Verbascum adzharicum* P. P. Gritz. (subalpine), *Veronica colchica* Kimer. (Scrophulariaceae); *Viola orthoceras* Ledeb. (Violaceae). MONOCOTS: *Allium albovianum* Vved. (Alliaceae); *Iris winogradowii* Fomin (Iridaceae).
Subalpine tall herb vegetation	DICOTS: *Angelica adzharica* Pimenov (Apicaceae); *Cirsium albowianum* Sommier & Levier, *Cirsium oblongifolium* K. Koch, *Senecio similiflorus* Kolak. (Asteraceae); *Delphinium elisabethae* N. Busch, *Delphinium ironorum* N. Busch, *Delphinium osseticum* N. Busch, *Delphinium thamarae* Kem.-Nath. (Ranunculaceae).
Subalpine forests and shrublands	DICOTS: *Betula megrelica* Sosn. (Betulaceae); *Rosa doluchanovii* Manden., *Rosa ermanica* Manden. (Rosaceae); *Verbascum adzharicum* P. P. Gritz. (Scrophulariaceae).
Montane forests	DICOTS: *Acer sosnowskyi* Doluch. (Aceraceae); *Cryptotaenia flahaultii*

HABITAT	SPECIES
Montane forests (continued)	(Woronow) Koso-Pol. (Apiaceae); *Anthemis saguramica* Sosn., *Cirsium imereticum* Boiss., *Cirsium kemulariae* Charadze, *Petasites georgicus* Manden. (Asteraceae); *Gymnospermium smirnovii* (Trautv.) Takht. (Berberidaceae); *Campanula suanetica* Rupr. (middle montane to subalpine, Campanulaceae); *Dianthus charadzeae* Gagnidze & Gvinian. (Caryophyllaceae); *Helianthemum georgicum* Juz. & Pozd. (lower montane, Cistaceae); *Swida armasica* (Sanadze) Gvin. (Cornaceae); *Onobrychis angustifolia* Chinth. (lower montane, Fabaceae); *Quercus imeretina* Steven ex Woronow (Fagaceae); *Alcea abchasica* Iljin (dry open slopes, forest edges), *Alcea transcaucasica* (Iljin ex Grossh.) Iljin (Malvaceae); *Paeonia lagodechiana* Kem.-Nath., *Paeonia majko* Ketsk., *Paeonia ruprechtiana* Kem.-Nath., *Paeonia steveniana* Kem.-Nath. (Paeoniaceae); *Cyclamen colchicum* (Albov) Albov, *Primula saguramica* Gavr. (Primulaceae); *Aquilegia colchica* Kem.-Nath.*, *Pulsatilla georgica* Rupr. (open forest areas in middle montane zone, Ranunculaceae); *Rosa irysthonica* Manden. (upper montane), *Rosa transcaucasica* Manden. (Rosaceae); *Thesium laxiflorum* Trautv. (Santalaceae); *Ulmus georgica* Schchian (Ulmaceae). MONOCOTS: *Allium gracilescens* Sommier & Levier (open areas in lower montane zone, Alliaceae); *Galanthus kemulariae* Kuth., *Galanthus ketzkhovelii* Kem.-Nath., *Galanthus schaoricus* Kem.-Nath. (Amaryllidaceae); *Ornithogalum imereticum* Sosn., *Scilla otschiauriae* Mordak (Hyacinthaceae).
Steppes and hemi-xerophilous scrub	DICOTS: *Echinops foliosus* Sommier & Levier, *Galatella eldarica* Kem.-Nath. (Asteraceae); *Campanula makaschvilii* E. A. Busch (Campanulaceae); *Onobrychis kachetica* Boiss. & Buhse, *Onobrychis meschetica* Grossh. (Fabaceae); *Paeonia carthalinica* Ketsk. (Paeoniaceae); *Amygdalus georgica* Desf. (Rosaceae).
Semi-deserts/saline areas	DICOTS: *Podospermum idae* Sosn. (Asteraceae); *Astragalus cyri* Fomin ex Grossh. (Fabaceae); *Reaumuria kuznetzovii* Sosn. & Manden. (Tamaricaceae).
Coastal sandy areas	DICOTS: *Paracynoglossum imeretinum* (Kusn.) Popov (Boraginaceae); *Verbascum sessiliflorum* Murb. (Scrophulariaceae).
Wetlands	DICOTS: *Achillea sedelmeyeriana* Sosn., *Solidago turfosa* Woronow ex Grossh. (Asteraceae); *Chrysosplenium albowianum* Kuth. (Saxifragaceae); *Hibiscus ponticus* Rupr. (Malvaceae); *Ficaria varia* Otsch. (Ranunculaceae); *Trapa colchica* Albov, *Trapa maleevii* V. N. Vassil. (Trapaceae). MONOCOTS: *Rhynchospora caucasica* Palla (Cyperaceae).

MAIN THREATS

The major threats to plant species in Georgia are:

1. Forest clear-cutting and selective logging
2. Overgrazing
3. Infrastructure development
4. Mining

Many non-endemic species are also threatened in Georgia, according to preliminary expert data. They still need conservation assessments and elaboration of relevant conservation actions.

LISTED IN THIS RED LIST 288

Extinct (EX) 0
Extinct in the Wild (EW) 5
Critically Endangered (CR) 6
Endangered (EN) 22
Vulnerable (VU) 79
Near Threatened (NT) 35
Least Concern (LC) 22
Data Deficient (DD) 119

CONSERVATION LEGISLATION

The Red List of Georgia was approved by the decree N 303, on 2 May 2006, by the President of Georgia. This Red List included 56 plant species (trees and shrubs) that were assessed, according to the IUCN Red List Categories and Criteria, Version 3.1 (2001), and the Guidelines for Application of IUCN Red List Criteria at Regional Levels (2003).

PRINCIPAL LAWS

Law of Georgia on Permits for Influence of the Environment. 29 December 1991.

Law of Georgia on Nature Protection. 10 December 1996, N 519-IS.

Law of Georgia on Protected Areas. 7 March 1996, N 136-IIS.

Law of Georgia on The Red List and The Red Data Book of Georgia. 6 June 2003, N 2356-IIS.

Georgia's Forest Codex. 22 June 2007, N 2124-IIS.

Law of Georgia on Ecological Expertise. 14 December 2007, N 5603-RS.

INTERNATIONAL CONVENTIONS

Convention on Access to Information, Public Participation in Decision-Making and Access to Justice in Environmental Matters done at Aarhus, Denmark, 25 June 1998.

Convention on International Trade in Endangered Species of Wild Fauna and Flora signed at Washington, D.C., on 3 March 1973; amended at Bonn, 22 June 1979.

United Nations Framework Convention on Climate Change, United Nations, 1992.

United Nations Convention to Combat Desertification in Countries Experiencing Serious Drought and/or Desertification, Particularly in Africa, United Nations, 14 October 1994.

The Ramsar Convention on Wetlands. Ramsar, Iran, 2 February 1971, with the Amendments of the Paris Protocol of 3 December 1982 and the Amendments of Regina Conference of the Participating Countries of 28 May 1987.

As previously indicated, the present Red List contains ca. 1330 Caucasian endemic plant species occurring in Georgia, 276 of which are endemic to the country. A summary of Georgia's endemic plant species assessments is given below.

TOP 50 NATIONAL CONSERVATION PRIORITIES FOR GEORGIA

AMARYLLIDACEAE

1. *Galanthus schaoricus* Kem.-Nath. **EN**

This is a rare species of snowdrop known from only one locality in the upper montane zone of the Racha-Lechkhumi region, western Georgia. Only a very limited number of individuals are known in the wild for this species.

APIACEAE

2. *Angelica adzharica* Pimenov **EN**

This is a rare species occurring only in Adjara, part of the Colchis region, in southwestern Georgia, bordering the Black Sea and Turkey to the south. It is found in a few locations in middle and upper montane zones in forest meadows and in tall herb communities. The habitat for this species is declining due to overgrazing and haymaking. The species is not protected at either the species or habitat level (Memiadze, 2003; Manvelidze et al., 2009).

3. *Astrantia colchica* Albov EN

This is an ornamental umbellate plant that occurs in subalpine or alpine meadows on limestone ridges of Abkhazeti and Samegrelo, western Georgia. The major threat to this species is overgrazing. It is not protected at either the species or habitat level.

4. *Chymsydia colchica* (Albov) Woronow ex Grossh. DD

This is an ornamental umbellate plant known from subalpine and alpine meadows on limestone ridges in Samegrelo, western Georgia. This species could reasonably be assessed as EN largely because its meadow habitat is declining due to overgrazing. Colchic *Chymsydia* Albov is not protected at either the species or habitat level.

5. *Heracleum egrissicum* Gagnidze EN

This cow parsnip, with the stems not exceeding 40 cm, is known from four or five locations in Racha-Lechkhumi and Samegrelo, western Georgia, where it is confined to subalpine and alpine limestone scree. Its major threats are overgrazing and road construction, and the known populations are fragmented and decreasing. No conservation measures are in place.

6. *Polylophium panjutinii* Manden. & Schischk. EN

This species occurs in meadows and rocky areas in the subalpine and alpine zones in Racha-Lechkhumi, Samegrelo, and Imereti, western Georgia. It is threatened mostly by habitat degradation from overgrazing and is not protected at either the species or habitat level.

ASTERACEAE

7. *Anthemis saguramica* Sosn. DD

The species is recorded from two localities in Kartli and Mtiuleti, in central and eastern Georgia, respectively. The Kartlian population is protected within the Tbilisi National Park, while the population in Mtiuleti is threatened by road construction, near the main motorway in eastern Georgia. The species needs further taxonomic investigation and was recently transferred to *Cota* J. Gay ex Guss. as *Cota saguramica* (Sosn.) Lo Presti & Oberpr. (2010).

8. *Cirsium oblongifolium* K. Koch VU

This species is a component of tall herb communities that occur in the upper montane to subalpine zones of Abkhazeti, Svaneti, and Racha-Lechkhumi, western Georgia. This unique habitat is continuing to decline because of haymaking and overgrazing. This species, as well as the entire plant community, needs habitat-based conservation action.

9. *Podospermum grigoraschvilii* Sosn. EN

This is a rare species known from two localities from alpine meadows in the mountains of Tush-Pshav-Khevsureti, eastern Georgia, where it is under threat from overgrazing.

10. *Scorzonera ketzkhowelii* Sosn. ex Grossh. CR

This species is a highly ornamental plant, distinguished by its broad leaves that are bright whitish silver, with a velvety texture, and by its large, lemon-yellow heads. It is known from a single locality, from a rocky ravine in the Tetrobi limestone area of southern Georgia. The Tetrobi Habitat/Species Management Area was established in 1995. Nevertheless, as the protected area borders hay and pasture grasslands, and this species is represented by only a small number of individuals, Ketzkhoveli's salsify is assessed as CR and requires species-based conservation action.

11. *Scorzonera kozlowskyi* Sosn. ex Grossh. CR

This species is another ornamental composite occurring in the Tetrobi limestone area of southern Georgia, at the same locality as *Scorzonera ketzkhowelii* Sosn. ex Grossh. As with the previous species, Kozlowsky's salsify needs species-based conservation action.

BERBERIDACEAE

12. *Gymnospermium smirnovii* (Trautv.) Takht. EN

This is a rare medicinal plant known from two isolated localities in the riparian forests of Kakheti, eastern Georgia. Its forest habitat is threatened by clear-cutting, which is particularly intense in the riparian areas. Another threat to this species is its collection for medicinal purposes. Smirnow's gymnospermium is not protected at either the species or habitat level.

BETULACEAE

13. *Betula megrelica* Sosn. VU

This is a rare species that forms elfin forests in the subalpine zone of limestone mountain ridges in Samegrelo, western Georgia. Megrelian birch grows with other short trees and shrubs at elevations from ca. 1700 to 2000 m, sometimes as low as 1250–1400 m. The species is included in the Red List of Georgia but is not protected at the habitat level.

14. *Corylus colchica* Albov VU

This species has some morphological characters that relate it to Chinese and North American representatives rather than to other hazels occurring in Georgia (Kemularia-Nathadze, 1975). Colchic hazel grows in the middle montane and subalpine zones in Abkhazeti and Samegrelo, western Georgia, where it is found mostly on limestone ridges in pure stands. The species is included in the Red List of Georgia but is not protected at the habitat level.

CAMPANULACEAE

15. *Campanula armazica* Charadze EN

This species consists of cushion-forming plants that grow in rock cracks and is found in the middle montane forest zone in Kartli, eastern Georgia, and Trialeti, southern Georgia. Its primary threats are forest cutting and road construction.

16. *Campanula dzaaku* Albov VU

This is an ornamental bellflower with a creeping rhizome and short stems that end with a single, dark sky-blue flower. Recorded from the alpine, rarely montane, forest zone of Abkhazeti, Racha-Lechkhumi, and Samegrelo, western Georgia, the species is confined to limestone rock cracks. The major threat to this species is road construction. It is not protected at either the species or habitat level.

17. *Campanula engurensis* Charadze VU

This is a rare bellflower, recorded from two localities on shaded rocks in deep ravines in the montane forest zone of Svaneti, western Georgia. The species is threatened by habitat degradation related to logging, which is a plausible threat to its population. It is not protected at either the species or habitat level.

18. *Campanula fondervisii* Albov VU

This bellflower is a rare species recorded from two localities at the subalpine zone of Racha-Lechkhumi and Samegrelo, western Georgia, where it grows on rocky limestone slopes. The major threats are road construction and climate change. The plants are not protected at either the species or habitat level.

19. *Campanula kachethica* Kantsch. CR

Two small subpopulations are known from Kiziki, eastern Georgia, where the species is confined to limestone rock cracks. A portion of the population is protected by the Artsivis Kheoba Natural Monument; however, the other portion is subjected to pressures from limestone quarrying. Based on recent observations, the viability of the overall population is estimated to be low. The causes for this decline need to be studied, particularly within the protected area.

20. ***Campanula kantschavelii*** Zagareli **DD**

This rare bellflower is recorded from a single locality in Kakheti, eastern Georgia, where the species is confined to rock cracks in the montane forest zone. Population trends are unknown, and threats are unclear. The species is not protected at either the species or habitat level.

21. ***Campanula kemulariae*** Fomin **VU**

This is a rare, highly ornamental bellflower recorded from a single locality in Imereti, western Georgia, where it is confined to limestone rock cracks in the lower and middle montane forest zones. Road construction is identified as the major plausible threat to Kemularia's bellflower. It is not protected at either the species or habitat level.

22. ***Campanula mirabilis*** Albov **DD**

This species was called by its author "florae Abchasiae regina," or the Queen of the flora of Abkhasia (Albov, 1895: 229). Indeed, one plant can produce up to 400 flowers, and the flowering period lasts for two or three (to four) months. The species is known in horticulture. However, in the wild it is recorded from two localities in Abkhazeti, western Georgia, where it is confined to limestone rocky areas in the montane and lower alpine zones. Current population trends, distributional range, and threats remain unknown and in need of study.

23. ***Campanula paradoxa*** Kolak. **VU**

Recorded from a single locality in Bzipi gorge, Abkhazeti, western Georgia, the paradox bellflower is confined to limestone rocky areas in the lower montane forest zone. Forest clear-cutting is the major plausible threat to this plant, which is not currently protected at either the species or habitat level.

24. ***Campanula radchensis*** Charadze **VU**

Recorded from four localities in Racha-Lechkhumi and Imereti, western Georgia, this species is confined to rocky limestone areas in the upper montane zone. Trampling by cattle and road construction are identified as plausible threats to this species. It is not protected at either the species or habitat level.

25. ***Campanula raddeana*** Trautv. **VU**

This species is known in horticulture. In the wild, it is recorded from a single location in rocky areas in the montane zone of the Borjomi area in Kartli, eastern Georgia. The species is threatened by human activity such as tourism, habitat degradation, overgrazing, and road construction. The species is not protected at either the species or habitat level.

26. ***Campanula suanetica*** Rupr. **VU**

This species is recorded from five localities in the montane forest zone of Svaneti, western Georgia, where the species is confined to shaded rock cracks. It is threatened by habitat degradation and road construction. The species is not protected at either the species or habitat level.

CARYOPHYLLACEAE

27. ***Cerastium svanicum*** Charadze **VU**

This species of snow-in-summer is known from several close localities in the subnival zone of Svaneti, western Georgia, at elevations from 2550 to 3100 m, where it is found mostly on rocky cliffs. Natural disasters such as landslides and avalanches as well as the negative effects of climate change are identified as the major plausible threats to this species.

28. *Dianthus azkurensis* Sosn. VU

The species is known from five localities in Meskheti, southern Georgia, where it is confined to rocky limestone areas in the middle montane zone. Road construction is the major plausible threat. Azkurian pink is not protected at either the species or habitat level.

CORNACEAE

29. *Swida armasica* (Sanadze) Gvin. CR

This is a very rare species in the dogwood family and is known from a single locality in the vicinity of Mtskheta, Kartli, eastern Georgia. Logging is the major threat to the forest habitat of Armazi dogwood, which is not protected at either the species or habitat level.

FABACEAE

30. *Astragalus cyri* Fomin ex Grossh. EN

This is a highly ornamental legume, with conspicuous purple flowers. The species is recorded from a single locality in Kartli, eastern Georgia, and at a few close localities in Meskheti, southern Georgia. Its arid habitat in the middle montane zone is declining due to extensive urbanization and infrastructure development.

31. *Astragalus meskheticus* Manden. VU

This is a very rare ornamental species recorded from a single locality on dry slopes in the middle montane zone in Meskheti, southern Georgia, where it is threatened by land development and road construction.

32. *Astragalus vardziae* Charadze & Chinth. VU

This is a rare ornamental species recorded from three localities on rocky slopes from middle montane to subalpine zones in Djavakheti and Meskheti, southern Georgia. It is threatened by human disturbance and over-collection related to tourism.

33. *Genista adzharica* Popov CR

This species is an ornamental subshrub occurring at a single locality, where it is found in rock crevices below 700 m in the montane forest zone near the settlement of Keda in Adjara, western Georgia. The species habitat is disturbed by forest clear-cutting, overgrazing, and infrastructure development. Adjaran broom is not protected at either species or habitat level (Memiadze, 2003; Manvelidze et al., 2009).

HYACINTHACEAE

34. *Muscari alpanicum* Schchian VU

This is a rare calciphilous species of lilyturf that is known from two localities on limestone mountain ridges of Imereti and Racha-Lechkhumi, western Georgia. Road construction and the over-collection of its attractive floral clusters for bouquets are identified as the major plausible threats to this species. It is not protected at either the species or habitat level.

IRIDACEAE

35. *Iris winogradowii* Fomin VU

This species had been previously indicated as Extinct in the Wild (EW) by the end of the 20th century. However, its introduction into horticulture as a garden plant and its subsequent reintroduction have reestablished plants in their native habitat on Gagripsha ridge in Abkhazeti, western Georgia, and on Mt. Lomtismta in Kartli, southeastern Georgia. It has been historically recorded from the Adjara, Meskheti, and Trialeti Floristic Regions. The locus classicus for the type of this species is part of the Borjomi Kharagauli National Park and is under protection.

LAMIACEAE

36. *Thymus ladjanuricus* Kem.-Nath. VU

This is a rare species known from five localities in Racha-Lechkhumi and Imereti, western Georgia, where it is confined to limestone rocky areas in the lower montane zone. Road construction is the major threat to Ladjanuri thyme, which is not protected at either the species or habitat level.

MALVACEAE

37. *Hibiscus ponticus* Rupr. CR

This species is the only Caucasus representative from this large genus of warm and tropical regions. It is a highly ornamental plant, with important medicinal uses. It was recorded from the wetland areas along the Black Sea coast of Abkhazeti, Samegrelo, Guria, and Adjara, western Georgia. However, its habitat has experienced severe human intrusions related to urbanization, infrastructure, and land development (drainage, irrigation) as well as fires, which has resulted in the observed population decline. At present, the species can be found only in the Imnati sphagnum wetland in Guria. There are no data on the separate population in Abkhazeti. Pontic hibiscus urgently needs in situ and ex situ conservation action. The species was formerly used as a source of fiber (Machutadze, 2011).

PAEONIACEAE

38. *Paeonia carthalinica* Ketsk. EN

This is an ornamental species with pinnatisect leaves composed of narrow, lanceolate lobes and with dark purple flowers. The species grows in steppe and thorny xerophilous scrub in the foothills of Shiga Kartli and Kiziki, eastern Georgia. The distributional area of this species coincides with pasturelands where it can be found in sheltered scrub patches.

39. *Paeonia majko* Ketsk. DD

This species is an ornamental plant with pinnatisect leaves composed of narrow, lanceolate lobes and with flowers that range from pinkish to reddish to violet shades. It is recorded from only one locality in Kartli, eastern Georgia. The species resembles *Paeonia hybrida* Pall. as well as a hybrid Crimean peony, *Paeonia tenuifolia* L. × *Paeonia triternata* Pall. ex DC. It needs further taxonomic study and is not protected at either the species or habitat level.

40. *Paeonia ruprechtiana* Kem.-Nath. EN

This is a highly ornamental plant with purple-brown emerging foliage (green after flowering) and large, bright pink-purple flowers. The species occurs in the montane forest zone of Racha-Lechkhumi, Imereti, western Georgia, and Kartli, eastern Georgia. Deforestation and over-collection are the major threats.

41. *Paeonia steveniana* Kem.-Nath. EN

This is a highly ornamental plant with large yellowish cream flowers. The species occurs at forest edges and in scrub in the high montane forest zone of Kartli, eastern Georgia. Deforestation and over-collection are the major threats to the species. It is not protected at either the species or habitat level.

PRIMULACEAE

42. *Cyclamen colchicum* (Albov) Albov VU

This is an ornamental species, with medicinal uses, occurring on limestone rocks from the lower montane to subalpine zones in Abkhazeti, Racha-Lechkhumi, and Samegrelo, western Georgia. Although population trends are unknown, the over-collection of flowers for bouquets and of tubers used in folk medicine are both plausible threats to the species.

RANUNCULACEAE

43. *Aquilegia colchica* Kem.-Nath. EN

This is a highly ornamental species with bicolored flowers that have blue perianth lobes and nectaries, but the upper portions of the blades are white. Three or four declining populations of the species are known, where they are restricted to rocky limestone areas in forests of the middle montane zone in Imereti, western Georgia. Population decline is caused by both deforestation and over-collection of the species, which is not protected at either the species or habitat level.

44. *Aquilegia gegica* Jabr.-Kolak. **VU**

This species is quite closely related to *Aquilegia colchica* but differs in its pale blue flowers that are not bicolored, as well as other morphological features (Kemularia-Nathadze, 1973). Four or five populations are known from rocky limestone areas in the middle montane zone in Abkhazeti, western Georgia. Over-collection is identified as the major threat to the species.

45. *Pulsatilla georgica* Rupr. **EN**

This species is recorded at two close localities near Tbilisi. This wonderful ornamental plant has large flowers that can vary in color from violet to sky-blue, with pinkish, reddish, yellowish, and even greenish shades. The flowering period (March to May) coincides with Easter, and over-collection is the major threat. It is not protected at either the species or habitat level.

ROSACEAE

46. *Amygdalus georgica* Desf. **VU**

This is a rare woody species known from only four localities in the steppe and hemi-xerophilous scrub zones of Kartli and Gare Kakheti, eastern Georgia, and Trialeti, southern Georgia. Land development and overgrazing are the major threats. The plants are not protected at the habitat level.

47. *Potentilla kemulariae* Kapeller & A. I. Kuth. **VU**

This is a very rare species known from only one locality on the limestone Nakerala ridge in Imereti, western Georgia. Plants are ornamental, with a dense white pubescence and sepals that are purple and contrast with the dark red petals. The species is intensively collected by local people. This over-collection and road construction are major threats to this plant, which is not protected at either the species or habitat level.

TRAPACEAE

48. *Trapa colchica* Albov **CR**

This rare species of water chestnut is recorded from four localities in wetland areas of Abkhazeti, Samegrelo, Guria, and Adjara, western Georgia. This wetland habitat is declining due to bog reclamation and land development. The current population does not exceed 250 mature individuals, with none of the existing subpopulations containing more than 50 mature individuals. Conservation action in situ is urgently needed (Machutadze, 2009).

49. *Trapa maleevii* V. N. Vassil. **EN**

Another rare species of water chestnut was recorded from three localities on wetland areas of Abkhazeti, Samegrelo, and Adjara, western Georgia. The wetland habitat continues to decline due to bog reclamation and land development. As a result, Maleev's water chestnut can now only be found at the Chrokhi River outflow and in freshwater ponds at Kvariati in Adjara. In situ conservation action for this species is urgently needed (Machutadze, 2009).

ULMACEAE

50. *Ulmus georgica* Schchian **DD**

This woody species is recorded from a single locality in Kiziki, eastern Georgia. It is selected as a conservation priority because of the need to investigate and clarify its taxonomic status. *Ulmus georgica* Schchian is separated from *Ulmus minor* Mill. principally by its elongate elliptic and rather elongate-tipped leaf blades that become leathery in autumn, in contrast to the finer texture of the ovate or obovate leaf blades of *Ulmus minor* Mill. (Shkhiani, 1975). A portion of the population grows in the protected areas in Kiziki, but the other portion is threatened by logging.

GEORGIA MAPS

1. *Galanthus schaoricus* Kem.-Nath.

2. *Angelica adzharica* Pimenov

3. *Astrantia colchica* Albov

4. *Chymsydia colchica* (Albov) Woronow ex Grossh.

5. *Heracleum egrissicum* Gagnidze

6. *Polylophium panjutinii* Manden. & Schischk.

7. *Anthemis saguramica* Sosn.

8. *Cirsium oblongifolium* K. Koch

9. *Podospermum grigoraschvilii* Sosn.

10. *Scorzonera ketzkhowelii* Sosn. ex Grossh.

11. *Scorzonera kozlowskyi* Sosn. ex Grossh.

12. *Gymnospermium smirnovii* (Trautv.) Takht.

13. *Betula megrelica* Sosn.

14. *Corylus colchica* Albov

15. *Campanula armazica* Charadze

16. *Campanula dzaaku* Albov

17. *Campanula engurensis* Charadze

18. *Campanula fondervisii* Albov

19. *Campanula kachethica* Kantsch.

20. *Campanula kantschavelii* Zagareli

21. *Campanula kemulariae* Fomin

22. *Campanula mirabilis* Albov

23. *Campanula paradoxa* Kolak.

24. *Campanula radchensis* Charadze

25. *Campanula raddeana* Trautv.

26. *Campanula suanetica* Rupr.

27. *Cerastium svanicum* Charadze

28. *Dianthus azkurensis* Sosn.

29. *Swida armasica* (Sanadze) Gvin.

30. *Astragalus cyri* Fomin ex Grossh.

31. *Astragalus meskheticus* Manden.

32. *Astragalus vardziae* Charadze & Chinth.

33. *Genista adzharica* Popov

34. *Muscari alpanicum* Schchian

35. *Iris winogradowii* Fomin

36. *Thymus ladjanuricus* Kem.-Nath.

37. *Hibiscus ponticus* Rupr.

38. *Paeonia carthalinica* Ketsk.

39. *Paeonia majko* Ketsk.

40. *Paeonia ruprechtiana* Kem.-Nath.

41. *Paeonia steveniana* Kem.-Nath.

42. *Cyclamen colchicum* (Albov) Albov

43. *Aquilegia colchica* Kem.-Nath.

44. *Aquilegia gegica* Jabr.-Kolak.

45. *Pulsatilla georgica* Rupr.

46. *Amygdalus georgica* Desf.

47. *Potentilla kemulariae* Kapeller & A. I. Kuth.

48. *Trapa colchica* Albov

49. *Trapa maleevii* V. N. Vassil.

50. *Ulmus georgica* Schchian

BIBLIOGRAPHY

Abashidze, I. (editor). 1981. Kartuli sabchota enciclopedia. [Georgian Soviet Encyclopedia.] Tbilisi [in Georgian].

Adzinba, Z. I. 1987. Endemy flory Abkhazii. [Endemic Flora of Abkhazia.] Metsniereba, Tbilisi [in Russian].

Agency of Protected Areas. 2011. Protected Areas. Ministry of Environment Protection of Georgia. <http://www.apa.gov.ge>, accessed 3 September 2013.

Akhalkatsi, M., J. Ekhvaia & Z. Asanidze. 2012. Diversity and genetic erosion of ancient crops and wild relatives of agricultural cultivars for food: implications for nature conservation in Georgia (Caucasus). Pp. 51–92 *in* J. Tiefenbacher (editor), Perspectives on Nature Conservation. InTech, Rijeka, Croatia

Albov, N. M. 1895. Nouvelles Contributions a la Flore de la Transcaucasie. 1. Une nouvelle Campanula remarquable. Bull. Herb. Boissier 3: 228–230.

Busch, N. A. 1935. Botaniko-geograficheskie issledovania Kavkaza. [Botanical-geographical Investigation of the Caucasus.] Izd Akad. Nauk SSSR, Moscow/Leningrad [in Russian].

Charadze, A. 1960. Endemichnyi gemikserofil'nyi element vysokogoriy Bol'shogo Kavkaza. [An endemic hemi-xerophilous element of the Greater Caucasus uplands.] Probl. Bot. 5: 115–126. Nauka, Moscow-Leningrad [in Russian].

Charadze, A. 1965. O subnival'noy zone Bol'shogo Kavkaza. On the subnival zone of the Greater Caucasus. Zametki Sist. Geogr. Rast. 25: 103–104 [in Russian].

Convention on Access to Information, Public Participation in Decision-Making and Access to Justice in Environmental Matters. 1998 [25 June]. Aarhus, Denmark. <http://www.unece.org/fileadmin/DAM/env/pp/documents/cep43e.pdf>, accessed 4 September 2013.

Convention on International Trade in Endangered Species of Wild Fauna and Flora (CITES). 1979 [22 June]. [Signed at Washington, DC, on 3 March 1973; amended at Bonn, 22 June 1979.] <http://www.cites.org/eng/disc/E-Text.pdf>, accessed 4 September 2013.

Dmitrieva, A. A. 1960. Opredelitel' rasteniy Adzharii. [Key to Plants of Adjara.] Publishing House of the Georgian SSR Academy of Sciences, Tbilisi [in Russian].

Dolukhanov, A. 1969. Flora i rastitel'nost' subnival'nykh landshaftov ush'el'ja reki Bolshaja Liakhvi i plato Keli (Bol'shoy Kavkaz). [Flora and vegetation of subnival landscapes of the Didi Liakhvi River and the Keli Upland (Greater Caucasus).] Bot. Zhurn. (Moscow & Leningrad) 54: 1662–1674 [in Russian].

Dolukhanov, A. 1980. Colchic Understory. Metsniereba, Tbilisi.

Dolukhanov, A. 2010. Lesnaya rastitel'nost' Gruzii. [Forest Vegetation of Georgia.] Universal, Tbilisi [in Russian].

Dolukhanov, A. & G. Nakhutsrishvili. 2003. Die natürlichen Vegetationsformationen Europas und ihre Untergliederung. Hygrophile thermophytische Laubmischwälder. Pp. 384–388 *in* U. Bohn, G. Gollub & Ch. Hettwer (editors), Karte der natürlichen Vegetation Europas. [Map of the Natural Vegetation of Europe.] Maßstab/Scale 1:2.500.000. Bundesamt für Natürschutz,
Federal Agency for Nature Conservation, Bonn.

Dolukhanov, A., M. Sakhokia & A. Charadze. 1942. O vysokogornykh rastitel'nykh pojasax Kavkaza. [On high mountain vegetation zones of the Caucasus.] Trudy Tbilissk. Bot. Inst. 8: 113–135 [in Russian].

Dolukhanov, A., M. Sakhokia & A. Charadze. 1946. Glavnye osobennosti rastitel'nogo pokrova Verkhvey Svaneti. [Main features of the vegetation cover of Upper Svaneti.] Trudy Tbilissk. Bot. Inst. 9: 79–130 [in Russian].

Gagnidze, R. 1974. Botaniko-geograficheskiy analiz florotsenoticheskogo kompleksa subal'piyskogo vysoskotrav'ja Kavkaza. [Botanical-geographical Analysis of the Floral Coenotic Complex of Subalpine Tall Herbaceous Vegetation of the Caucasus.] Metsniereba, Tbilisi [in Russian].

Gagnidze, R. 1977. Ekologo-cenoticheskaja kharakteristika i analiz vertical'nogo rasprostranenija vidov vysoskotrav'ja Kavkaza. [Ecological-cenotic characteristic and analysis of the vertical distribution of tall herbaceous species in the Caucasus.] Zametki Sist. Geogr. Rast. 34: 41–83 [in Russian].

Gagnidze, R. 2005. Vascular Plants of Georgia. A Nomenclatural Checklist. Universal, Tbilisi.

Gagnidze, R. & N. Margalitadze. 2004. Holocene history of the high-mountain regions of botanical-geographical provinces of Georgia (according to pollen analysis). Caucasian Geographical Review 4: 63–67.

Gamtsemlidze, Z. G. 1977. Bioekologicheskie I cenoticheskie osobennosti subnival'nykh rasteniy Kazbegi. [Bioecological and cenotic peculiarities of Kazbegi subnival plants.] Pp. 44–49 *in* G. Nakhutsrishvili (editor), Vysokogornye ekosistemy Kazbegi. [The High Mountain Ecosystems of Kazbegi.] Metsniereba, Tbilisi [in Russian].

Golitsin, S. V. 1939. Shkeriani – kustarnikovye fitocenozy vlazhnykh lesistykh gor. [Shkeriani – shrub communities of humid forested mountains of Adjara.] Trudy Voronezsk. Gosud. Univ. 11(2): 141–175 [in Russian].

Grossheim, A. A. 1936. Analiz flory Kavkaza. [Analysis of the flora of the Caucasus.] Trudy Bot. Inst. Azerb. Fil. AN SSSR, Baku [in Russian].

Grossheim, A. A. 1948. Rastitel'nyi pokrov Kavkaza. [Vegetation cover of the Caucasus.] Byull. Moskovsk. Obshch. Isp. Prir., Moscow [in Russian].

Gulisashvili, V., L. Makhatadze & L. Prilipko. 1975. Rastitel'nost' Kavkaza. [Vegetation of the Caucasus.] Nauka, Moscow (in Russian).

International Union for Conservation of Nature and Natural Resources (IUCN). 2001. IUCN Red List Categories and Criteria, Version 3.1. IUCN, Gland, Switzerland, and Cambridge, United Kingdom. <http://www.iucn.org>, accessed 4 September 2013.

International Union for Conservation of Nature and Natural Resources (IUCN). 2003. Guidelines for Application of IUCN Red List Criteria at Regional Levels. Version 3.0. IUCN Species Survival Commission, Gland, Switzerland, and Cambridge, United Kingdom. <http://life.bio.sunysb.edu/ee/akcakayalab/RegionalGuidelines.pdf>, accessed 4 September 2013.

Ivanishvili, M. A. 1973. Flora formaciy koljucheastragalovykh tragakantnukov severnogo sklona Bol'shogo Kavkaza. [Flora of Thorny Astragal Tragacanth Formation of the Northern Slope of the Greater Caucasus.] Metsniereba, Tbilisi [in Russian].

Kemularia-Nathadze, L. 1973. Ojakhi 28. [Family 28.] Ranunculaceae Juss. Pp. 38–39 *in* N. Ketskoveli, A. Charadze & R. Gagnidze (editors), Sakartvelos Flora. [Flora of Georgia.] Second edition, Vol. 2. Metsniereba, Tbilisi [in Georgian].

Kemularia-Nathadze, L. 1975. Ojakhi 37. [Family 37.] Corylaceae Mirb. Pp. 39–40 *in* N. Ketskoveli, A. Charadze & R. Gagnidze (editors), Sakartvelos Flora. [Flora of Georgia.] Second edition, Vol. 3. Metsniereba, Tbilisi [in Georgian].

Ketskhoveli, N. N. 1960. Saqartvelos mcenareuli sapari. [Vegetational Cover of Georgia.] Metsniereba, Tbilisi [in Georgian].

Ketskhoveli, N. N., A. Charadze & R. Gagnidze (editors). 1971–2011. Sakartvelos Flora. Me-2 gamocema, tom. 1–16. [Flora of Georgia. Second edition, Vols. 1–16.] Metsniereba, Tbilisi [in Georgian].

Kimeridze, K. R. 1992. Bolotnaja rastitel'nost' Gruzii. [Bog Vegetation of Georgia.] Tbilisi [unpubl. ms. in Russian].

Kolakovsky, A. A. 1961. Rastitel'nyi mir Kolkhidy. [Flora and vegetation of Colchis.] Moscow University, Moscow [in Russian].

Kolakovsky, A. A. 1964. Pliocenovaja flora Kodora. [Pliocene Flora of the Kodori River.] Metsniereba, Sokhumi [in Russian].

Kolakovsky, A. A. 1974. Vertikal'nye pojasa rastitel'nosti kolkhidskikh lesov Tretichnogo perioda. [Vertical zones of Colchis forest vegetation in the Tertiary.] Trudy Tbilissk. Inst. Lesa 21: 98–116 [in Russian].

Kolakovsky, A. A. 1980–1986. Flora Abkhazii. 2-oe izd., tom. 1–4. [Flora of Abkhazia. Second edition, Vols. 1–4.] Metsniereba, Tbilisi [in Russian].

Komarov Botanical Institute. 1996. Map of the reconstructed vegetation of Central and Eastern Europe. Scale 1:2,500,000. (6 sheets).

Kvachakidze, R. K. 2009a. Sakartvelos mcenareuloba. [Vegetation of Georgia.] Tbilisi [in Georgian].

Kvachakidze, R. K. 2009b. Sakartvelos tkeebi. Dziritadi asociaciebi (tkis tipebi). [Forests of Georgia. Principal Associations (Forest Types).] Metsniereba. Publishing House of the Georgian National Academy of Sciences, Tbilisi [in Georgian with summary in English].

Lachashvili, I. & M. Khachidze. 2010. Sakartvelos udabnos flora da mcenareuloba. [Georgia's Desert Flora and Vegetation.] Universal, Tbilisi [in Georgian].

Lachashvili, I., N. Lachashvili & M. Khachidze. 2004. Synopsis of Flora of the Vashlovani State Reserve (East Georgia). Universal, Tbilisi.

Lo Priesti, R. M., S. Oppolzer & C. Oberprieler. 2010. A molecular phylogeny and a revised classification of the Mediterranean genus Anthemis s.l. (Compositae, Anthemideae) based on three molecular markers and micromorphological characters. Taxon 59(5): 1441–1456.

Lortkipanidze, O. 2002. Kartuli civilizaciis sataveebtan. [At the Sources of the Georgian Civilization.] Tbilisi [in Georgian].

Machutadze, I. 2011. Kolkhetis dablobis charbteniani habitatebis ishviati da qrobadi sakheobebi. [Rare and threatened species of wetland habitats of the Colchis lowland.] Pp. 45–47 *in* G. Nakhutsrishvili (editor), Saqartvelos biomravalferovneba. [Biodiversity of Georgia.] Georgian National Academy of Sciences, Tbilisi [in Georgian].

Makashvili, A., D. Sosnovsky, N. Ketskoveli & A. Charadze (editors). 1941–1952. Sakartvelos Flora. 1-li gamocema, tom. 1–8. [Flora of Georgia, First edition, Vols. 1–8.] Metsniereba, Tbilisi [in Georgian].

Mandenova, I. 1975. Ojakhi 109. [Family 109.] Umbelliferae Juss. P. 259 *in* N. Ketskoveli, A. Charadze & R. Gagnidze (editors), Sakartvelos flora. [Flora of Georgia], Second edition, Vol. 9. Metsniereba, Tbilisi [in Georgian].

Manvelidze, Z., O. Eminagaoglu, N. Memiadze & D. Kharazishvili. 2009. Conservation of Endemic Plant Species of Georgian-Turkish Transboundary Area. World Wildlife Fund Caucasus office, Tbilisi.

Memiadze, N. 2003. Botanical and geographic review of Adjara-Shavsheti endemics. Bull. Georgian Acad. Sci. 168(3): 62–64.

Menabde, V. 1948. Pshenitsi Gruzii. [Wheats of Georgia.] Izdatel'stvo Akademii nauk Gruzii [Georgian Academy of Sciences Press], Tbilisi [in Russian].

Menabde, V. 1961. Kartuli khorblebi da mati roli khorblis saerto evolutsiashi. [Georgian wheat cultivars and their role in wheat evolution.] Proc. Tbilisi Bot. Inst. 21(1): 229–259 [in Georgian].

Mikheev, A. D. 2006. The genus *Iris* L. P. 105 *in* A. L. Takhtajan (editor), Konspekt Flory Kavkaza [Caucasus Flora Conspectus], Vol. 2. St. Petersburg University Press, St. Petersburg [in Russian].

Nakhutsrishvili, G. S. 1999. The Vegetation of Georgia (Caucasus). Braun-Blanquetia, Vol. 15. Dipartimento di Botanica ed Ecologia dell'Università, Camerino.

Nakhutsrishvili, G. S. 2003. High mountain vegetation of the Caucasus region. Pp. 93–103 *in* L. Nagy, G. Grabherr, C. Körner & D. B. A. Thompson (editors), Alpine Biodiversity in Europe. Ecological Studies 167. Springer, Berlin-Heidelberg.

Nakhutsrishvili, G. S. 2004. Kaukasus. Pp. 124–134 *in* C. A. Burga, F. Klötzli & G. Grabherr (editors), Gebirge der Erde (Landschaft, Klima, Pflanzenwelt). Ulmer Verlag, Stuttgart.

Nakhutsrishvili, G. S. 2005. Gliederung der Hochgebirgsvegetation des Kaukasus (im Vergleichzu den Alpen). [Caucasian high mountain vegetation (in comparison with that of the Alps).] Pp. 335–345 *in* U. Böhn, C. Hettwer & G. Gollub (editors), Anwendung und Auswretung der Karte der natürlichen Vegetation Europas. Bundesamt für Naturschutz, Bonn.

Nakhutsrishvili, G. S. & Z. Gamtsemlidze. 1984. Zhizn' rasteniy v ekstremal'nykh uslovijakh vysokogoriy. [Plant Life in Extreme Environment of High Mountains.] Nauka, Leningrad [in Russian].

Nakhutsrishvili, G. S., S. Sikharulidze & O. Abdaladze. 2006. Bakuriani – Natural and Cultural Resources of the Borjomi Region. Institute of Botany, Tbilisi.

Nakhutsrishvili, G. S., N. Zazanashvili & K. Batsatsashvili. 2010. Regional profile: Colchic and Hyrcanic temperate rainforests of the western Eurasian Caucasus. Pp. 214–221 *in* D. DellaSala (editor), Temperate and Boreal Rainforests of the World. Ecology and Conservation, Island Press, Washington, DC.

Nakhutsrishvili, Sh. 1966. Osnovnye osobennosti rastitel'nogo pokrova juzhnogo vysokogor'ja Gruzii. [Main features of the vegetation cover of the southern upland of Georgia.] Probl. Bot. 8: 106–111. Nauka, Moscow-Leningrad [in Russian].

Ramsar Convention on Wetlands (RCW). 1971 [2 Feb.]. Convention on Wetlands of International Importance especially as Waterfowl Habitat. Ramsar (Iran). UN Treaty Series No. 14583. As amended by the Paris Protocol, 3 December 1982, and Regina Amendments, 28 May 1987. <http://www.ramsar.org/cda/en/ramsar-about-about-ramsar/main/ramsar/1-36%5E7687_4000_0__>, accessed 4 September 2013.

Sakhokia, M. F. 1958. Botanicheskie ekskursii po Gruzii. [Botanical excursions over Georgia.] Botanicheskie ekskursii po Gruzii. Izd. AN Gruzinskoi SSR [Publishing House of the Acad. Sci. Georgian SSR], Tbilisi [in Russian].

Shatilova, I., N. Mchedlishvili, L. Rukhadze & E. Kvavadze. 2011. The History of the Flora and Vegetation of Georgia (South Caucasus). Georgian National Museum, Tbilisi.

Shetekauri, S. & M. Jakoby. 2009. Mountain Flowers and Trees of Caucasia. Bunebaprinti, Istanbul.

Tumajanov, I. I. 1938. Lesa gornoy Tusheti. [Forests of mountainous Tusheti.] Trudy Tbilissk. Bot. Inst. 5: 105–248 [in Russian].

United Nations. n.d. United Nations Convention to Combat Desertification (UNCCD). <http://www.unccd.int/en/Pages/default.aspx>, accessed 4 September 2013.

United Nations. 1992. United Nations Framework Convention on Climate Change. FCCC/INFORMAL/84. GE.05-6222 (E) 200705. <http://unfccc.int/resource/docs/convkp/coveng.pdf>, accessed 4 September 2013.

United Nations. 1994 [14 October]. 10. United Nations Convention to Combat Desertification in Those Countries Experiencing Serious Drought and/or Desertification, Particularly in Africa. United Nations Treaty Series, 1954: 3. [Depositary notification C.N.176.1995.] <http://treaties.un.org/pages/ViewDetails.aspx?src=TREATY&mtdsg_no=XXVII-10&chapter=27&lang=en>, accessed 4 September 2013.

Urushadze, T. F. 1997. Sakartvelos dziritadi niadagebi. [The Main Soils of Georgia.] Metsniereba, Tbilisi [in Georgian].

Zazanashvili, N., R. Gagnidze & G. Nakhutsrishvili. 1995. High mountain vegetation on the new vegetation map of Georgia. J. Veget. Sci. 6: 157–158.

IRAN

ایران

جلیل نوروزی

ناحیه قفقاز، از نظر تاریخی به منطقه بین دریای سیاه و خزر گفته می‌شود که وسعت آن ۵۸۰،۰۰۰ کیلومترمربع بوده و در شش کشور گسترش دارد. این ناحیه به دو بخش شمالی و جنوبی تقسیم می‌شود که قفقاز شمالی بخشهایی از جنوب غرب روسیه، و قفقاز جنوبی کل جمهوری آذربایجان، ارمنستان، گرجستان، شمال‌شرق ترکیه و شمال‌غرب ایران را در بر می‌گیرد.

منطقه زیستی قفقاز، یک ناحیه بسیار متنوع از نظر گیاهی و جانوری بوده و از درصد بومزادی بسیار بالایی برخوردار است. بصورت گذرگاه زیستی است که گونه‌های اروپای مرکزی و شمالی، آسیای مرکزی، خاورمیانه و شمال آفریقا در آن پراکنده‌اند که با گونه‌های انحصاری آن درآمیخته و ناحیه‌ای با تنوع زیستی بسیار بالا حاصل شده است. بیش از ۶،۵۰۰ گونه گیاهی آوندی در این ناحیه یافت می‌شود که حداقل یک چهارم آنها انحصاری قفقاز هستند که بیشترین نرخ بومزادی در نواحی معتدله نیمکره شمالی است.

طرح بررسی گونه‌های گیاهی بومی و در معرض انقراض ناحیه قفقاز، توسط IUCN با همکاری باغ گیاهشناسی میسوری آمریکا (DGM)، WWF و گیاهشناسان شش کشور قفقازی (آذربایجان، ارمنستان، گرجستان، روسیه، ترکیه و ایران)، با هدف شناسایی همه گونه‌های انحصاری و همینطور در معرض انقراض قفقازی، بین سالهای ۲۰۰۶ تا ۲۰۰۹ میلادی به اجرا درآمد. نتیجه این طرح، تهیه لیستی از تمام گونه‌های بومی موجود در این ناحیه و پراکنش آنها در این شش کشور بوده است، که گیاهشناسان هر کشوری بطور هماهنگ با دیگر کشورها، همه گونه‌های قفقازی موجود در منطقه خود را شناسایی و در نهایت با هم ادغام نمودند. از گونه‌های قفقازی هر کشور، ۵۰ مورد بسیار نادر و در معرض انقراض، انتخاب و معرفی شدند. چندین کارگاه علمی، برای هماهنگی بیشتر بین همه محققان این طرح در کشورهای آلمان، گرجستان و آمریکا برگزار شد.

بخش ایران این طرح توسط نگارنده انجام شد که در مراحل پایانی به این همکاری پیوست. در این پژوهش، همه گیاهان آوندی محدود به ناحیه قفقاز ایران از منابع گیاهشناسی استخراج و با همکاری گیاهشناسان کشورهای دیگر، گونه‌های محدود به ناحیه قفقاز معین شد. بدین منظور همه ۷،۳۰۰ گونه گیاهی موجود در ایران بررسی، و موارد محدود به ناحیه قفقاز مشخص شدند. همچنین لیستی از کل گونه‌هایی که در کشورهای دیگر در محدوده قفقازی بودند (نزدیک به ۲،۰۰۰ گونه) نیز بررسی شد تا وجود یا عدم وجودشان در ایران و همچنین محدود بودنشان به مرزهای قفقاز مشخص شود. سپس برای بخش قفقاز ایران متنی در مورد اقلیم، پستی و بلندی، فلور و پوشش‌های گیاهی ناحیه نوشته شد، و همه گونه‌های بومی قفقازی ایران بر اساس زیستگاههای آنها طبقه‌بندی، و ۵۰ گونه بسیار نادر و در معرض خطر انقراض با نقشه پراکنش معرفی شدند.

ناحیه قفقازی ایران، شمال غرب کشور یعنی استانهای اردبیل و گیلان را بطور کامل، و بخش عمده آذربایجان شرقی، شمال آذربایجان غربی و غرب مازندران را دربر می‌گیرد. ناحیه کاملا کوهستانی بوده و دامنه ارتفاعی از ۲۶ متر زیر سطح دریا تا نزدیک ۴۸۰۰ متر بالای سطح دریا می‌باشد. غرب کوههای البرز، تالش، سبلان، سهند، بزقوش، میشوداغ و کیامکی داغ از جمله کوههای بلند این ناحیه هستند. میزان بارندگی از ۳۰۰ میلی-متر در سال در بعضی نواحی آذربایجان تا ۲۰۰۰ میلی‌متر در سال در نواحی غربی ناحیه خزری متغیر است.

این ناحیه در ایران بصورت گذرگاه زیستی بین البرز، زاگرس، آناتولی و کوههای قفقاز است و از نظر جغرافیای گیاهی در دو منطقه ایران-تورانی و اروپا-سیبری قرار گرفته است. پستی و بلندیهای زیاد، تنوع اقلیمی، قرار گرفتن در بین چندین رشته کوه و چندین ناحیه جغرافیای گیاهی باعث تنوع بالای گونه‌ای و پوششهای گیاهی آن شده است. تعداد گیاهان موجود در این ناحیه حدودا ۴،۰۰۰ گونه تخمین زده می‌شود. نزدیک به ۴۲۰ گونه گیاه آوندی منحصر به قفقاز در ایران وجود دارد که ۱۸۱ مورد آنها فقط در ایران پراکنش دارند. بسیاری از این گونه، پراکنش جغرافیایی بسیار محدودی داشته، در گروه گیاهان در معرض خطر انقراض قرار می‌گیرند.

از پوششهای گیاهی مهم این ناحیه می‌توان به جنگلهای خزری یا هیرکانی، درختچه‌زارهای استپی، نواحی نیمه‌آلپی، آلپی و نیمه‌برفی-برفی اشاره کرد. جنگلهای خزری، دامنه‌های شمالی البرز و تالش را تا ارتفاع ۲،۸۰۰ متر از سطح دریا می‌پوشانند و جزو جنگلهای معتدله برگریز محسوب می‌شوند. درختچه‌زارهای استپی، نواحی کوهستانی پایین‌دست آذربایجان (حدودا ۱،۰۰۰ تا ۲،۲۰۰ متر) را می‌پوشاند که متشکل از پوششهای استپی با درختها و درختچه‌های پراکنده در آن است. نواحی نیمه‌آلپی حدودا بین ۲،۲۰۰ تا ۲،۸۰۰ متر متمرکز شده‌اند که گونه‌هایی از خانواده چتریان، فرم رویشی غالب را دارند. نواحی آلپی بین ۳،۰۰۰ تا ۳،۶۰۰ متر متمرکز هستند و بیشتر از بالشتکی‌های خاردار و چمنزارها پوشیده شده‌اند. بالاتر از نواحی آلپی، نواحی نیمه برفی-برفی است که پوششها تنکتر بوده و بستر، بیشتر از سنگریزه و صخره تشکیل شده است. حد بالایی گیاهان آوندی در شمال غرب ایران به ارتفاع حدود ۴،۴۰۰ متر می‌رسد.

امید است این پژوهش مورد توجه و استفاده محققان علوم زیستی کشور قرار گرفته، آنها را تشویق به مطالعات و بررسی‌های بیشتر و دقیق‌تر در این ناحیه نموده، به دانش ما در مورد تنوع زیستی منطقه بیافزاید، و مسوولان حفاظت محیط زیست و منابع طبیعی کشور را به توجه بیشتر به جنبه‌های حفاظتی این ناحیه ترغیب نموده، از تخریب زیستگاهها جلوگیری، و به گسترش مناطق حفاظت شده منجر شود.

IRAN

Jalil Noroozi

CAPITAL
Tehran, the largest city.

IMPORTANT REGIONAL CITIES
Tabriz, Ardabil, Rasht.

LANGUAGE
Farsi and Turkish.

FLORA
About 4000 (total plant species); 181 (total national endemics in the Caucasus region); ca. 420 (Caucasian region endemics).

The *Red Data Book of Iran* (Jalili & Jamzad, 1999) enumerated 1727 endemic plant taxa for the entire country.

BOTANICAL INSTITUTIONS
The Botanical Garden of Tabriz covers an area of ca. 18 hectares. It is especially designed to represent the different vegetation types of the region as well as those of other parts of Iran and adjacent countries. This botanical garden has a herbarium with mostly native northwestern Iranian plants. It is one of the provincial branches of the National Herbarium of Iran (TARI). Collections for most of the plant species of the Caucasian part of Iran, as in other parts of the country, have been assembled in TARI, which is part of the Research Institute of Forests and Rangelands of Iran (RIFR). Address: Khalaban Sq., Tabriz, East Azerbaijan, Iran; Tel: +98-411 2611601-5.

PROTECTED AREAS
The protected areas include one national park, five wildlife refuges, five national natural monuments, and nine additional protected areas.

National park (3175 hectares): Bojagh.

Wildlife refuges (99,822 hectares): Kiamaki, Selke, Sarkhangol, Amir Kalaye, Londovil.

National natural monuments (12,191 hectares): Khoshkedaran, Alamkuh, Susan-e-sefid (*Lilium ledebourii* (Baker) Boiss. protection area), Yakhkan Cave, Sabalan Peak.

Protected areas (340,744 hectares): Marakan, Arasbaran, Lisar, Siah Kashim, Ghasht Rudkhan and Siahmazghi, Siahrud Rudbar, Sarvelat and Javaherdasht, Belskuh, Chahar Bagh.

The Caucasian part of Iran is in northwestern Iran and encompasses the provinces of Ardabil, Gilan, the main portion of East Azerbaijan, northern West Azerbaijan, and western Mazandaran. This area is bounded on the west by Turkey, on the north by Armenia and the Republic of Azerbaijan, on the northeast by the Caspian Sea, and reaches to the Chalus valley at its easternmost limit. The entire area lies between 36°15′–39°47′N and 44°03′–51°25′E.

The topography is completely mountainous and ranges in elevation from 26 m below sea level on the shores of the Caspian Sea to 4850 m at the summit of Alamkuh, in western Alborz, and this peak is the second highest in Iran after Mount Damavand, at 5671 m. Sabalan, Sahand, Talysh, Bozqush, Misho Dagh, and western Alborz are the major mountains within the Caucasian territory of Iran. Sabalan (4811 m) and Sahand (3707 m) are two isolated, inactive volcanic mountains in northwestern Iran. Mount Sabalan is the third highest mountain of Iran with a beautiful permanent crater lake on its summit.

CLIMATE

Caucasian part of Iran is located in two completely different Zonobiomes (according to Breckle, 2002): the warm-temperate humid climate (Zonobiome V) to the southwest of the Caspian region, including the northern side of the Talysh and Alborz mountains to the Chalus valley, and the transition between the Mediterranean Zonobiome with winter rain and arid summer (Zonobiome IV) and the Zonobiome of an arid temperate climate (continental with cold winters; Zonobiome VII) in the northwestern part of the country. Based on the Global Bioclimatic Classification System (GBC) developed by Rivas-Martínez et al. (1997, 1999), this region can be classified into two different macrobioclimates. The southwestern Caspian region (Caucasian part of the Hyrcanian forests) represents the Temperate Macrobioclimate (temperate oceanic bioclimate) with the absence or very short duration of a dry season, while the high plateau of northwest Iran represents the Mediterranean Macrobioclimate (Mediterranean xeric-continental bioclimate) with a long summer drought and cold winters (Djamali et al., 2011).

High precipitation along the southern Caspian Sea results in Hyrcanian deciduous forests, forming an arc along the southern shore of the Caspian Sea and the adjacent lower zones of the northern slopes of the Alborz and Talysh mountain ranges. The climate of this area is mesic and warm, with rainy summers and mild winters. The mean annual precipitation for the Caucasian portion of these forests ranges from 1200 mm in the eastern part to ca. 2000 mm in the western part. From west to east the amount of precipitation decreases (Khalili, 1973; Domoers et al., 1998; Djamali et al., 2011).

The rainfall regime in northwestern Iran, with the exception of the Arasbaran forest in East Azerbaijan Province, is a semi-Mediterranean type. This regime is characterized by a dry period of ca. three to six months, which coincides with the warm season of the year. The summer is arid, hot, and sunny, with minimum rainfall, and most precipitation is during late autumn, winter, and early spring. Mean annual precipitation in areas of the plain (1100–1500 m above sea level) is ca. 280–330 mm. With increasing elevation, precipitation increases and temperature and aridity decrease. At high elevations, annual and diurnal amplitudes of temperature can be very high (Noroozi et al., 2008, 2010). The analytical data show that these higher altitudes are affected by the northwesterly flow of polar air masses (Khalili, 1973).

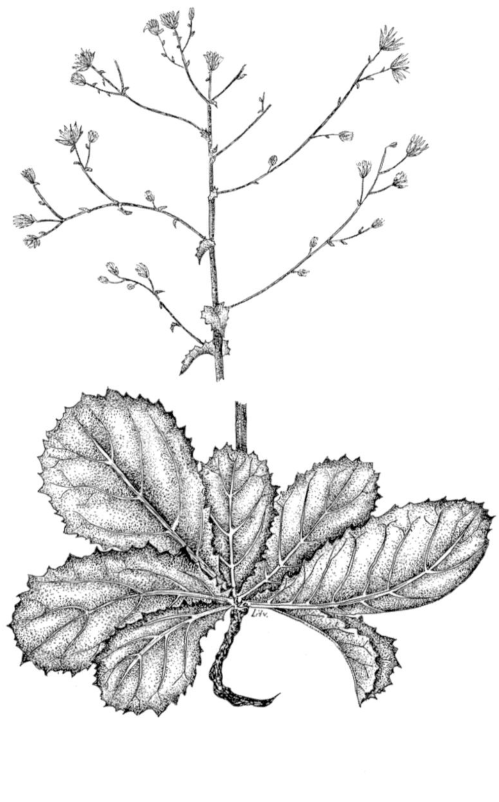

Cephalorrhynchus takhtadzhianii (Sosn.) Kirp., endemic to the southeastern Caucasus region.

Erythronium caucasicum Woronow, endemic to the western Caucasus region.

FLORA AND VEGETATION

The Caucasian part of Iran is a transitional corridor between the Alborz, Zagros, Anatolia, and Caucasus mountain ranges. This territory belongs to two different phytogeographic regions. The subtropical Hyrcanian forests are included in the Euxino-Hyrcanian province of the Euro-Siberian region and northwest of Iran (Azerbaijan plateau), while the high elevations of the Alborz Mountains are included in the Armeno-Iranian province of the Irano-Turanian region (Zohary, 1973; Takhtajan, 1986). As an aid to identifying the Caucasian plant species of Iran, *Flora Iranica* (Rechinger, 1963–2012) and *Flora of Iran* (Assadi et al., 1988–2012) were used as the main sources, along with other taxonomic monographs and new species records from the Caucasian part of Iran. Almost 4000 vascular plant species are living in this region of which ca. 420 species are endemic to the Caucasian region and 181 species are endemic to the Caucasian part of Iran. Some genera with a high number of endemic species in the Caucasian part of Iran are *Astragalus* L. (38 species, Fabaceae), *Cousinia* Cass. (14 species, Asteraceae), *Alchemilla* L. (seven species, Rosaceae), *Dianthus* L. (six species, Caryophyllaceae), and *Acantholimon* Boiss. (five species, Plumbaginaceae). Due to the diverse orography and climate, different vegetation types have developed in the region. The main vegetation types are the Hyrcanian forests and montane steppe shrublands, as well as the subalpine zone, alpine zone, and subnival-nival zone.

HYRCANIAN FORESTS

The Hyrcanian forests occupy the southern Caspian lowlands and extend on the northern slopes of the Alborz and Talysh mountains up to almost 2800 m. Arasbaran is an isolated part of this forest in northwestern Iran, East Azerbaijan Province. The Hyrcanian subprovince belongs to the Euro-Siberian phytogeographical region and is characterized by deciduous forests and in peripheral areas by deciduous scrub (Zohary, 1973). This includes remnants of a former Arcto-Tertiary flora that includes *Celtis australis* L. (Ulmaceae), *Parrotia persica* C. A. Mey. (Hamamelidaceae), *Pterocarya fraxinifolia* (Poir.) Spach (Juglandaceae), and *Zelkova carpinifolia* (Pall.) K. Koch (Ulmaceae) (Zohary, 1973), and also includes an Indo-Malaysian flora, with *Albizia julibrissin* Durazz. (Fabaceae), *Buxus hyrcana* Pojark. (Buxaceae), *Diospyros lotus* L. (Ebenaceae), and *Nelumbo nucifera* Gaertn. (Nelumbonaceae), which survived the ice age in this region (Frey et al., 1999). These forests can be altitudinally divided into three vegetation types characterized as lowland, montane, and subalpine deciduous forests (Bobek, 1951; Zohary, 1973; Rastin, 1983; Klein & Lacoste,

1989; Frey et al., 1999; Akhani et al., 2010).

The lowland deciduous forests are distributed from the Caspian coast up to 700(–1000) m, where it merges with montane forests that are determined by the upper limit of frost-sensitive tree species (Frey et al., 1999). The dominant and characteristic arboreal species of this zone are *Acer cappadocicum* Gled. (Aceraceae), *Albizia julibrissin* Durazz. (Fabaceae), *Alnus glutinosa* (L.) Gaertn. (Betulaceae), *Buxus hyrcana* Pojark. (Buxaceae), *Danae racemosa* (L.) Moench (Ruscaceae), *Ficus carica* L. (Moraceae), *Gleditsia caspica* Desf. (Caesalpiniaceae), *Melia azedarach* L. (Meliaceae), *Mespilus germanica* L. (Rosaceae), *Parrotia persica* C. A. Mey. (Hamamelidaceae), *Populus caspica* (Bornm.) Bornm. (Salicaceae), *Pterocarya fraxinifolia* (Poir.) Spach (Juglandaceae), *Quercus castaneifolia* C. A. Mey. (Fagaceae), *Tilia platyphyllos* Scop. (Tiliaceae), and *Zelkova carpinifolia* (Pall.) K. Koch (Ulmaceae) (Frey et al., 1999; Hamzeh'ee et al., 2008).

The montane deciduous forests range from 700 to 1800(to 2000) m and are dominated by beech (*Fagus orientalis* Lipsky, Fagaceae). The other dominant arboreal species are *Acer platanoides* L. and *Acer velutinum* Boiss. (both Aceraceae); *Alnus subcordata* C. A. Mey. and *Carpinus betulus* L. (both Betulaceae); *Diospyrus lotus* L. (Ebenaceae); *Fraxinus excelsior* L. (Oleaceae); *Ilex spinigera* (Loes.) Loes. (Aquifoliaceae); *Ruscus hyrcanus* Woronow (Ruscaceae); *Sorbus graeca* (Spach) Lodd. ex S. Schauer and *Sorbus torminalis* (L.) Crantz (both Rosaceae); and *Ulmus glabra* Huds. (Ulmaceae) (Zohary, 1973; Frey et al., 1999).

The most frequent undergrowth taxa of the lowland and montane zones in the Caucasian part are *Actaea spicata* L., *Aquilegia olympica* Boiss., and *Ranunculus repens* L. (these three, Ranunculaceae); *Alcea hyrcana* (Grossh.) Grossh. (Malvaceae); *Cardamine bulbifera* (L.) Crantz (Brassicaceae); *Carex remota* L. subsp. *remota* and *Carex strigosa* Huds. (both Cyperaceae); *Corydalis angustifolia* (M. Bieb.) DC. (Fumariaceae); *Crocus caspius* Fisch. & C. A. Mey. ex Hohen. (Iridaceae); *Cyclamen coum* Mill. and *Primula macrocalyx* Bunge (both Primulaceae); *Epimedium pinnatum* Fisch. ex DC. (Berberidaceae); *Euphorbia amygdaloides* L. (Euphorbiaceae); *Galanthus transcaucasicus* Fomin and *Leucojum aestivum* L. (both Amaryllidaceae); *Geranium pyrenaicum* Burm. f. (Geraniaceae); *Lathyrus vernus* (L.) Bernh. (Fabaceae); *Parietaria officinalis* L. (Urticaceae); *Phyllitis scolopendrium* (L.) Newman (Aspleniaceae); *Poa nemoralis* L. (Poaceae); *Polypodium vulgare* L. (Polypodiaceae); *Pteridium aquilinum* (L.) Kuhn (Dennstaedtiaceae); *Pteris cretica* L. (Pteridaceae); *Rumex sanguineus* L. (Polygonaceae); *Scutellaria tournefortii* Benth. and *Teucrium hircanicum* L. (both Lamiaceae); *Silene schafta* J. G. Gmel. ex Hohen. (Caryophyllaceae); and *Smilax excelsa* L. (Smilaceae) (Zohary, 1973; Hamzeh'ee et al., 2008).

The subalpine deciduous forests cover the uppermost zone of the Hyrcanian forests at elevations from 1800 to 2500(to 2800) m and are dominated by *Quercus macranthera* Fisch. & C. A. Mey. ex Hohen. (Fagaceae) (Klein & Lacoste, 1989; Akhani et al., 2010). Other characteristic species of this zone are *Acer hyrcanum* Fisch. & C. A. Mey. (Aceraceae); *Cervaria cervariifolia* (C. A. Mey.) Pimenov and *Chaerophyllum aureum* L. (both Apiaceae); *Crataegus pentagyna* Waldst. & Kit. ex Willd. and *Pyrus boissierana* Buhse (both Rosaceae); *Dryopteris filix-mas* (L.) Schott (Dryopteridaceae); *Festuca drymeja* Mert. & W. D. J. Koch (Poaceae); *Fragaria vesca* L. and *Geum urbanum* L. (both Rosaceae); *Hesperis hyrcana* Bornm. & Gauba (Brassicaceae); *Hypericum hirsutum* L. (Hypericaceae); *Lathyrus roseus* Steven and *Vicia truncatula* Fisch. ex M. Bieb. (both Fabaceae); and *Lonicera caucasica* Pall. and *Viburnum lantana* L. (both Caprifoliaceae) (Klein & Lacoste, 1989). In contrast to the Euxinian region, Hyrcanian forests do not have any species of *Rhododendron* L. (Ericaceae) or the three conifers *Abies* Mill., *Picea* A. Dietr., and *Pinus* L. (all Pinaceae), which are usually timberline trees. New paleobotanical research shows that the aforementioned coniferous trees have not been present in this region for more than 200,000 years BP (see Djamali et al., 2008).

MONTANE STEPPE SHRUBLANDS

The montane zone of the Azerbaijan plateau in northwestern Iran, at elevations between ca. 1000 and 2200 m, is covered by steppe shrublands. Knowledge about the vegetation complexes of this area is limited. Sparse trees and shrubs belonging to *Amygdalus* L., *Cotoneaster* Medik., *Crataegus* L., and *Rosa* L. (all Rosaceae), as well as *Pistacia* L. (Anacardiaceae), among others, are scattered in this steppe vegetation, and the understory is covered with herbaceous, graminoid and thorn-cushion plants (Zohary, 1973; Frey & Probst, 1986). Many endemic species of *Astragalus* L. (Fabaceae) and *Cousinia* Cass. (Asteraceae), with narrow distributions, are restricted to these habitats (Table 1). Some of the arboreal species in these habitats are *Amygdalus lycioides* Spach var. *lycioides* and *Amygdalus nairica* Fed. & Takht. (both Rosaceae); *Berberis integerrima* Bunge (Berberidaceae); *Cerasus incana* (Pall.) Spach, *Cerasus microcarpa* (C. A. Mey.) Boiss. subsp. *microcarpa*, *Cotoneaster nummularioides* Pojark., *Cotoneaster nummularius* Fisch. & C. A. Mey., *Crataegus curvisepala* Lindm., *Crataegus meyeri* Pojark., *Pyrus salicifolia* Pall., and *Rosa canina* L. (these eight, Rosaceae); *Pistacia atlantica* Desf. (Anacardiaceae); *Rhamnus kurdica* Boiss. & Hohen. and *Rhamnus pallasii* Fisch. & C. A. Mey. (both Rhamnaceae) (Zohary, 1973; Noroozi, pers. obs.). Ephemeral plants (annuals and geophytes) are also common. Some of the more notable understory species are *Aethionema grandiflorum* Boiss. & Hohen. (Brassicaceae); *Artemisia fragrans* Willd., *Artemisia sieberi* Besser subsp. *sieberi, Gundelia tournefortii* L., *Scariola orientalis* (Boiss.) Soják, and *Xeranthemum longepapposum* Fisch. & C. A. Mey. (these five, Asteraceae); *Astragalus microcephalus* Willd. (Fabaceae); *Bellevalia longistyla* (Miscz.) Grossh. (Hyacinthaceae); *Bromus tomentellus* Boiss., *Stipa arabica* Trin. & Rupr., and *Stipa hohenackeriana* Trin. & Rupr. (these three, Poaceae); *Iris reticulata* M. Bieb. (Iridaceae); *Muscari caucasicum* (Griseb.) Baker (Hyacinthaceae); *Pterocephalus canus* Coult. ex DC. (Dipsacaceae); and *Stachys inflata* Benth. (Lamiaceae).

SUBALPINE ZONE

The umbelliferous large herbs are the most remarkable vegetation type in the subalpine zone of the Irano-Turanian parts of Iran (Klein, 1987, 1988; Noroozi et al., 2008). This vegetation type is usually found at elevations from 2000 to 3000 m in the Caucasian portion of Iran. Klein (1987) introduced the vegetation classification of the Prangetea ulopterae community for the subalpine zone of the Irano-Turanian parts of the Alborz Mountains. Different tall umbelliferous genera like *Prangos* Lindl., *Ferula* L., and *Diplotaenia* Boiss. (all Apiaceae) are conspicuously distributed in this zone of the Azerbaijan and Alborz mountains. The most dominant species with high coverage and abundance in this vegetation is *Prangos uloptera* DC. Other community-dominant species are *Asperula setosa* Jaub. & Spach and *Cruciata taurica* (Pall. ex Willd.) Ehrend. (both Rubiaceae); *Astragalus aegobromus* Boiss. & Hohen. (Fabaceae); *Diplotaenia cachrydifolia* Boiss. (Apiaceae); *Fibigia suffruticosa* (Vent.) Sweet (Brassicaceae); *Helichrysum oligocephalum* DC. (Asteraceae); *Hypericum scabrum* L. (Hypericaceae); *Marrubium astracanicum* Jacq., *Nepeta fissa* C. A. Mey., and *Stachys lavandulifolia* Vahl (these three, Lamiaceae); *Mesostemma kotschyanum* (Fenzl ex Boiss.) Vved. (Caryophyllaceae); *Papaver bracteatum* Lindl., *Papaver lasiothrix* Fedde, and *Papaver orientale* L. (these three, Papaveraceae); as well as *Rheum ribes* L. and *Rumex elbrusensis* Boiss. (both Polygonaceae).

The subalpine zone on the northern side of the Alborz Mountains occurs at elevations from ca. 2800 to 3200 m, just above the Quercus macranthera communities and is covered by a graminoid vegetation type. Klein and Lacoste (1994) classified this vegetation under Festuco-Brometea community type, which belongs to the Euro-Siberian region. The dominant species of this subalpine community are *Campanula glomerata* L. (Campanulaceae); *Centaurea hyrcanica* Bornm., *Dolichorrhiza persica* (Boiss.) B. Nord., *Hieracium caucasicum* Nägeli & Peter, and *Tanacetum coccineum* (Willd.) Grierson (these four, Asteraceae); *Cephalaria kotschyi* Boiss. & Hohen. (Dipsacaceae); *Gentiana septemfida* Pall. (Gentianaceae); *Hypericum hirsutum* L. (Hypericaceae); *Juniperus communis* L. and *Juniperus sabina* L. (both Cupressaceae); *Phleum phleoides* (L.) G. Karst. and *Poa pratensis* L. (both Poaceae); *Polylophium involucratum* (Pall.) Boiss. (Apiaceae); *Ranunculus buhsei* Boiss. (Ranunculaceae); and *Trifolium canescens* Willd. (Fabaceae) (Klein & Lacoste, 1994).

Windswept areas of the subalpine to alpine zones in northwestern Iran and the Alborz Mountains in the Irano-Turanian region are covered by thorn-cushion steppe vegetation types. The most important plant species here belong to *Astragalus* L. (Fabaceae), *Acantholimon* Boiss. (Plumbaginaceae), and *Onobrychis* Mill. (Fabaceae), among other genera, with *Onobrychis cornuta* (L.) Desv. as the most dominant species (Klein, 1987; Noroozi et al., 2008, 2010).

ALPINE ZONE

The alpine flora of the Iranian mountains has an Irano-Turanian origin, with a continental climate and Mediterranean precipitation pattern (Zohary, 1973; Klein, 1982; Noroozi, 2008, 2010). This zone in the Caucasian part of Iran is almost entirely above 3000 m, with a high degree of endemism. The growing period is mostly restricted to July and August, and alpine areas have been used as summer pastures. The main vegetation types in this zone are snowbed, thorn-cushion grasslands, and scree.

The snowbed communities are mostly dwarf, rosette plants with low taxonomic diversity. They are usually dominated by *Alopecurus aucheri* Boiss. and *Catabrosella parviflora* (Boiss. & Buhse) E. B. Alexeev ex R. R. Mill (both Poaceae); *Carum caucasicum* (M. Bieb.) Boiss. (Apiaceae); *Cerastium cerastoides* (L.) Britton (Caryophyllaceae); *Gnaphalium supinum* L. and *Taraxacum brevirostre* Hand.-Mazz. (both Asteraceae); *Plantago atrata* Hoppe (Plantaginaceae); *Sibbaldia parviflora* Willd. (Rosaceae); and *Veronica gentianoides* Vahl (Scrophulariaceae).

The thorn-cushion grasslands are widespread in the alpine habitat, and their plant diversity is rich in comparison to the other alpine zone vegetation types. The most remarkable taxa found here are *Alopecurus textilis* Boiss., *Poa araratica* Trautv., and *Koeleria eriostachya* Pančić (these three, Poaceae); *Asperula prostrata* (Adams) K. Koch and *Galium decumbens* (Ehrend.) Ehrend. & Schönb.-Tem. (both Rubiaceae); *Astragalus aureus* Willd. and *Astragalus sahendi* Buhse ex Fisch. (both Fabaceae); *Campanula stevenii* M. Bieb. subsp. *beauverdiana* (Fomin) Rech. f. & Schiman-Czeica (Campanulaceae); *Dianthus erythrocoleus* Boiss., *Herniaria caucasica* Rupr., *Minuartia glandulosa* (Boiss. & A. Huet) Bornm., and *Silene marschallii* C. A. Mey. subsp. *sahendica* (Boiss. & Buhse) Melzh. (these four, Caryophyllaceae); *Erigeron uniflorus* Boiss. subsp. *elbrusensis* (Boiss.) Rech. f., *Helichrysum psychrophilum* Boiss., and *Tragopogon kotschyi* Boiss. (these three, Asteraceae); and *Veronica orientalis* Mill. (Scrophulariaceae).

Scree vegetation is usually found on

steep slopes with either moving or stabilized scree and is dominated by *Alyssum condensatum* Boiss. & Hausskn. and *Physoptychis gnaphalodes* (DC.) Boiss. (both Brassicaceae); *Bromus tomentosus* Trin. (Poaceae); *Jurinella moschus* (Habl.) Bobrov (Asteraceae); *Minuartia recurva* (All.) Schinz & Thell. subsp. *oreina* (Mattf.) McNeill (Caryophyllaceae); *Pedicularis caucasica* M. Bieb. (Scrophulariaceae); and *Ziziphora clinopodioides* Lam. (Lamiaceae). These species are also common in the subnival zone (Noroozi et al., 2011).

SUBNIVAL-NIVAL ZONE

Habitats in the subnival-nival zone have a highly scattered distribution over the Caucasian portion of Iran and are restricted to the highest elevations of the Sahand, Sabalan (between 3600 and 4400 m), and western Alborz mountains (between 3900 and 4800 m), above the alpine zone. In spite of the low species diversity of this zone the degree of local endemism is quite high in comparison to lower elevations (Noroozi et al., 2011). The ground is mainly covered with scree and rocks, and plant coverage and abundance are mostly low (Klein & Lacoste, 1999). The most characteristic taxa of this zone are *Achillea aucheri* Boiss., *Artemisia melanolepis* Boiss., *Erigeron caucasicus* Steven subsp. *venustus* (Botsch.) Grierson, *Erigeron hyrcanicus* Bornm. & Vierh., *Senecio taraxacifolius* (M. Bieb.) DC., and *Tripleurospermum caucasicum* (Willd.) Hayek (these six, Asteraceae); *Alopecurus dasyanthus* Trautv. (Poaceae); *Arabis caucasica* Willd., *Didymophysa aucheri* Boiss., *Draba bruniifolia* Steven, and *Erysimum gelidum* Bunge (these four, Brassicaceae); *Campanula karakuschensis* Grossh. (Campanulaceae); *Dracocephalum aucheri* Boiss. (Lamiaceae); *Kobresia humilis* (C. A. Mey. ex Trautv.) Serg. (Cyperaceae); *Oxyria digyna* (L.) Hill (Polygonaceae); *Potentilla porphyrantha* Juz. (Rosaceae); *Saxifraga sibirica* L. (Saxifragaceae); and *Veronica aucheri* Boiss. (Scrophulariaceae) (Noroozi et al., 2011).

MAIN THREATS

All of the vegetation types are affected by human activities. However, the impact differs considerably between the different habitats. Overgrazing, agriculture, forestry, fire, overexploitation, urbanization, littering, invasive species, extensive dam building, and infrastructure developments are the main human impacts on the natural habitats of the Caucasian territory of Iran, as in other parts of the country. All habitats are under high grazing pressure. This overgrazing by livestock, especially sheep and goats, leads to soil degradation and changes in plant composition with a higher portion of spiny and/or poisonous, inedible plants such as the fern *Pteridium aquilinum* (L.) Kuhn (Dennstaedtiaceae), as well as *Colchicum trigynum* (Steven ex Adams) Stearn (Colchiaceae) in the Hyrcanian forests and *Gundelia tournefortii* L. (Asteraceae) and species of *Euphorbia* L. (Euphorbiaceae) in the mountains of northwestern Iran. Some introduced plant species have invaded natural habitats and threaten native species and communities. For instance, *Azolla filiculoides* Lam. (Salviniaceae), an introduced aquatic fern species from southeastern Asia, is the most harmful invasive species in the Hyrcanian region and represents an extreme threat to natural wetlands (Hashemian & Ataei Azimi, 2009). Some other threats to the Hyrcanian forests include tourism, fire, extensive road-building, and cutting of trees for timber (Akhani et al., 2010). The cold-adapted plant species and vegetation of the alpine and subnival-nival zones are highly vulnerable to global warming effects due to the absence of alternative cold habitats in the Iranian mountains (Noroozi et al., 2011).

Table 1. Distribution of endemic species within different vegetation types.

HABITAT	SPECIES
Hyrcanian forests	*Acantholimon esfandiarii* Rech. f. & Schiman-Czeika, *Alchemilla amardica* Rothm., *Alchemilla citrina* A. Froehner, *Alchemilla condensa* A. Froehner, *Alchemilla farinosa* A. Froehner, *Alchemilla pectiniloba* A. Froehner, *Alchemilla surculosa* A. Froehner, *Allium esfandiarii* Matin, *Allium subvineale* Wendelbo, *Aristolochia hyrcana* P. H. Davis & M. S. Khan, *Asperula sherardioides* Jaub. & Spach, *Astragalus acutifolius* Bunge, *Astragalus ammodendroides* Bornm. ex Podlech & Zarre, *Astragalus arasbaranensis* Maassoumi & Ranjbar, *Astragalus leushanensis* Maassoumi, *Astragalus pinetorum* Boiss. subsp. *declinatus* Podlech, *Bupleurum wolffianum* Bornm. ex H. Wolff, *Colutea uniflora* Beck, *Cousinia alfredii* Bornm. & Gauba, *Cousinia hablitzii* C. A. Mey. ex DC., *Crepis alfredii* Bornm., *Crocus gilanicus* B. Mathew, *Crucianella gilanica* Trin. subsp. *gilanica*, *Dianthus agrostolepis* Rech. f., *Dianthus hyrcanicus* Rech. f., *Dianthus rudbaricus* Assadi, *Heracleum rechingeri* Manden., *Johrenia ramosissima* Mozaff., *Lactuca azerbaijanica* Rech. f., *Lindelofia kandavanensis* Bornm. & Gauba, *Myosotis anomala* Riedl, *Phlomis ghilanensis* K. Koch, *Rindera regia* (S. G. Gmel.) Kusn., *Satureja isophylla* Rech. f., *Scrophularia oxysepala* Boiss., *Silene lineata* Boiss. & Buhse, *Thlaspi maassoumii* Mozaff., *Tulipa wendelboi* Matin & Iranshahr
Lowland dry hills and saline soils	*Achillea moghanica* Iranshahr, *Alcea assadii* Pakravan, *Astragalus birangiae* Maassoumi, *Atraphaxis aucheri* Jaub. & Spach, *Atraphaxis suaedifolia* Jaub. & Spach, *Centaurea congesta* Wagenitz, *Cousinia hypochionea* Bornm., *Gamanthus pilosus* (Pall.) Bunge, *Halanthium mamanense* Bunge, *Heliotropium gypsaceum* Rech. f. & Riedl., *Hyacinthella persica* (Boiss. & Buhse) Chouard, *Jurinea leptoloba* DC., *Leutea polyscias* (Boiss.) Pimenov, *Salvia oligophylla* Aucher ex Benth., *Tanacetum salsugineum* Podlech, *Taraxacum ruberuliforme* Soest, *Tulipa urumiensis* Stapf
Montane steppe shrublands	*Acantholimon atropatanum* Bunge, *Acantholimon densiflorum* Assadi, *Acantholimon tragacanthinum* (Jaub. & Spach) Boiss., *Acinos arvensis* (Lam.) Dandy, *Alcea ghahremanii* Pakravan & Assadi, *Alcea leariloba* I. Riedl, *Alcea wilhelminae* I. Riedl, *Allochrusa bungei* Boiss., *Allochrusa persica* (Boiss.) Boiss., *Alococarpum erianthum* (DC.) Riedl & Kuber, *Anthemis triumfettii* (L.) DC. subsp. *decumbens* Iranshahr, *Asperula sherardioides* Jaub. & Spach, *Astragalus aharicus* Maassoumi & Podlech, *Astragalus basilicus* Maassoumi & Podlech, *Astragalus canosus* Maassoumi, *Astragalus choicus* Bunge, *Astragalus hendelanicus* Maassoumi, *Astragalus horridissimus* Širj. & Bornm., *Astragalus inoblitus* Maassoumi, *Astragalus khajiboulaghensis* Maassoumi, *Astragalus mozaffarianii* Maassoumi, *Astragalus pseudomacrostachys* Maassoumi, *Astragalus rubrocalycinus* Maassoumi & Podlech, *Astragalus zarjabadensis* Ranjbar, *Bellevalia assadii* Wendelbo, *Bellevalia tabriziana* Turrill, *Centaurea aziziana* Rech. f., *Cirsium iranicum* Petr., *Cousinia adnata* Bunge, *Cousinia atropatana* Bunge, *Cousinia azerbaijanica* Djavadi, Attar

HABITAT	SPECIES
Montane steppe shrublands (continued)	& Najafi, *Cousinia boissieri* Buhse, *Cousinia calolepis* Boiss., *Cousinia erinacea* Jaub. & Spach, *Cousinia gilanica* Bornm., *Cousinia gilliatii* Rech. f., *Cousinia grandis* C. A. Mey. ex DC., *Cousinia microcephala* C. A. Mey. ex DC., *Cousinia prasina* Jaub. & Spach, *Cousinia tabrisiana* Bunge, *Cousinia urumiensis* Bornm., *Cousinia wilhelminae* Rech. f., *Deyeuxia parsana* Bor, *Eremostachys azerbaijanica* Rech. f., *Haplophyllum buhsei* Boiss., *Hieracium azerbaijanense* Lack, *Hieracium hoppeanum* Schult., *Nepeta makuensis* Jamzad & Mozaff., *Onobrychis aucheri* Boiss. subsp. *aucheri*, *Onobrychis depauperata* Boiss., *Onobrychis szovitsii* Boiss., *Peltariopsis drabicarpa* (Boiss.) N. Busch, *Peucedanum knappii* Bornm., *Pseudocamelina szowitsii* (Boiss.) N. Busch, *Ranunculus dolosus* Fisch. & C. A. Mey., *Rindera media* (Turrill) Riedl, *Salsola persica* Bunge ex Boiss., *Salvia hypochionaea* Boiss., *Sameraria nummularia* Bornm., *Satureja atropatana* Bunge, *Scabiosa schimperiana* Boiss. & Buhse, *Scutellaria bornmuelleri* Hausskn. & Bornm. subsp. *mianensis* Rech. f., *Sedum elburzense* Akhani & Assadi, *Sterigmostemum contortuplicatum* (Boiss.) Kuntze, *Tragopogon rezaiyensis* Rech. f., *Typha azerbaijanensis* Hamdi & Assadi
Subalpine zone	*Acantholimon gilliatii* Turrill, *Acantholimon mishaudaghense* Mobayen, *Acantholimon wilhelminae* Rech. f. & Schiman-Czeika, *Achillea millefolium* L. subsp. *elbursensis* Hub.-Mor., *Alchemilla rechingeri* Rothm., *Arenaria szowitsii* Boiss., *Astragalus hymenocystis* Fisch. & C. A. Mey., *Astragalus hymenostegis* Fisch. & C. A. Mey., *Astragalus juladakensis* Maassoumi, *Astragalus karl-heinzii* Maassoumi, *Astragalus neomobayenii* Maassoumi, *Astragalus pinetorum* Boiss. subsp. *pinetorum*, *Bupleurum ghahremanii* Mozaff., *Campanula gilliatii* Milne-Redh. & Turrill, *Centaurea incanescens* (Fisch. & C. A. Mey. ex DC.) Sch. Bip., *Crucianella suaveolens* C. A. Mey., *Dianthus sahandicus* Assadi, *Eritrichium gracillimum* Rech. f., *Euphorbia sahendi* Bornm., *Ferulago trifida* Boiss., *Gypsophila bazorganica* Rech. f., *Gypsophila wilhelminae* Rech. f., *Hedysarum papillosum* Boiss., *Lathyrus alamutensis* Mozaff., Ahvazi & Charkhch., *Minuartia acuminata* Turrill, *Onobrychis talagonica* Rech. f., *Onosma sharifii* Riedl, *Ranunculus sahendicus* Boiss. & Buhse, *Salvia sahendica* Boiss. & Buhse, *Saxifraga mazanderanica* Rech. f., *Silene gynodioica* Ghaz. subsp. *gynodioica*, *Stenotaenia nudicaulis* Boiss., *Thymus marandensis* Jamzad, *Veronica chionantha* Bornm., *Vicia rechingeri* Chrtková
Alpine zone	*Alchemilla fluminea* A. Froehner, *Astragalus azizii* Maassoumi, *Astragalus beckii* Bornm., *Astragalus herbertii* Maassoumi, *Astragalus parvulus* Bornm., *Astragalus pauperiflorus* Bornm., *Astragalus pinetorum* Boiss. subsp. *alamutensis* Maassoumi & Podlech, *Astragalus sahendi* Buhse ex Fisch., *Astragalus savellanicus* Podlech, *Campanula savalanica* Fed., *Centaurea elbrusensis* Boiss. & Buhse, *Cousinia gmelinii* C. Winkl., *Cousinia pterocaulos* (C. A. Mey.) Rech. f., *Crucianella gilanica* Trin. subsp. *suleimanica* Ehrend. & Schönb.-Tem., *Dianthus seidlitzii* Boiss., *Dolichorrhiza persica* (Boiss.) B. Nord., *Festuca rechingeri* E. B. Alexeev, *Festuca sabalanica* E. B. Alexeev, *Gypsophila*

HABITAT	SPECIES
Alpine zone (continued)	*iranica* Barkoudah, *Nepeta pogonosperma* Jamzad & Assadi, *Nepeta sahandica* Noroozi & Ajani, *Nonea anchusoides* Boiss. & Buhse, *Onosma bilabiata* Boiss. & Buhse, *Phlomis ghilanensis* K. Koch, *Ranunculus renzii* Iranshahr & Rech. f., *Thlaspi tenue* (Boiss. & Buhse) Hedge, *Veronica euphrasiifolia* Link
Subnival-nival zone	*Achillea aucheri* Boiss. subsp. *aucheri*, *Achillea aucheri* Boiss. subsp. *glabra* Hub.-Mor., *Astragalus rudimentus* Maassoumi, *Campanula karakuschensis* Grossh., *Oxytropis takhti-soleimanii* Vassilcz., *Potentilla porphyrantha* Juz., *Saxifraga iranica* Bornm., *Saxifraga ramsarica* Jamzad

TOP 50 NATIONAL CONSERVATION PRIORITIES FOR IRAN

The selected species are only known from the type location or are distributed in a very narrow geographic region. Two species, *Phlomis ghilanensis* K. Koch and *Veronica euphrasiifolia* Link, do not have maps.

ALLIACEAE

1. *Allium aznavense* R. M. Fritsch **CR**

This species was recently discovered in north Khalkhal, on the western slopes of the Talysh Mountains near Aznav spring, among rubble and marble rocks at an elevation of ca. 1800 m.

2. *Allium esfandiarii* Matin **CR**

This species is restricted to one site in the montane zone near Kaleybar, in East Azerbaijan Province.

APIACEAE

3. *Bupleurum wolffianum* Bornm. ex H. Wolff **EN**

This species has only been recorded from the type locality in the Arasbaran forests in northwestern Iran.

4. *Johrenia ramosissima* Mozaff. **EN**

This species is only known from the type locality in the montane zone (1500–1700 m) of the Hyrcanian forests between Mazanabad and Kelardasht, in Mazandaran Province.

ASTERACEAE

5. *Centaurea incanescens* (Fisch. & C. A. Mey. ex DC.) Sch. Bip. **CR**

This species has only been recorded from the subalpine vegetation belt of the Qotur region in northern Khoy.

6. *Cirsium iranicum* Petr. **EN**

This species is only known from the type locality in the mountains between Tabriz and Ahar, from an elevation of 1800 m.

7. *Cousinia atropatana* Bunge **CR**

This species has only been recorded from the type location on mountains near Tabriz.

8. *Cousinia gilliatii* Rech. f. **EN**

This species has been recorded from the mountains of east Tabriz at elevations from 1600 to 1700 m.

9. *Cousinia hablitzii* C. A. Mey. ex DC. **VU**

This species is restricted to the Samam Mountains in Gilan Province.

10. *Cousinia wilhelminae* Rech. f. **VU**

This species has only been recorded from the montane zone (1300–1500 m) of the Qotur region in northern Khoy.

11. *Lactuca azerbaijanica* Rech. f. **CR**

This species has only been recorded from south-facing slopes of *Quercus* L. forests around Arasbaran region at ca. 1800 m.

BORAGINACEAE

12. *Heliotropium gypsaceum* Rech. f. & Riedl **CR**

This species grows only on the gypsum

hills between Khoy and Marand at an elevation of ca. 1000 m.

BRASSICACEAE

13. *Pseudocamelina szowitsii* (Boiss.) N. Busch **CR**

This species grows only in the montane zone of Misho Dagh near Marand.

14. *Thlaspi maassoumii* Mozaff. **VU**

This species has only been recorded from the lowlands of the Hyrcanian forests near Ramsar, in Mazandaran Province, and could be threatened by urbanization in the region.

CAMPANULACEAE

15. *Campanula savalanica* Fed. **CR**

This species is restricted to the alpine zone of the Sabalan Mountains, where it is under high pressure from recreational trampling and overgrazing. It could also be threatened by climate warming.

CARYOPHYLLACEAE

16. *Dianthus seidlitzii* Boiss. **CR**

This species has only been recorded from the alpine zone of the Sahand Mountains and could be threatened from overgrazing and climate change effects. Sahand is one of the most over-grazed mountains in Iran due to its famous cheese production.

17. *Dianthus talyschensis* Boiss. & Buhse **CR**

This species is only known from the type locality in montane forests of the western Hyrcanian region near the border of Iran and the Republic of Azerbaijan.

18. *Gypsophila bazorganica* Rech. f. **CR**

This species is restricted to Gogeh Dagh Mountain, west of Bazorgan and near the border of Iran and Turkey.

19. *Gypsophila iranica* Barkoudah **CR**

This species is known from the alpine zone of Sabalan Mountains, where it is under high pressure from overgrazing due to the large herds moved by nomads to the alpine regions in summer.

20. *Minuartia sabalanica* Assadi & Mostafavi **CR**

This is a rare species known from the alpine zone of the Sabalan Mountains, where it is under threat from over-grazing.

EUPHORBIACEAE

21. *Euphorbia sahendi* Bornm. **CR**

This species has only been recorded from the high alpine vegetation belt of the Sahand Mountains, in northwestern Iran, and could be threatened by climate change effects.

FABACEAE

22. *Astragalus aharicus* Maassoumi & Podlech **EN**

This species has a narrow distribution in the mountains between Ahar and Tabriz.

23. *Astragalus arasbaranensis* Maassoumi & Ranjbar **EN**

This species is only known from the type locality in the Arasbaran forests.

24. *Astragalus birangiae* Maassoumi **CR**

This species is only known from the type locality and is restricted to the lowlands of Dashte Mogan at an elevation of 160 m.

25. *Astragalus horridissimus* Širj. & Bornm. **EN**

This species has only been recorded from the rocky slopes of Eskandar village (1700–2100 m) in northeastern Tabriz.

26. *Astragalus ilachchiensis* Ranjbar & Ghahrem. **CR**

This recently discovered species is from a locality between Tabriz and Azarshar in the northwestern Sahand Mountains at elevations from 1800 to 2200 m.

27. *Astragalus montismishoudaghi* Sheikh Akbari Mehr, Ghorbani & Maassoumi **CR**

This newly discovered species was found in the area of Misho Dagh Mountain at 1300 m, in northwestern Iran.

28. *Hedysarum marandense* Mozaff., Akrami & Maassoumi **EN**

This species was discovered and described recently from the mountain area between Zunuzagh and Kuhkamar, northwest of

Marand, at elevations from 2080 to 3300 m.

29. *Hedysarum orumiehense* Ranjbar **VU**

This recently discovered species comes from the eastern slope of the Silvaneh and Razhan mountains, in western Orumieh, at ca. 1500 m.

30. *Onobrychis depauperata* Boiss. **CR**

This species is only known from the type locality in the mountains near Khoy, West Azerbaijan Province.

31. *Onobrychis szovitsii* Boiss. **CR**

This species has only been recorded from the type locality on southern slopes in the montane zone of Misho Dagh.

32. *Oxytropis takhti-soleimanii* Vassilcz. **CR**

This is a species from the subnival-nival zone that is only known from the type locality on Takhtesoleyman Mountain in western Alborz. It is threatened by climate change effects.

HYACINTHACEAE

33. *Bellevalia assadii* Wendelbo **CR**

This species is only known from the type locality in the montane zone, at 1700 m, in the mountains between Tabriz and Marand.

LAMIACEAE

34. *Nepeta makuensis* Jamzad & Mozaff. **CR**

This species is only known from the type locality in the montane zone, at 1700 m, near Maku, West Azerbaijan Province.

35. *Nepeta sahandica* Noroozi & Ajani **CR**

The species is restricted to unstable talus in the Sahand Mountains at an elevation of ca. 3500 m. It is critically endangered due to the absence of alternative cold habitats for any upward shift of species distribution under the impact of global warming.

36. *Phlomis ghilanensis* K. Koch **EN**

This species is restricted to the Gambu Mountains in Gilan Province.

37. *Salvia hypochionaea* Boiss. **EN**

This species is only known from the type locality in northwestern Iran, on the southern slopes of Misho Dagh.

38. *Thymus marandensis* Jamzad **VU**

This species is only known from the type locality in the subalpine zone (2000–2600 m) of the mountains near Marand, East Azerbaijan Province.

LILIACEAE

39. *Tulipa wendelboi* Matin & Iranshahr **CR**

This species is only known from the type locality in the montane zone of the Hyrcanian forests near Marzanabad and could be threatened by urbanization.

MALVACEAE

40. *Alcea assadii* Pakravan **CR**

This species is only found in the lowlands (500 m) between Kaleybar and Aslanduz, near the border of Iran and the Republic of Azerbaijan.

PLUMBAGINACEAE

41. *Acantholimon atropatanum* Bunge **CR**

This species has only been recorded from the montane zone on the southern slopes of the Bozqush Mountains, where it is under high pressure from overgrazing.

42. *Acantholimon esfandiarii* Rech. f. & Schiman-Czeika **CR**

This species is only known from the type location in the western Alborz Mountains, from Amarlu region, Gilan Province.

43. *Acantholimon wilhelminae* Rech. f. & Schiman-Czeika **CR**

This species is only known from the type locality in the montane-subalpine zone, at elevations from 2100 to 2500 m, on Gogeh Dagh Mountain, near Maku.

POACEAE

44. *Festuca sabalanica* E. B. Alexeev **EN**

This species is restricted to the high alpine and subnival zone of the Sabalan Mountains, where it occurs on northern slopes.

POLYGONACEAE

45. *Atraphaxis suaedifolia* Jaub. & Spach **CR**

This species has only been recorded from the type locality in the mountains near Tabriz.

POLYPODIACEAE

46. *Polypodium iranicum* Mazooji **CR**

This is an epiphytic or epilithophytic fern recorded from the Arasbaran Protected Area. It is presently known from only two rainforest locations at 1320 and 1450 m.

RANUNCULACEAE

47. *Ranunculus renzii* Iranshahr & Rech. f. **CR**

This alpine species is only known from Khalil Kuh near Orumieh, West Azerbaijan Province. It could be threatened by overgrazing and global warming effects.

RUBIACEAE

48. *Asperula microphylla* Boiss. **CR**

This species is only recorded from the Julfak Mountains, Gilan Province.

SCROPHULARIACEAE

49. *Veronica euphrasiifolia* Link **CR**

This species is only known from the type locality in the alpine regions of Gilan.

TYPHACEAE

50. *Typha azerbaijanensis* Hamdi & Assadi **VU**

This species has only been found in the montane zone, at 1100 m, near Khoy, in West Azerbaijan Province.

IRAN MAPS

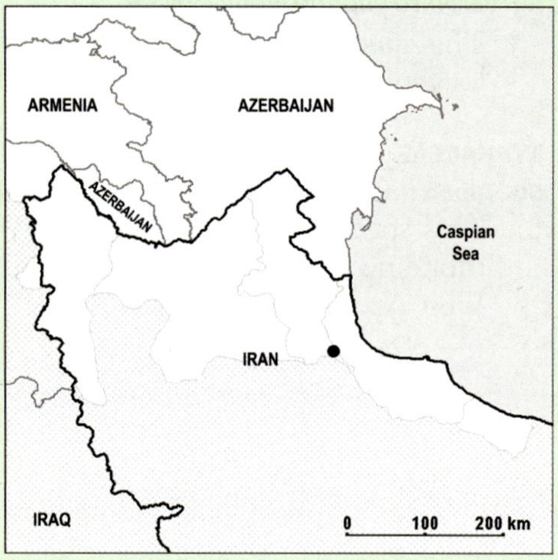

1. *Allium aznavense* R. M. Fritsch

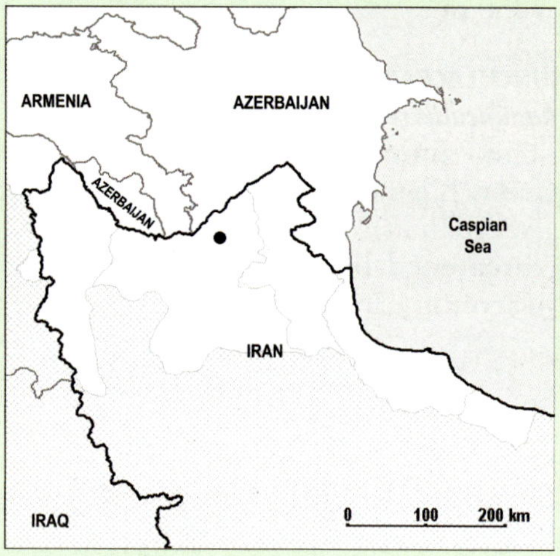

2. *Allium esfandiarii* Matin

3. *Bupleurum wolffianum* Bornm. ex H. Wolff

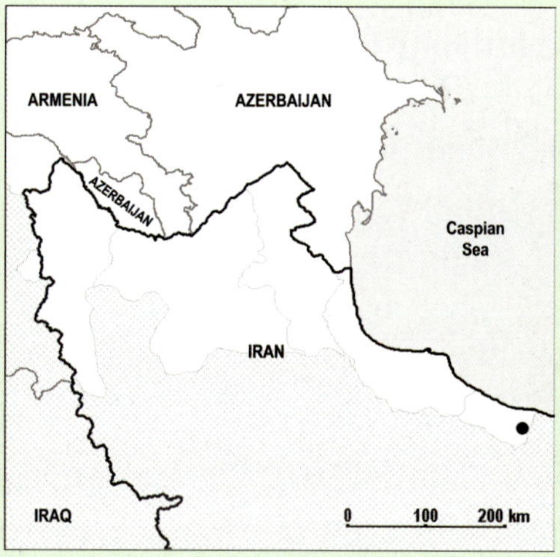

4. *Johrenia ramosissima* Mozaff.

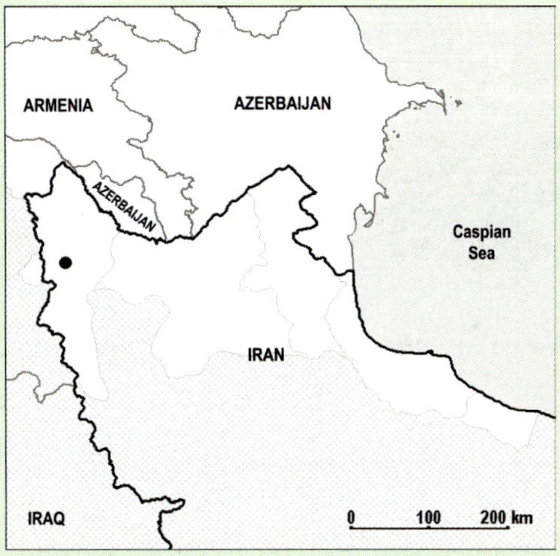

5. *Centaurea incanescens* (Fisch. & C. A. Mey. ex DC.) Sch. Bip.

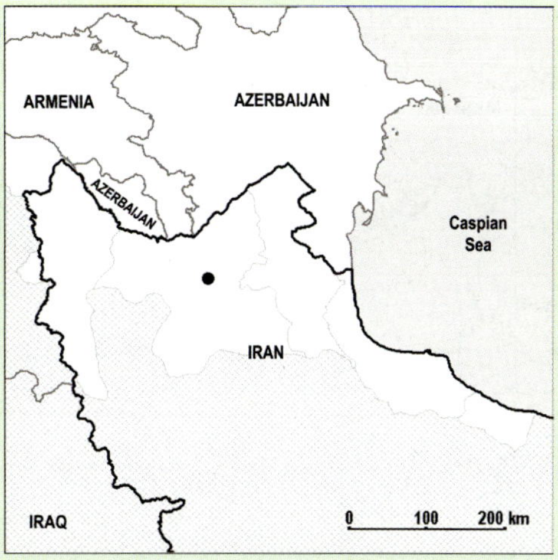

6. *Cirsium iranicum* Petr.

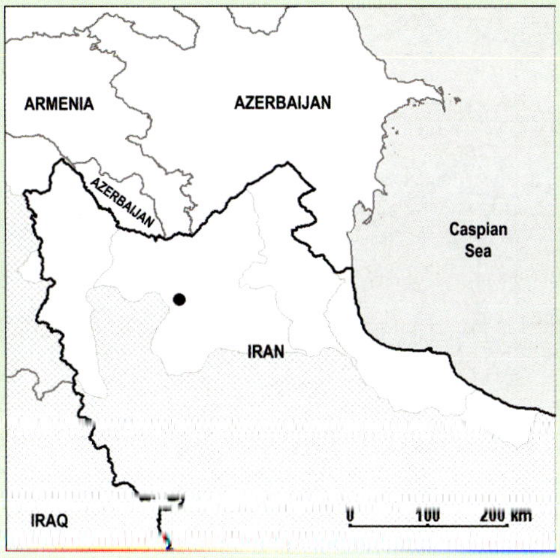

7. *Cousinia atropatana* Bunge

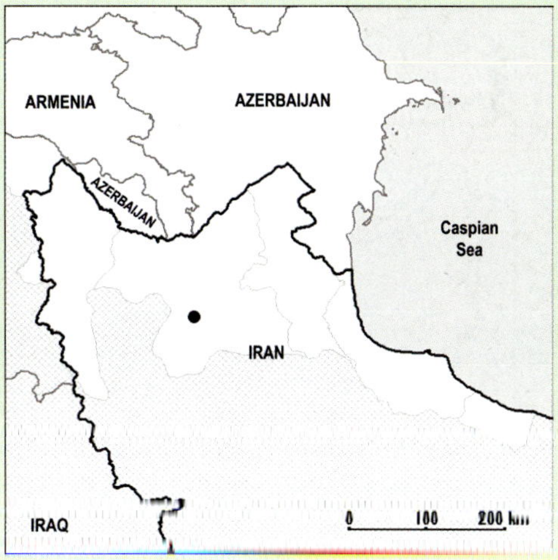

8. *Cousinia gilliatii* Rech. f.

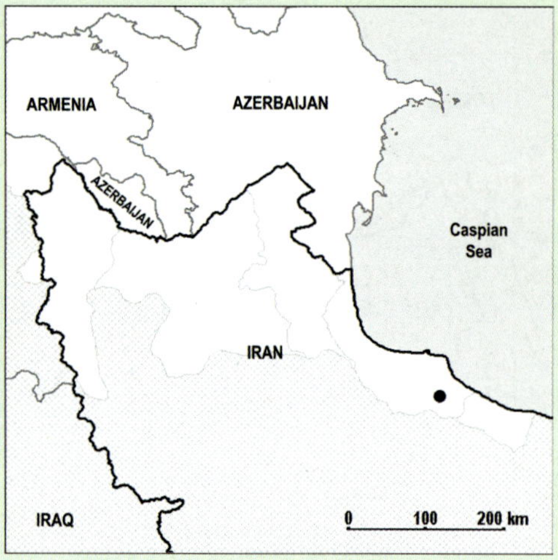

9. *Cousinia hablitzii* C. A. Mey. ex DC.

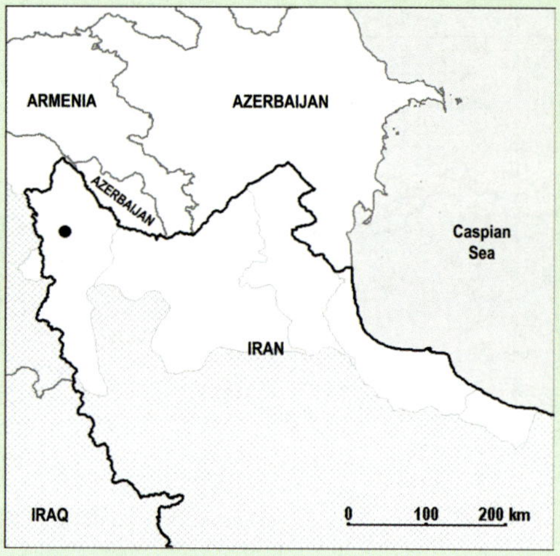

10. *Cousinia wilhelminae* Rech. f.

11. *Lactuca azerbaijanica* Rech. f.

12. *Heliotropium gypsaceum* Rech. f. & Riedl

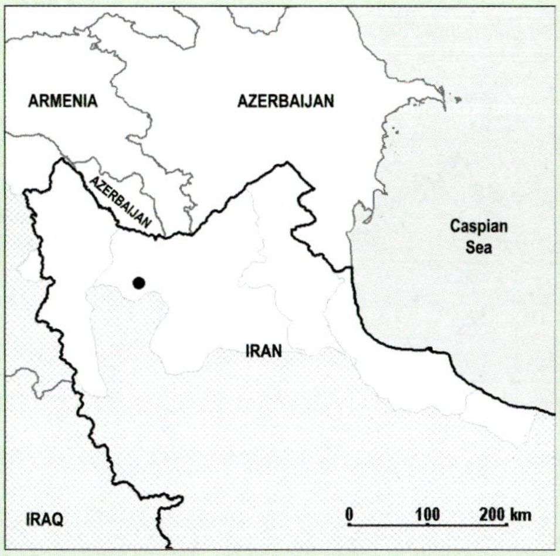

13. *Pseudocamelina szowitsii* (Boiss.) N. Busch

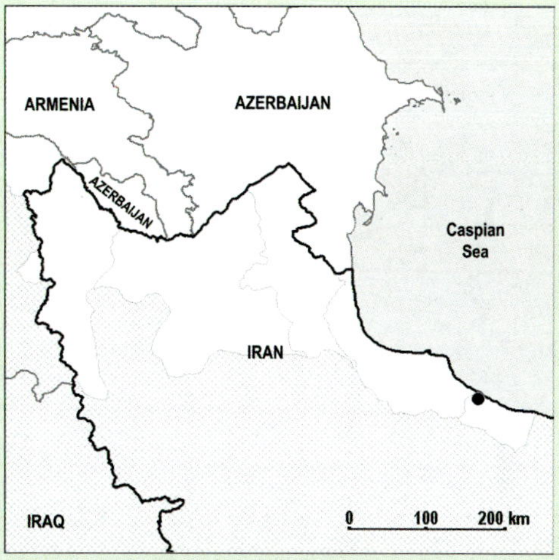

14. *Thlaspi maassoumii* Mozaff.

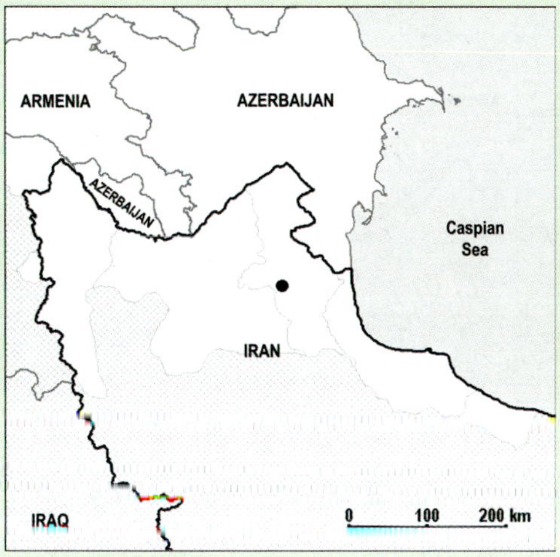

15. *Campanula savalanica* Fed.

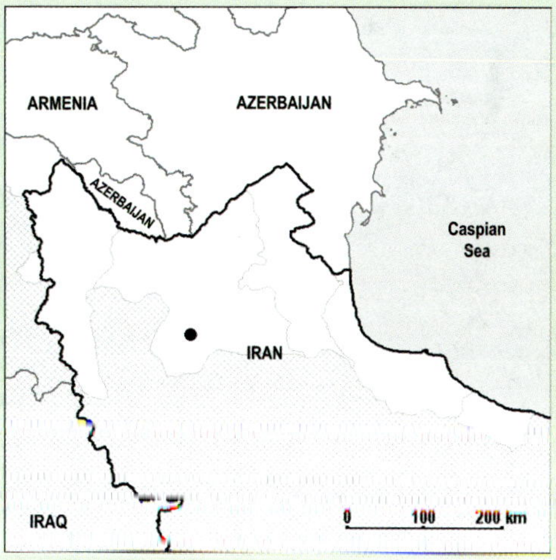

16. *Dianthus seidlitzii* Boiss.

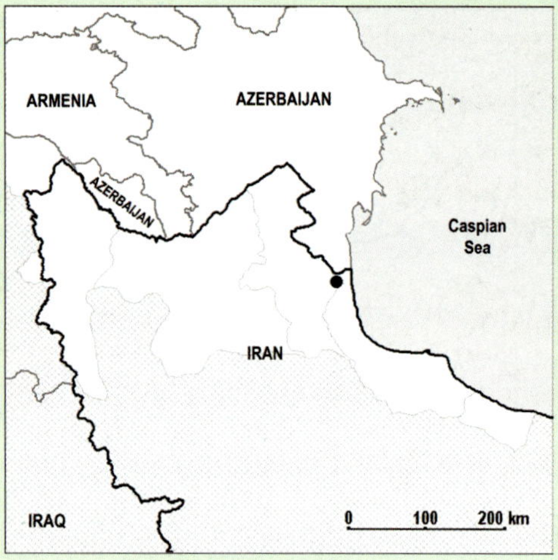

17. *Dianthus talyschensis* Boiss. & Buhse

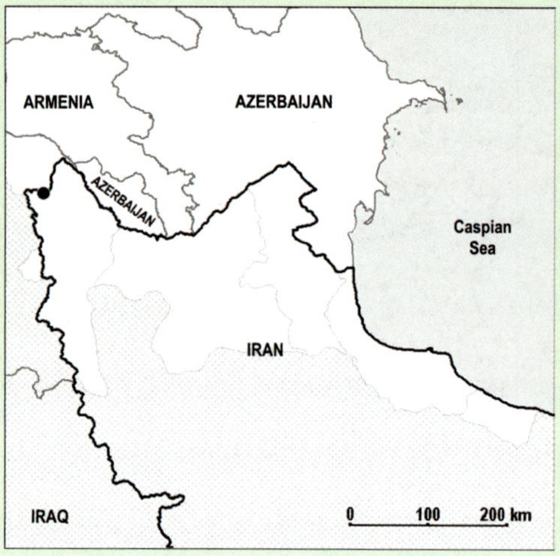

18. *Gypsophila bazorganica* Rech. f.

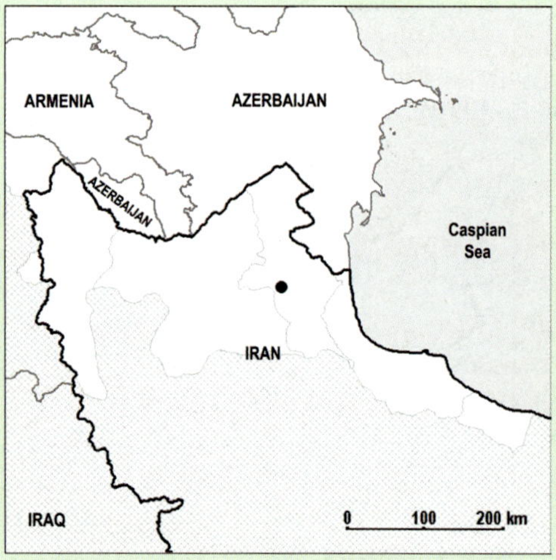

19. *Gypsophila iranica* Barkoudah

20. *Minuartia sabalanica* Assadi & Mostafavi

21. *Euphorbia sahendi* Bornm.

22. *Astragalus aharicus* Maassoumi & Podlech

23. *Astragalus arasbaranensis* Maassoumi & Ranjbar

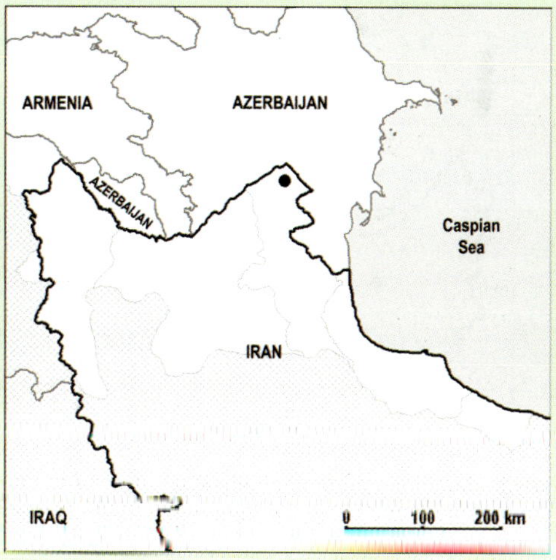

24. *Astragalus birangiae* Maassoumi

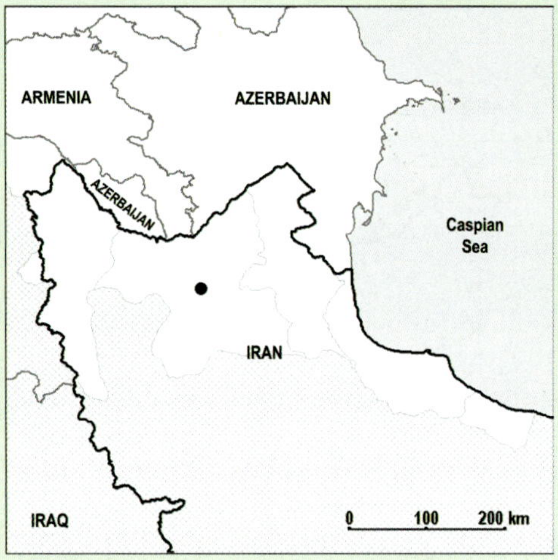

25. *Astragalus horridissimus* Širj. & Bornm.

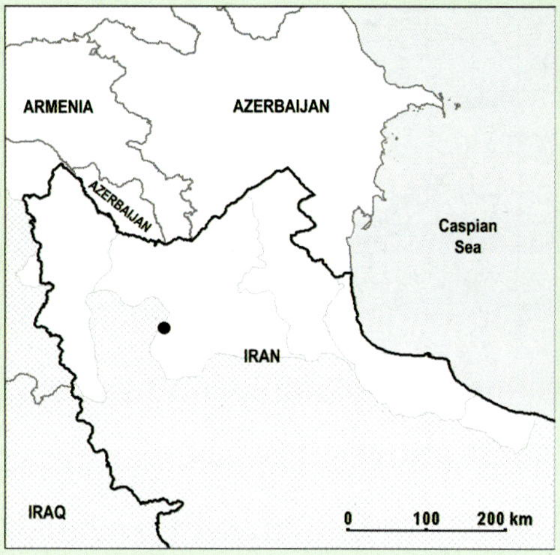

26. *Astragalus ilachchiensis* Ranjbar & Zarin

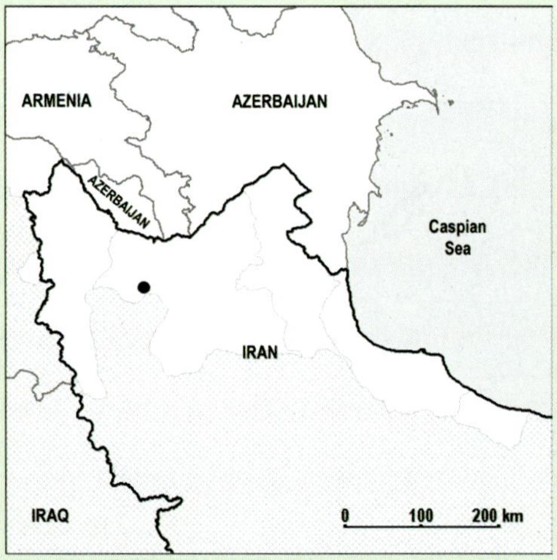

27. *Astragalus montismishoudaghi* Sheikh Akbari Mehr, Ghorbani & Maassoumi

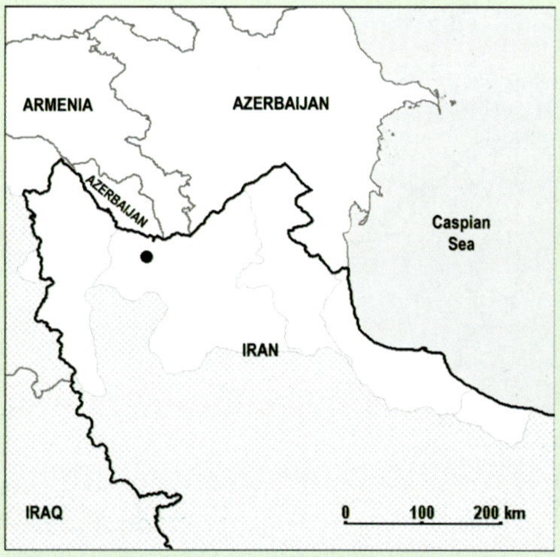

28. *Hedysarum marandense* Mozaff., Akrami & Maassoumi

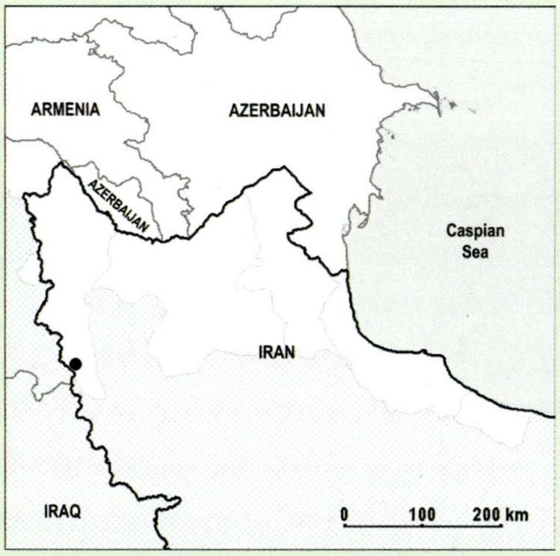

29. *Hedysarum orumiehense* Ranjbar

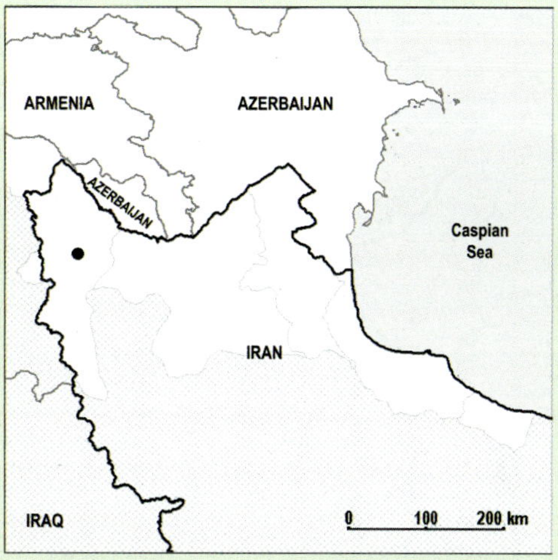

30. *Onobrychis depauperata* Boiss.

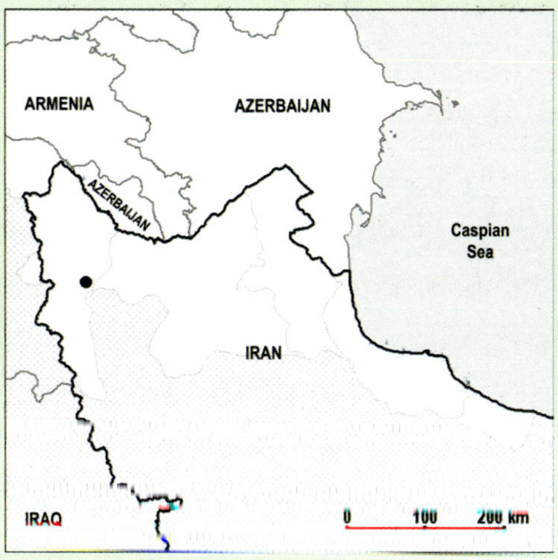

31. *Onobrychis szovitsii* Boiss.

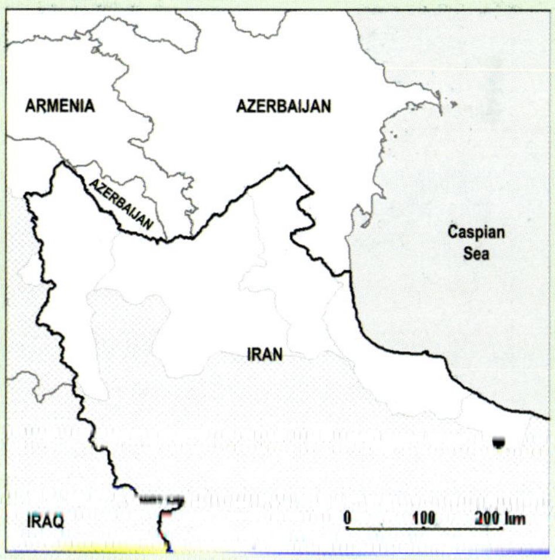

32. *Oxytropis takhti-soleimanii* Vassilcz.

33. *Bellevalia assadii* Wendelbo

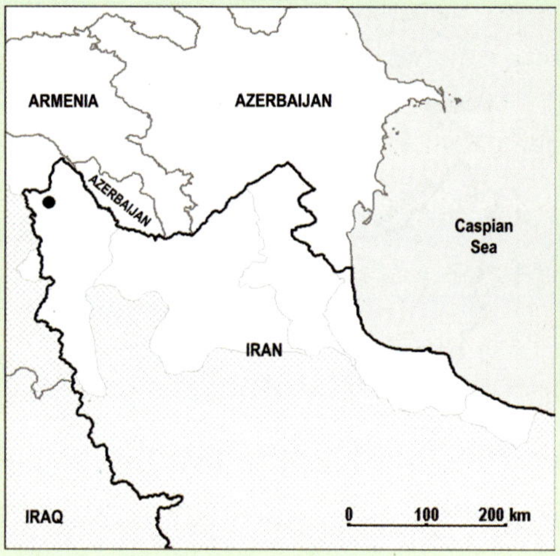

34. *Nepeta makuensis* Jamzad & Mozaff.

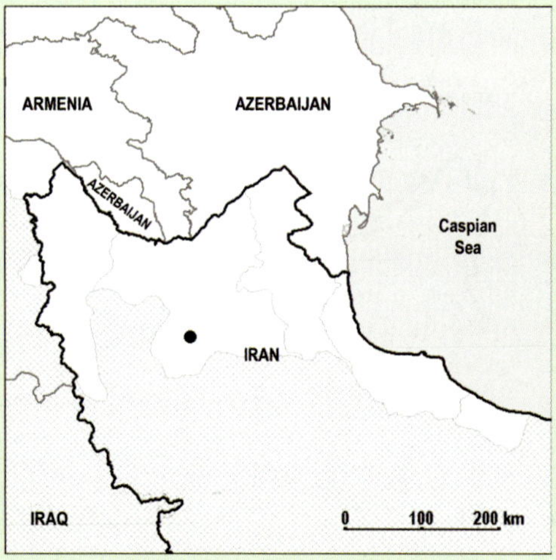

35. *Nepeta sahandica* Noroozi & Ajani

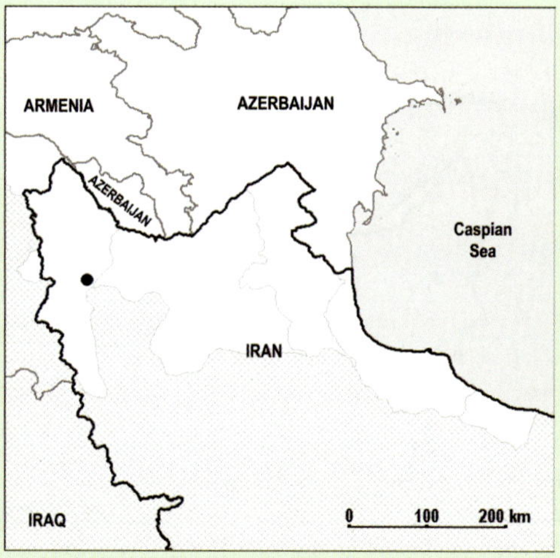

37. *Salvia hypochionaea* Boiss.

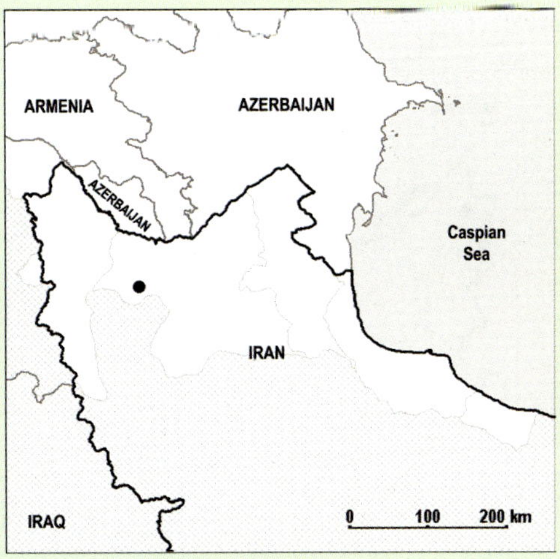

38. *Thymus marandensis* Jamzad

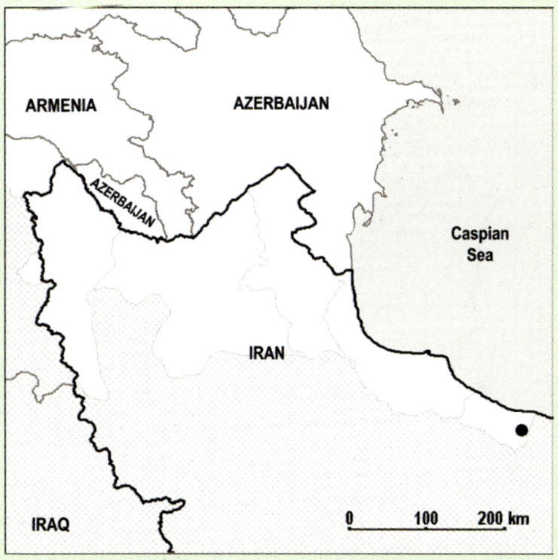

39. *Tulipa wendelboi* Matin & Iranshahr

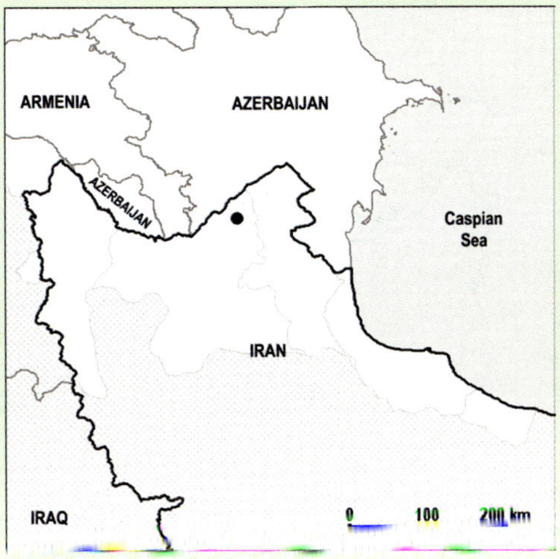

40. *Alcea assadii* Pakravan

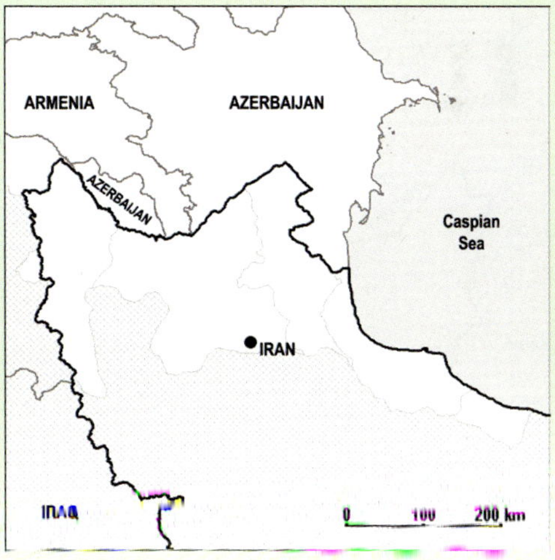

41. *Acantholimon atropatanum* Bunge

42. *Acantholimon esfandiarii* Rech. f. & Schiman-Czeika

43. *Acantholimon wilhelminae* Rech. f. & Schiman-Czeika

44. *Festuca sabalanica* E. B. Alexeev

45. *Atraphaxis suaedifolia* Jaub. & Spach

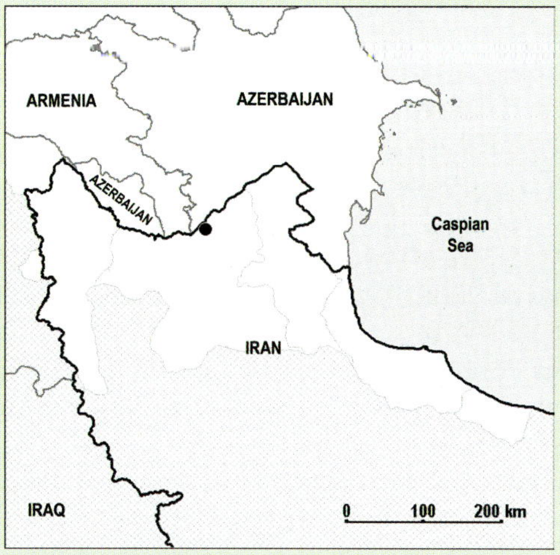

46. *Polypodium iranicum* Mazooji

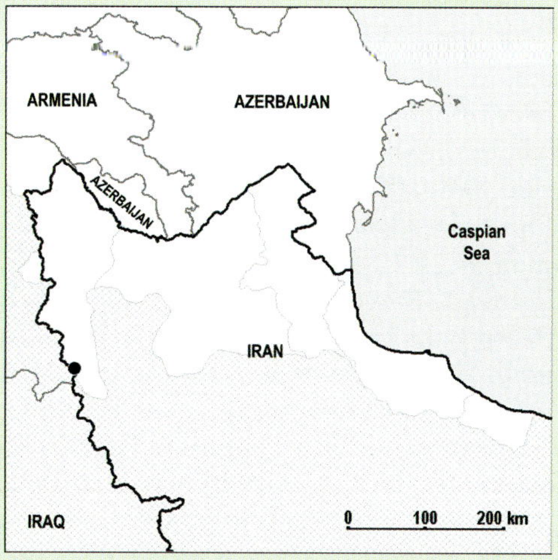

47. *Ranunculus renzii* Iranshahr & Rech. f.

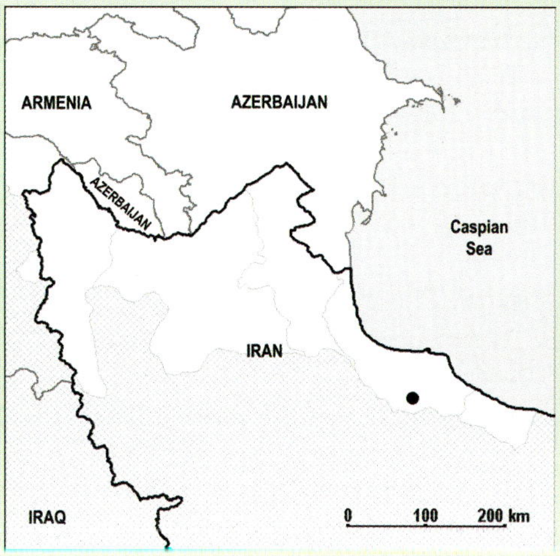

48. *Asperula microphylla* Boiss.

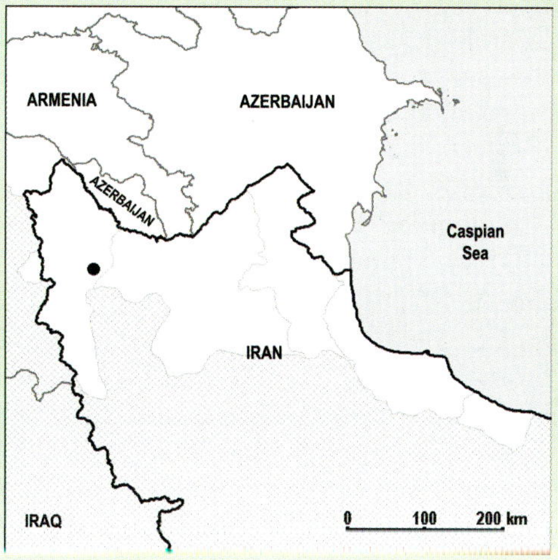

50. *Typha azerbaijanensis* Hamdi & Assadi

BIBLIOGRAPHY

Akhani, H., M. Djamali, A. Ghorbanalizadeh & E. Ramezani. 2010. Plant biodiversity of Hyrcanian relict forests, N. Iran: An overview of the flora, vegetation, palaeoecology and conservation. Pakistan J. Bot. 42: 231–258.

Assadi, M., A. A. Massoumi, M. Khatamsaz & V. Mozaffarian (editors). 1988–2012. Flora of Iran. Research Institute of Forests and Rangelands of Iran (RIFR), Tehran [in Farsi].

Bobek, H. 1951. Die natürlichen Wälder und Gehölzfluren Irans. Bonner Geogr. Abh. 8. Bonn.

Breckle, S. W. 2002. Walter's Vegetation of the Earth. The Ecological Systems of the Geo-Biosphere. Springer, Heidelberg.

Djamali, M., J. L. de Beaulieu, M. Shah-Hosseini, V. Andrieu-Ponel, A. Amini, H. Akhani, S. A. G. Leroy, L. Stevens, H. Lahijani, P. Ponel & S. Brewer. 2008. A late Pleistocene long pollen record from Lake Urmia, NW Iran. Quatern. Res. 69: 413–420.

Djamali, M., H. Akhani, R. Khoshravesh, V. Andrieu-Ponel, P. Ponel & S. Brewer. 2011. Application of the global bioclimatic classification to Iran: Implications for understanding the modern vegetation and biogeography. Ecol. Medit. 37(1): 91–114.

Domoers, M., M. Kaviani & D. Schaefer. 1998. An analysis of regional and intra-annual precipitation variability over Iran using multivariate statistical methods. Theor. Appl. Climatol. 61: 151–159.

Frey, W. & W. Probst. 1986. A synopsis of the vegetation of Iran. Pp. 9–43 *in* H. Kürschner (editor), Contributions to the Vegetation of Southwest Asia. Beih. Tübinger Atlas Vorderen Orients, Reihe A (Naturwiss.), Nr. 24. Wiesbaden.

Frey, W., H. Kürschner & W. Probst. 1999. Flora and vegetation, including plant species and larger vegetation complexes in Persia. Pp. 46–63 *in* E. Yarshater (editor), Encyclopedia Iranica, Vol. X, Fasc. 1. Bibliotheca Persia Press, New York.

Hamzeh'ee, B., A. Naqinezhad, F. Attar, A. Ghahreman, M. Assadi & N. Prieditis. 2008. Phytosociological survey of remnant *Alnus glutinosa* ssp. *barbata* communities in the lowland Caspian forests of northern Iran. Phytocoenologia 38(1-2): 117–132.

Hashemian, B. D. & A. Ataei Azimi. 2009. Alien and exotic *Azolla* in northern Iran. African J. Biotech. 8(2): 187–190.

Jalili, A. & Z. Jamzad. 1999. Red Data Book of Iran: A Preliminary Survey of Endemic, Rare and Endangered Plant Species in Iran. Research Institute of Forests and Rangelands of Iran (RIFR), Tehran.

Khalili, A. 1973. Precipitation patterns of Central Elburz. Theor. Appl. Climat. 21: 215–232.

Klein, J. C. 1982. Les groupements chionophiles de l'Alborz central (Iran). Comparaison avec leurs homologues d'Asie centrale. Phytocoenologia 10: 463–486.

Klein, J. C. 1987. Les Pelouses xérophiles d'altitude du franc sud de l'Alborz ventral (Iran). Phytocoenologia 15(2): 253–280.

Klein, J. C. 1988. Les groupements à grandes ombellifères et à xérophytes orophiles: Essai desynthèse à l'échelle de la région irano-touranienne. Phytocoenologia 16(1): 1–36.

Klein, J. C. & A. Lacoste. 1989. Les chênaies à *Quercus macranthera* F. et M. dans le massif de l'Alborz (Iran) et les chaînes limitrophes (grand et petit Caucase). Ecol. Medit. 15(3/4): 65–93.

Klein, J. C. & A. Lacoste. 1994. Les pelouses subalpines (Alchemilletum plicatissimae ass. nov.) de l'Alborz central (Iran): ultime avance´e sud-orientale de l'aire des Festuco-Brometea Br.-Bl. et Tx. 1943. Phytocoenologia 24: 401–421.

Klein, J. C. & A. Lacoste. 1999. Observation sur la végétation des éboulis dans les massifs irano-touraniens: Le Galietum aucheri ass. nov. de l'Alborz central (N-Iran). Doc. Phytosociol., n.s.

19: 219–228.
Noroozi, J., H. Akhani & S. W. Breckle. 2008. Biodiversity and phytogeography of alpine flora of Iran. Biodivers. & Conservation 17: 493–521.
Noroozi, J., H. Akhani & W. Willner. 2010. Phytosociological and ecological study of the high alpine vegetation of Tuchal Mountains (Central Alborz, Iran). Phytocoenologia 40: 293–321.
Noroozi, J, H. Pauli, G. Grabherr & S. W. Breckle. 2011. The subnival-nival vascular plant species of Iran: A unique high-mountain flora and its threat from climate warming. Biodivers. & Conservation 20: 1319–1338.
Rastin, N. 1983. Vegetationskundliche Untersuchungen in Hochwaldresten der Kaspischen Ebene. Phytocoenologia 11(2): 245–289.
Rechinger, K. H. (editor). 1963–2012. Flora Iranica. Akademische Druck-u. Verlagsanstalt und Naturhistorisches Museum Wien, Graz and Wien.
Rivas-Martínez, S., D. Sánchez-Mata & M. Costa. 1997. Syntaxonomical synopsis of the potential natural plant communities of North America, Vol. I. Itinera Geobot. 10: 5–148.
Rivas-Martínez, S., D. Sánchez-Mata & M. Costa. 1999. North American boreal and western temperate forest vegetation. Syntaxonomical synopsis of the potential natural plant communities of North America, Vol. II. Itinera Geobot. 12: 3–311.
Takhtajan, A. L. 1986. Floristic Regions of the World. University of California Press, Los Angeles.
Zohary, M. 1973. Geobotanical Foundations of the Middle East, Vol. 2. Fischer, Stuttgart.

RUSSIAN FEDERATION

Российская Федерация

Д. В. Гельтман, С. А. Литвинская, А. Д. Михеев,
Р. А. Муртазалиев, Н. Н. Портениер, В. В. Шванова

Флора Кавказского региона отличается большим разнообразием и насчитывает около 6300 видов сосудистых растений. Эта книга посвящена тем из них, которые встречаются только в этом регионе и нигде более; ботаники называют такие виды *эндемичными*, или *эндемиками*. Цель работы, результаты которой представлены в данной книге — выяснить, в каких странах региона встречаются эти виды, а также оценить, находятся ли они под угророзой исчезновения в связи с деятельностью человека. Оценка степени угрозы осуществлялась согласно системе категорий и критериев, принятой Международным союзом по охране природы (МСОП): исчезнувшие (EX), находящиеся в критическом состоянии (CR), находящиеся в опасном состоянии (EN), уязвимые (VU), находящиеся в состоянии, близком к угрожаемому (NT), вызывающие наименьшие опасения (LC), недостаток информации (DD).

Эта книга — первая сводка по охране растений Кавказского региона на английском языке; в ней приняли участие представители всех стран этого региона, специалисты других стран, представители международных природоохранных организаций.

Российская Федерация (Россия) на Кавказе занимает территорию, находящуюся к северу от Главного Кавказского хребта, а также узкую полосу северной части Черноморского побережья (Северо-Западное и частично Западное Закавказье). Здесь полностью расположены 9 и частично — 1 субъект федерации, входящие в Северо-Кавказский и Южный федеральные округа.

Флора России насчитывает около 12 тыс. видов сосудистых растений, из них на Кавказе встречается около 3700, 1255 видов являются эндемиками Кавказа, а 280 — эндемиками Российского Кавказа.

Растительность Российского Кавказа очень разнообразна и отражает разнообразие природных условий и климата этой территории. Северную, равнинную часть занимают степи, значительная часть которых в настоящее время распахана и используется под посевы различных культур. В засушливой северо-восточной части распространены полупустыни. В горах наблюдается высотная поясность. Широколиственные леса (различные виды дубов, бук, каштан) сменяются хвойными (пихта, ель, сосна), а затем — субальпийскими лугами и зарослями, альпийскими коврами. На северо-западе по берегу Черного моря встречаются сообщества, сложенные жестколистными кустарниками, напоминающие растительность Средиземноморья.

Российский Кавказ (кроме высокогорных районов) — густонаселенная территория с развитым сельским хозяйством и несколькими крупными промышленными центрами. Здесь проживает около 16 млн человек. Черноморское побережье, Кавказские Минеральные Воды и Приэльбрусье (Центральный Кавказ), некоторые другие районы являются популярными курортами общероссийского значения, привлекающими все более многочисленных туристов. Леса активно используются для заготовки древесины. Различные виды человеческой деятельности оказывают негативное воздействие на растительный покров и приводят к исчезновению отдельных видов растений. Главными угрозами являются рубка лесов, распашка степей, активное строительство, особенно в курортной зоне Черноморского побережья, прокладка дорог, газопроводов и нефтепроводов, чрезмерный выпас скота, сбор населением отдельных видов декоративных, лекарственных и пищевых растений. В последнее время отрицательное воздействие оказывает также активное распространение чужеродных инвазионных видов, которые вытесняют редкие растения природной флоры.

Богатейшая флора Кавказа, включая его северную часть, издавна привлекала внимание ученых. Первые экспедиции ботаников на Северный Кавказ состоялись еще в XVIII веке. С тех здесь работали многие ученые, накоплены значительные коллекционные материалы, однако флора Кавказа еще не может считаться полноценно изученной.

В настоящее время ведущим центром по изучению флоры Российского Кавказа является Ботанический институт им. В. Л. Комарова Российской академии наук в Санкт-Петербурге. В состав этого института входит гербарий, где хранятся важнейшие коллекции, имеющие огромное значения для изучения флоры Кавказа, в том числе собранные в XVIII — начале XX века. Институт имеет эколого-ботаническую станцию «Пятигорск» в Ставропольском крае, где в условиях культуры сохраняются редкие растения флоры Кавказа.

Специалисты по флоре Кавказа работают также в ботанических учреждениях Москвы, Ростова-на-Дону, Краснодара, Ставрополя, Махачкалы, Пятигорска, Сочи, Нальчика, Грозного и других городов; эти учреждения, как правило, имеют собственные гербарные коллекции, а иногда и ботанические сады.

Охрана природы в Российской Федерации регулируется целым рядом законодательных актов, главнейшие из которых — Федеральный закон «Об охране окружающей среды» (от 10.01.2002 № 7-ФЗ), с изменениями в 2004-2012 г.г. и «Об особо охраняемых природных территориях» (от 14.03.1995 № 33-ФЗ, с изменениями в 2001–2012 г.г.). Россия подписала и ратифицировала основные международные соглашения, регулирующие охрану растений. Подготовка, составление и ведение Красной книги Российской Федерации, в которую заносятся виды растений и животных, подлежащие особой охране, регулируется Постановлением Правительства Российской Федерации (от 19.02.1996 № 158). В последнее издание этой книги (2008 г.) внесены 185 сосудистых растений, встречающихся на Кавказе. Подготовлены и изданы также Красные книги Краснодарского и Ставропольского краев, Кабардино-Балкарии, Дагестана.

На территории Российского Кавказа расположены 6 государственных заповедников, 3 национальных парка и ряд особо охраняемых природных территорий иного статуса (заказники и памятники природы). Старейшей охраняемой природной территорией является Кавказский государственный биосферный заповедник, созданный в 1924 г. Здесь исчезающие растения флоры Кавказа сохраняются в природных условиях.

Всего в данную книгу включено 3017 таксонов (видов, подвидов и разновидностей) сосудистых растений, из которых на территории России зафиксировано 1346. Целый ряд этих видов требует охраны в масштабе всего Кавказа: 23 находятся в критическом состоянии, 77 – в опасном состоянии, 51 — уязвимые и 78 — в состоянии, близком к угрожаемому. Все эти растения нуждаются в тех или иных мерах охраны, в первую очередь, в организации в местах их обитания тех или иных особо охраняемых природных территорий — заповедников, заказников, памятников природы. Создание таких территорий — лучший вклад России в дело сохранения биологического разнообразия растений Кавказа.

В то же время следует учитывать, что на территории России отдельные виды — эндемики Кавказа могут требовать специальных мер охраны даже в том случае, если их состояние в соседних странах не внушает опасения. Поэтому всем кавказским эндемикам должно уделяться особое внимание при планировании природоохранных мероприятий и подготовке Красных книг различного уровня.

Выход данной книги будет способствовать усилению практических мер по охране растений флоры Кавказа, сохранению его уникальной природы, укреплению международного сотрудничества в этой области.

RUSSIAN FEDERATION

Dmitry Geltman, Anatoly Mikheev†, Svetlana Litvinskaya, Ramazan Murtazaliev, Nikolay Portenier†, and Valeria Shvanova

ADMINISTRATIVE DIVISION

The Russian Federation is represented in the Caucasian region by 10 states of the federation: the Republic of Adygea (capital, Maykop), the Republic of Dagestan (capital, Makhachkala), the Republic of Ingushetia (capital, Magas), the Kabardino-Balkaria Republic (capital, Nal'chik), the Karachayevo-Circassian Republic (capital, Cherkessk), the Republic of North Ossetia (capital, Vladikavkaz), the Chechen Republic (capital, Grozny), Krasnodar Territory (administrative center, Krasnodar), Stavropol Territory (administrative center, Stavropol), and part of Rostov Oblast (administrative center, Rostov-on-Don). Most of these states of the federation belong to the North Caucasian Federal District, with the exception of Krasnodar Territory and Rostov Oblast, which belong to the South Federal District.

LARGEST CITIES IN THE CAUCASUS AREA

In order of population size (population estimate provided in parentheses): Rostov-on-Don (1,050,000), Krasnodar (709,000), Makhachkala (650,000), Stavropol (366,000), Sochi (334,000; with suburbs 406,000), Vladikavkaz (312,000), Novorossiysk (284,000), Nal'chik (270,000), Grozny (250,000), Pyatigorsk (207,000), Armavir (189,000), Maykop (153,000).

POPULATION

The republics and autonomous regions differ in size. There are very small republics such as Adygea, with a population of ca. 444,000, and North Ossetia, with ca. 702,000. At the same time, Krasnodar and Stavropol territories are both very large, with populations of ca. 5 million and ca. 2 million, respectively. The total number of people living in the Russian portion of the Caucasus region is ca. 16 million.

LANGUAGE

Russian is the common language across the region, but people also speak local languages from numerous North Caucasian ethnic groups.

FLORA

In the Russian part of the Caucasus: ca. 3700 (total plant species); 280 (total national endemics); 1300 (Caucasian region endemics).

Of the species that occur in the Russian Caucasus region, 185 are included in the *Red Data Book of Russia* (Iliashenko & Iliashenko, 2000).

BOTANICAL INSTITUTIONS AND COLLECTIONS

The Komarov Botanical Institute of the Russian Academy of Sciences in St. Petersburg is a leading institution studying the flora of the Russian portion of the Caucasus. Its affiliated herbarium (LE), with ca. 6 million specimens, has a special Caucasian section that contains ca. 800,000 specimens, including numerous type specimens and historical collections. The Komarov Botanical Institute, in cooperation with colleagues in other countries, produces and publishes the *Caucasian Flora Conspectus*, with three volumes published to date (Takhtajan, 2003–2012). Address: 2 Prof. Popov St., St. Petersburg 197376, Russia; Tel./FAX: +7 (812) 372-54-43 (director's office), +7 (812) 372-54-57 (herbarium); email: binadmin@binran.ru (director's office), Ivantatanov@mail.ru and Shvanova@yandex.ru (Caucasian Department), herbarium@binran.ru (herbarium).

The Komarov Institute is affiliated with the Ecology-Botanical Station in the Caucasian region. This is located in the Stavropol

Territory, in the city of Pyatigorsk, with a special living collection of endangered plant species of the area and a reference herbarium. Address: Energetik, Pyatigorsk, Stavropol Territory, 357506, Russia; email: stbot@yandex.ru.

The herbarium of the Moscow State University (MW) also has important collections for the Caucasian region. The Main Botanical Garden of the Russian Academy of Sciences in Moscow maintains the MHA herbarium, as well as living plant collections. The herbarium of the Moscow State Pedagogical University (MOSP) is also of interest for this area.

Northern Caucasian herbaria include the Rostov State University (RV) in Rostov-on-Don, the Kuban Agricultural University (KBAI) in Krasnodar, the Kabardino-Balkaria State University (AUBSN) in Nal'chik, the Dagestan State University (LENUD) and the Mountain Botanical Garden (DAG) in Makhachkala, the Stavropol State University (SPI) and the Stavropol Botanical Garden (SBG) in Stavropol, and the Pyatigorsk State Pharmaceutical Academy (PGFA) in Pyatigorsk.

The Kuban State University in Krasnodar deals with rare Caucasian endemics and publishes a series of red data books and endangered plant lists. Address: 96/35 Stavropolskaia St., Krasnodar, Krasnodar Territory, 350033, Russia; email: litvinsky@yandex.ru.

The Mountain Botanical Garden in Makhachkala has collections of living plants and provides intensive research on the endangered plants of Dagestan. Address: 45 M. Gadjiev St., Makhachkala, Republic of Dagestan, 367000, Russia; email: pibreklab@yahoo.com.

PROTECTED AREAS

There are seven nature reserves (zapovednik) in the Russian Caucasus region: Caucasian Biosphere Nature Reserve, extending across Krasnodar Territory, the Republic of Adygea, and the Karachayevo-Circassian Republic; Utrish Nature Reserve in Krasnodar Territory; Teberda Nature Reserve in the Karachayevo-Circassian Republic; Kabardino-Balkaria High Mountain Nature Reserve in the Kabardino-Balkaria Republic; Severo-Osetinsky Nature Reserve in the Republic of North Ossetia; Ersi Nature Reserve in the Republic of Ingushetia; and Dagestan Nature Reserve in the Republic of Dagestan. There are also three national parks, which include the Sochi National Park in Krasnodar Territory, the Prielbrusye National Park in the Kabardino-Balkaria Republic, and the Alania National Park in the Republic of North Ossetia. In addition, there are numerous other protected areas that include sanctuaries (zakaznik), natural monuments, and more.

Russia occupies the northern portion of the Caucasian region, mainly to the north of the Greater Caucasus Range and its foothills, bounded to the south by Georgia and Azerbaijan. It has access to both the Black Sea and Caspian Sea. Much of the northern part of the territory is fertile plain. It is essentially an agricultural region, producing wheat, barley, rice, sunflowers, vegetables, tobacco, and other crops. The central portion is occupied by the Stavropol Heights. The famous health resort with mineral springs is situated to the south of the Stavropol Heights in the Greater Caucasus foothills. It was established in 1803 as Mineralnye Vody (Mineral Waters) and includes the resort towns of Essentuki, Kislovodsk, Pyatigorsk, and Zheleznovodsk. Oil fields are situated in the eastern part of this northern Caucasus region, mostly in the Chechen Republic and Dagestan, and partly in the west near Maykop.

OROGRAPHY

The Russian portion of the northern Caucasus includes the northern slopes of the Greater Caucasus Range that can be divided into four orographic zones: 1) the Main or Watershed (Glavny) Range, 2) the Lateral (Bokovoi) Range, 3) the Rocky (Skalisty) Range, and 4) the Chalk (Melovoy) Range. Among these ranges, intermountain depressions are intersected by transverse valleys formed by montane rivers.

At its extreme western extent, the Main (Glavny) Range is a small montane upheaval, consisting mainly of marls and limestone, and is composed of several closely parallel ranges of complex configurations. In its central portion, it is composed of crystalline rocks, with its highest altitudinal point on Mount Shkhara, at 5098 m. In the east, this range is formed by sedimentary metamorphosed rocks or shales and is much lower than in the central part.

The Lateral (Bokovoi) Range is located 10–20 km north of the Main Range and joins in many places with it. Almost throughout its extent, the Lateral Range is higher than the Main Range and in its central portion reaches 5633 m at Mount Elbrus, which is the principal peak in the Caucasus and the highest in Europe. To the north, the Lateral Range descends in altitude to merge with the tectonic-erosive valley that is a Jurassic depression (broadest at its western extent), separating the high-mountain Caucasus with a system of limestone cuesta mountains. This limestone system is located to the north of the Jurassic depression, where it is represented by two or three mountain ranges. The highest and southernmost of them, the Rocky (Skalisty) Range, has an asymmetric profile, with a gentle northeastern slope but steep and precipitous southwestern slopes, and is composed mainly of limestone and dolomites. Northward, several cuesta ridges in the Chalk (Melovoy) Range have a gentle outline and are sometimes separated from the Rocky Range by noticeable depressions. North of the Chalk Range, the lowest mountain range or hill system comprises the central portion of Ciscaucasia in the Stavropol Heights. To the north of these ranges and uplands, the area is represented by plains and lowlands that are sometimes referred to as Pre-Caucasia.

CLIMATE

The climate of the Russian Caucasus region is diverse, varying from very wet and subtropical in the western portion to very dry and continental semi-desert at the eastern extent. Generally, the Russian portion is colder than the southern Caucasus because it is mostly open to the influence of cold, windy air masses from the north. The only exception is a narrow belt between the mountains and the Black Sea, known as northwestern and western Transcaucasia, which is protected from this northern continental influence by the western portion of the Greater Caucasus.

The warmest month in the plains is July, with a mean temperature of 21–25°C. In the eastern part, the summers are generally warmer and drier, and there the temperature can sometimes reach 40°C. The coldest month is January, with the mean temperature around –2 to –5°C. The maximum low temperature can sometimes reach –30°C, though usually only for a short period. The mean annual precipitation varies from 200 to 800 mm.

The climate along the Black Sea coast is relatively mild. The mean January temperature in Novorossiysk and Sochi is around 3°C. The maximum high temperature in summer is 36°C (less than the plains). Despite the milder climate, on some days in December it could be as high as 19°C and in February as low as –23°C. The mean annual precipitation is 700–800 mm along the northern part of the Black Sea coast and 1000–1500 mm in the south. Some locations along the Black Sea are characterized by very strong winds (bora) during the winter months. Thus, in Novorossiysk the wind is so strong that it can overturn trains, break electric wires, and cause much physical damage.

In the mountainous areas, the climate is very diverse and is highly dependent on the altitude and orientation of the valleys. The mean January temperature at 2000 m elevation is –8°C, while in August, the warmest month, it is 13°C. The amount of precipitation varies considerably. Snow cover in winter is much deeper in the mountains of the western Caucasus compared to the eastern portion. In summer, due to the specific circulation of air masses, a rain shadow precipitation effect is formed in some places.

Arabis nordmanniana Rupr., endemic to the western Caucasus region.

Campanula daghestanica Fomin, endemic to the eastern Caucasus region.

Erodium stevenii M. Bieb., an endangered endemic in the Greater Caucasus region.

Sorbus velutina (Albov) C. K. Schneid., endemic to the western Caucasus region.

FLORA AND VEGETATION

According to Takhtajan (1978), the Russian part of the Caucasus falls into five floristic provinces: Eastern European, Crimean-Novorossiysk, Caucasian, Euxine, and Turanian (Aralo-Caspian). Some botanists estimate there are 3900 species of vascular plants in the Russian Caucasus (Galushko, 1978–1981). New data show that the number of species is closer to 3700, and among them are 1300 Caucasian endemics that include ca. 280 national endemics. Some of these are narrow, local endemics. Thus the Fisht-Oshten mountain system alone has 120 endemic species (Alper, 1960). More than 70 species are narrow endemics in Dagestan, in the eastern part of the North Caucasus, and ca. 200 species have distributional areas that include Dagestan and other territories along the border.

The northern border of the Caucasian hot spot runs along the steppe region. In the northwestern part, steppes are found up to 600–800 m in the Markhot, Navagirsky, and Kozekhur ranges. These steppes feature relictual taxa such as the feather grass *Stipa pulcherrima* K. Koch (Poaceae) and many other perennial herbs, mostly of Mediterranean origin: e.g., foxtail lily, *Eremurus spectabilis* M. Bieb. (Asphodelaceae); onosma, *Onosma polyphylla* Ledeb. (Boraginaceae); peony, *Paeonia tenuifolia* L. (Paeoniaceae); and tulip, *Tulipa gesneriana* L. (Liliaceae), the ancestor of many cultivated tulip varieties. Perennial herbs alternate with short semishrubs such as broom, *Genista humifusa* L. (Fabaceae); *Onosma polyphylla* Ledeb. (Boraginaceae); and thyme, *Thymus markhotensis* Maleev (Lamiaceae). Steppes in the northeastern portion transition to semi-deserts and consist of graminoid forbs, e.g., *Agropyron fragile* (Roth) P. Candargy (Poaceae) and summer cypress, *Kochia prostrata* (L.) Schrad. (Cyperaceae). The steppe zone in Dagestan extends up to 600–1000 m in altitude.

The foothills of the slopes that belong to the Greater Caucasus have a very rich vegetation. Drier places often support steppes or, according to Russian vegetation classification, meadow-steppes that are composed mainly of grasses, e.g., *Bothriochloa ischaemum* (L.) Keng, *Koeleria cristata* (L.) Pers., *Phleum phleoides* (L.) H. Karst., and *Poa pratensis* L. (all Poaceae) and legumes such as *Astragalus bungeanus* Boiss. (Fabaceae). These meadow-steppes are used as permanent pasture. Some areas on relatively dry soils are covered with oak forests. There are five species of oaks in the North Caucasus: *Quercus robur* L., with two subspecies; *Quercus petraea* L. ex Liebl., with three subspecies; *Quercus hartwissiana* Steven; *Quercus pubescens* Willd.; and *Quercus macranthera* Fisch. & C. A. Mey. ex Hohen. (all Fagaceae). Each forest-forming oak species has its own altitudinal zone and is linked with a definite ecosystem. The more xerophytic *Quercus petraea* forms the second oak belt in the mountains. Together with oak, other trees include *Acer campestre* L., *Acer laetum* C. A. Mey., and *Acer platanoides* L. (all Aceraceae) and *Fraxinus excelsior* L. (Oleaceae). The understory includes Cornelian cherry, *Cornus mas* L. (Cornaceae); privet, *Ligustrum vulgare* L. (Oleaceae); and bladdernut, *Staphylea pinnata* L. (Staphyleaceae). These forests are very beautiful in early spring because of the many perennial geophytes that come into flower, e.g., sowbread, *Cyclamen coum* Mill. (Primulaceae); dog tooth violet, *Erythronium caucasicum* Woronow (Liliaceae); snow drop, *Galanthus alpinus* Sosn. (Amaryllidaceae); hellebore, *Helleborus caucasicus* A. Braun (Ranunculaceae); and species of *Corydalis* DC. (Fumariaceae) and *Scilla* L. (Hyacinthaceae). Oak forests occupy considerable territory in the forest-steppe portion of the Stavropol Heights. The conditions for forest growth are sometimes not favorable because of moisture deficits. In these areas, forests are confined to deep ravines. Above 400 m in elevation, forests appear as islands in the open steppe.

The next tree formation is dominated by beech, *Fagus orientalis* Lipsky (Fagaceae). Beech requires moist and rich soil to grow and is accompanied by many other deciduous trees and shrubs. The upper limit of the vertical distribution of beech depends on temperature and winter precipitation, while the lower limit is controlled by air humidity. The beech forest belt starts in the west by the Pshish River, where forest occurs at an elevation of ca. 600 m, and spreads to the east, where the forest boundary rises to ca. 1800 m. There are many other tree species that grow together with beech and form various kinds of beech forest. One of the most characteristic species of the beech forest is hornbeam, *Carpinus betulus* L. (Betulaceae), while in the western regions chestnut, *Castanea sativa* Mill. (Fagaceae), is more common. Among the other species that grow with beech are oaks, linden (*Tilia* L., Tiliaceae), and mountain ash (*Sorbus* L., Rosaceae). Beech is the most shade-tolerant tree in the Caucasus region. This is why the number of understory plants is limited. Among endemics that grow in clearings and along the forest margins are many perennials such as *Geranium gracile* Ledeb. ex Nordm. (Geraniaceae) and *Lilium caucasicum* (Miscz. ex Grossh.) Grossh. (Liliaceae).

The mountain ranges that are situated in the central Caucasus are covered by pine–birch forests with *Pinus kochiana* Klotzsch ex K. Koch (Pinaceae) and *Betula litwinowii* Doluch. and *Betula pendula* Roth. (both Betulaceae). Dark coniferous forests are confined to an elevation of 1000–2000 m and consist mostly of fir, *Abies nordmanniana* (Steven) Spach, and spruce, *Picea orientalis* (L.) Peterm. (both Pinaceae).

Above the timberline, at an elevation of 1800–2000 m up to 2400–2800 m, the forest is replaced by subalpine meadows. Woody plants, if present, are crook-stemmed and short in stature. Subalpine meadows are very rich and include many tall perennials that reach 1.5–2(–3) m. They are very picturesque in the midsummer when most of the species come into flower and one can see the numerous blue flowers of *Gadellia lactiflora* (M. Bieb.) Shulkina (Campanulaceae), the bright yellow inflorescences of showy *Telekia speciosa* (Schreb.) Baumg. and *Inula magnifica* Lipsky (both Asteraceae), and the dark blue flowers of monkshood, *Aconitum nasutum* Fisch. ex Rchb., *Aconitum orientale* Mill., and *Delphinium pyramidatum* Albov (these three, Ranunculaceae). There are other types of subalpine meadows that consist of grasses and dicots but do not contain gigantic herbs. Some of the characteristic species are *Arrhenatherum elatius* (L.) J. Presl & C. Presl, *Bromopsis variegata* (M. Bieb.) Holub, *Calamagrostis arundinacea* (L.) Roth, and *Milium schmidtianum* K. Koch (these four, Poaceae), as well as forbs such as *Anemonastrum fasciculatum* (L.) Holub and *Aquilegia olympica* Boiss. (both Ranunculaceae); *Astragalus freynii* Albov (Fabaceae); and *Geranium ruprechtii* (Woronow) Grossh. (Geraniaceae). When overgrazed, species of *Calamagrostis* Adans. are usually replaced by other grasses, mostly fescues in *Festuca* L. (Poaceae). These meadows occupy huge slopes of the subalpine zone. Meadows with different species of fescue are spread throughout the northern Caucasus.

Above elevations of 2300–2800 m, alpine meadows or alpine carpets develop. They consist of low perennial plants with various special life forms such as carpet-like, cushion-like, or prostrate. Snow covers the ground during most of the year, and the active growing period is very short. Perennials usually have shortened stems with leaves in rosettes and short reproductive stems often with single flowers. Some shrubs and semishrubs can also grow in this belt. A dominant species is *Rhododendron caucasicum* Pall. (Ericaceae), which develops thickets. Several species of juniper (*Juniperus* L., Cupressaceae), dryad (*Dryas* L.), and sibaldia (*Sibbaldia* L. [both Rosaceae]) with prostrate stems develop a carpet-like form. These carpet plants with small evergreen leaves usually grow among large rocks where they find protection from severe conditions in winter. Among the perennials are some very attractive plants such as *Campanula argunensis* Rupr. (Campanulaceae), with its large flower and very tiny vegetative body, and *Crocus scharojanii* Rupr. (Iridaceae), which develops its orange-yellow flowers in late summer. At higher elevations of 2700–3000 m, mosses and lichens are prevalent. Among the characteristic herbaceous perennials are carpet saxifrage (*Saxifraga juniperifolia* Adams, Saxifragaceae), with bright yellow flowers, the dense cushions of *Draba bryoides* DC. (Brassicaceae), and endemic species such as *Arabis farinacea* Rupr. (Brassicaceae) and *Nonea daghestanica* Kusn. (Boraginaceae). Alpine carpets with species of *Sibbaldia* L. (Rosaceae) and *Euphrasia* L. (Scrophulariaceae), as well as *Crocus vallicola* Herb. (Iridaceae), are of wide distribution as well.

At higher elevations in the drier, easternmost parts of Dagestan, the high mountain vegetation is composed of xerophytes such as *Astragalus beckerianus* Trautv. (Fabaceae), *Anthemis fruticulosa* M. Bieb. (Asteraceae), *Convolvulus ruprechtii* Boiss. (Convolvulaceae), and *Salvia canescens* C. A. Mey. and *Thymus daghestanicus* Klokov & Des.-Shost. (both Lamiaceae).

The southwestern slopes of the Greater Caucasus and their high-elevation meadows that fall in the Russian territories along the Black Sea are also xeromorphic communities. This region is the most xeric in the western Caucasus region. Vegetation here is very similar to that of dry regions of the Mediterranean. Open juniper woodlands in mid-montane zones are replaced by herb communities lower on the slopes. There are numerous species with very restricted distribution that are listed in local red data books, and among them are *Arabis sachokiana* N. Busch (Brassicaceae), *Astrantia pontica* Albov (Apiaceae), *Campanula komarovii* Maleev (Campanulaceae), and *Corylus colchica* Albov (Betulaceae).

Table 1. Distribution of endemic species within vegetation types.

HABITAT	SPECIES
Alkali soils	*Euphorbia normannii* Schmalh. ex Lipsky, *Taraxacum ciscaucasicum* Schischk.
Arid depressions	*Alchemilla tephroserica* (Buser) Juz., *Astragalus flexicaulis* Sosn., *Astragalus haesitabundus* Lipsky, *Dianthus elbrusense* Charadze, *Tragopogon leonidae* Kuth.
Flood plain terrace (pebbly)	*Picris canescens* (Steven) V. N. Vassil., *Symphytum podcumicum* Yu. M. Frolov
Forests	*Allium grande* Lipsky, *Cerastium holosteum* Fisch. & Hornem. var. *meyerianum* (Rupr.) I. V. Sokolova ex Czerep., *Corydalis tarkiensis* Prokh., *Galanthus angustifolius* Koss, *Galanthus bortkewitschianus* Koss, *Epimedium circinatocucullatum* Sosn., *Scrophularia mollis* Sommier & Levier
Glacial moraines	*Campanula besenginica* Fomin, *Campanula songutica* Amirkh., *Cerastium polymorphum* Rupr., *Festuca inguschetica* E. B. Alexeev, *Oxytropis kubanensis* Leskov, *Papaver lisae* N. Busch, *Psephellus salviifolius* Boiss., *Rosa brotherorum* Chrshan.
Littoral	*Astragalus hyrcanus* Pall., *Cladochaeta caspica* Sosn. ex Grossh.
Meadows (incl. subalpine)	*Alchemilla elisabethae* Juz., *Amoria raddeana* (Trautv.) Roskov, *Astragalus ciceroides* Sosn., *Cirsium balkharicum* Charadze, *Cirsium czerkessicum* Charadze, *Cirsium dealbatum* M. Bieb., *Crepis caucasigena* Czerep., *Delphinium crispulum* Rupr., *Dianthus kubanensis* Schischk., *Euphrasia daghestanica* Juz., *Gentiana owerinii* (Kusn.) Grossh., *Gentianella lipskyi* (Kusn.) Holub, *Gypsophila meyeri* Rupr., *Hieracium rigidellum* Litv. & Zahn, *Hyalopoa lakia* (Woronow) Tzvelev, *Onobrychis hamata* Vassilcz.
Petrophilous (rocky and stony) situations	*Allium daghestanicum* Grossh., *Allium gunibicum* Miscz. ex Grossh., *Allium mirzojevii* Tscholok., *Allium psebaicum* Mikheev, *Allium samurense* Tscholok., *Artemisia daghestanica* Krasch. & Poretzky, *Asphodeline tenuior* (Fisch. ex M. Bieb.) Ledeb., *Astragalus beckerianus* Trautv., *Astragalus daghestanicus* Grossh., *Astragalus fissuralis* T. Alex., *Astragalus lasioglottis* Steven ex M. Bieb., *Astragalus salatavicus* Bunge, *Atraphaxis daghestanica* (Lovelius) Lovelius, *Axyris caucasica* (Sommier & Levier) Lipsky, *Campanula ardonensis* Rupr., *Campanula autraniana* Albov, *Campanula czerepanovii* Fed., *Campanula kadargavanica* Amirkh. & Komzha, *Campanula kryophila* Rupr., *Campanula sclerophylla* (Kolak.) Czerep., *Centaurea avarica* Tzvelev, *Centaurea ruprechtii* (Boiss.) Wagenitz, *Cicer balcaricum* Galushko, *Cotoneaster nefedovii* Galushko,

HABITAT	SPECIES
Petrophilous (rocky and stony) situations (continued)	*Daphne baksanica* Pobed., *Daphne circassica* Woronow ex Pobed., *Dianthus awaricus* Charadze, *Dianthus daghestanicus* Charadze, *Dianthus vladimiri* Galushko, *Draba elisabethae* N. Busch, *Eremogone holostea* (M. Bieb.) Rupr., *Erigeron schalbusi* Vierh., *Erysimum brevistylum* Sommier & Levier, *Erysimum meyerianum* (Rupr.) N. Busch, *Euphorbia buschiana* Grossh., *Euphorbia daghestanica* Geltman, *Festuca elbrusica* E. B. Alexeev, *Galium oshtenicum* Ehrend. & Schanzer ex Mikheev, *Genista angustifolia* Schischk., *Gypsophila capitata* M. Bieb., *Gypsophila glauca* Steven ex Ser., *Haplophyllum ciscaucasicum* (Rupr.) Grossh. & Vved., *Helianthemum dagestanicum* Rupr., *Heracleum nanum* Satzyp., *Hypericum asperuloides* Czern. ex Turcz., *Jurinea alata* (Desf.) Cass., *Jurinea annae* Sosn., *Jurinea bellidioides* Boiss., *Jurinea brachypappa* Nemirova, *Jurinea coronopifolia* Sommier & Levier, *Jurinea dolomitica* Galushko, *Jurinea galushkoi* Nemirova, *Jurinea iljinii* Grossh., *Jurinea prokhanovii* Nemirova, *Jurinea pseudoiljinii* Galushko & Nemirova, *Jurinea ruprechtii* Boiss., *Jurinea sosnowskyi* Grossh., *Jurinea woronowii* Iljin, *Limoniopsis owerinii* (Boiss.) Lincz., *Medicago daghestanica* Rupr. ex Boiss., *Medicago gunibica* Vassilcz., *Minuartia inamoena* (C. A. Mey.) Woronow ex Grossh., *Muehlbergella oweriniana* (Rupr.) Feer, *Myosotis schistosa* A. P. Khokhr., *Onobrychis daghestanica* Grossh., *Onosma levinii* T. N. Popova, *Paederotella daghestanica* (Trautv.) Kem.-Nath., *Pedicularis balkharica* E. A. Busch, *Poa primae* Tzvelev, *Potentilla alexeenkoi* Lipsky, *Potentilla sphenophylla* Th. Wolf, *Primula renifolia* Volgunov, *Psathyrostachys dagestanica* (T. Alex.) Nevski, *Psephellus boissieri* (Sosn.) Sosn., *Psephellus circassicus* (Albov) Galushko, *Psephellus galushkoi* Alieva, *Psephellus hymenolepis* (Trautv.) Boiss., *Psephellus maleevii* Sosn., *Psephellus troitzkyi* Sosn., *Pulsatilla andina* (Rupr.) Woronow, *Pyrethrum tricholobum* Sosn. ex Manden., *Rosa balcarica* Galushko, *Rosa irinae* Demurova, *Rosa prokhanovii* Galushko, *Salsola daghestanica* (Turcz. ex Bunge) Lipsky, *Satureja subdentata* Boiss., *Saxifraga carinata* Oett., *Saxifraga charadzeae* Otsch., *Saxifraga dinnikii* Schmalh. ex Akinf., *Scabiosa gumbetica* Boiss., *Scrophularia charadzeae* Kem.-Nath., *Seseli alexeenkoi* Lipsky, *Silene daghestanica* Rupr., *Silene ikonnikovii* Lazkov, *Silene kubanensis* Sommier & Levier, *Silene lipskyi* Lazkov, *Tanacetum akinfiewii* (T. Alex.) Tzvelev, *Thymus pulchellus* C. A. Mey., *Tripleurospermum nathaliae* Zernov, *Veronica bogosensis* Tumadz.
Sands	*Asperula diminuta* Klokov, *Centaurea scripczinskyi* Mikheev, *Jurinea ciscaucasica* (Sosn.) Iljin
Steppes	*Centaurea czerkessica* Dobrocz. & Kotov, *Cephalaria dagestanica* Bobrov, *Echinops viridifolius* Iljin, *Erodium stevenii* M. Bieb., *Fumaria daghestanica* Mikhailova, *Hieracium adenobrachion* (Litv. & Zahn) Uksip, *Iris notha* M. Bieb., *Onobrychis dielsii* (Sirj.) Vassilcz., *Psephellus leucophyllus* (M. Bieb.) C. A. Mey., *Scorzonera schischkinii* Lipsch. & Vassilcz., *Vincetoxicum stauropolitanum* Pobed.

HABITAT	SPECIES
Submediterranean ecosystems (dry slopes)	*Agropyron pinifolium* Nevski, *Asperula lipskyana* V. I. Krecz., *Centaurea vicina* Lipsky, *Cleome circassica* Tzvelev, *Galatella pontica* (Lipsky) Novopokr. & Bogdan, *Scutellaria novorossica* Juz., *Seseli ponticum* Lipsky, *Veronica filifolia* Lipsky
Submediterranean ecosystems (woodlands, light forest, shibliak, and montane steppes)	*Astracantha arnacanthoides* (Boriss.) Podlech, *Astragalus circassicus* Grossh., *Campanula komarovii* Maleev, *Carduus novorossicus* Porten., *Dianthus acantholimonoides* Schischk., *Erysimum callicarpum* Lipsky, *Euphorbia dubovikiae* Oudejans, *Thymus markhotensis* Maleev
Phrygana (tomillares)	*Genista lipskyi* Novopokr. & Schischk., *Scorzonera kubanica* (Krasch. & Lipsch.) Dubovik, *Thymus helendzhicus* Klokov & Des.-Shost.
Wet places near springs	*Cirsium elbrusense* Sommier & Levier
Wet rocks and gorges	*Asplenium daghestanicum* Christ, *Galium buschiorum* Mikheev, *Petrocoma hoefftiana* (Fisch. ex C. A. Mey.) Rupr., *Saxifraga dinnikii* Schmalh. ex Akinf.
Wetlands (incl. salt marshes)	*Juncus elbrusicus* Galushko
Woodland margins, bush thickets	*Galatella fominii* (Kem.-Nath.) Czerep., *Hypericum maleevii* Zernov & Elenevsky, *Rosa galushkoi* Demurova, *Rosa zaramagensis* Demurova

TOP 50 CONSERVATION PRIORITIES FOR NORTH CAUCASUS, RUSSIAN FEDERATION

ALLIACEAE

1. *Allium circassicum* Kolak. **CR**

This species is endemic to the Russian portion of western Transcaucasia, in the area of Sochi. It occurs in subalpine and alpine belts in glades surrounded by crooked beech or maple trees, as well as meadows and scree slopes. This species could become extinct because of intensive land transformation in the Sochi area and from its over-collection as an ornamental plant.

2. *Allium daghestanicum* Grossh. **CR**

Endemic to Dagestan, this species is known from a single locality. It occurs on stony slopes and rocks and could become extinct from various kinds of land development.

3. *Allium grande* Lipsky **EN**

Endemic to Dagestan, this species occurs in oak and hornbeam woodlands. It is disappearing because of lumbering, overgrazing, and its over-collection as an ornamental plant.

4. *Allium mirzojevii* Tscholok. **CR**

Endemic to Dagestan, this species occurs on scree and rocky slopes. It could become extinct due to the low number of individuals and from various land transformations.

5. *Allium psebaicum* Mikheev **EN**

Endemic to the western Caucasus region, this species is known from a single locality on dry shrubby slopes. It could become extinct due to various kinds of land transformation, especially overgrazing and gypsum mining.

6. *Allium samurense* Tscholok. CR

Endemic to Dagestan, this species occurs on shale screes and rocks. Due to the very low number of individuals, it could become extinct from land transformation.

AMARYLLIDACEAE

7. *Galanthus bortkewitschianus* Koss EN

Endemic to the central Caucasus region, this species occurs in evergreen woods. It is a very attractive spring plant but is disappearing due to the low number of individuals, lack of seed production, and effects of lumbering, collecting for bouquets, and excavation of bulbs for transplanting into gardens (including illegal export). It is also endangered by various kinds of land transformation.

ASPHODELACEAE

8. *Asphodeline tenuior* (Fisch. ex M. Bieb.) Ledeb. VU

Endemic to the western and central Caucasus region, this species occurs on gypsum outcrops and stony steppes. It is disappearing due to overgrazing, gypsum mining, collecting for bouquets, and excavation of bulbs for transplanting into gardens.

ASPLENIACEAE

9. *Asplenium daghestanicum* Christ EN

Endemic to Dagestan, this species occurs on wet shale rocks in the middle mountain belt, at elevations of ca. 2000 m. Its main threats include narrow ecological amplitude, low number of individuals, and construction activities.

ASTERACEAE

10. *Centaurea avarica* Tzvelev CR

Endemic to Dagestan, this species occurs on limestone outcrops or stony slopes in the middle mountain belt. Its main threats include narrow ecological amplitude, low number of individuals, and construction activities.

11. *Centaurea ruprechtii* (Boiss.) Wagenitz EN

Endemic to Dagestan, this species occurs on limestone outcrops and dry slopes. It is disappearing due to the low number of individuals and land tranformation.

12. *Psephellus galushkoi* Alieva CR

Endemic to Dagestan, this species occurs on dry mountain slopes with southern and southeastern orientations, at elevations from 1200 to 1600 m. It could become extinct due to the low number of individuals, overgrazing, and road construction.

13. *Tanacetum akinfiewii* (T. Alex.) Tzvelev CR

Known from a single locality in Dagestan near Tsudakhar village, this species occurs on limestone rocks. It could become extinct due to construction work.

BRASSICACEAE

14. *Crambe cordifolia* Steven EN

Endemic to eastern Pre-Caucasia and the central Caucasus region, this species is already extirpated in many localities due to the plowing of steppes and collection of roots, which are used as a substitute for horseradish.

15. *Hornungia angustilimbata* V. I. Dorof. CR

Endemic to Dagestan, this species is known from only two localities. It occurs on limestone rocks, with the number of individuals very low, and could become extinct due to stone mining.

CAMPANULACEAE

16. *Campanula andina* Rupr. VU

Endemic to the eastern Caucasus region, this species occurs on limestone rocks, cliffs, and in cracks, mainly in the mid-mountain zone from 700 to 1800 m in elevation. There are approximately 10 known populations. Its major threats are climate change and its highly specific habitat requirements.

17. *Campanula autraniana* Albov EN

Endemic to the western Caucasus region, this species occurs mainly on limestone rocks and outcrops. It is under threat because of narrow ecological amplitude and over-collection as an ornamental plant.

18. *Campanula besenginica* Fomin EN

Endemic to the central Caucasus region, this species occurs on rocks near glaciers or the snow line. Known from a few localities, it could be accidentally destroyed in the process of land transformation.

19. *Campanula czerepanovii* Fed. CR

Endemic to Dagestan, this species occurs on rocks and rocky slopes, where it is known from only two localities. It could become extinct due to the low number of individuals and overgrazing.

20. *Campanula dolomitica* E. A. Busch VU

Endemic to the central Caucasus region, this species is disappearing because of overgrazing, road construction, and over-collection for bouquets.

21. *Campanula komarovii* Maleev EN

Endemic to northwestern Transcaucasia, this species is restricted to the northern portion of the Black Sea shore. It occurs on disturbed slopes and in montane steppes in areas that resemble Mediterranean-like vegetation. In some localities, *Campanula komarovii* Maleev is still numerous but is rapidly disappearing due to resort construction, terracing of montane slopes with pine plantings, and collection as an ornamental plant.

22. *Campanula kryophila* Rupr. EN

Endemic to the central Caucasus region, this species occurs in the high mountain belt on granitic and other acidic rocks. Its main threats include low number of individuals, narrow ecological amplitude, and tourist activity.

23. *Muehlbergella oweriniana* (Rupr.) Feer CR

Endemic to Dagestan, this cushion-like plant occurs on limestone rocks. The number of individuals is very low, and the species is disappearing due to overgrazing and road construction.

CARYOPHYLLACEAE

24. *Dianthus acantholimonoides* Schischk. VU

Endemic to northwestern Transcaucasia, this species is restricted to the northern part of the Black Sea shore. It occurs on marl-like outcrops, mainly in the low mountain belt in areas that resemble Mediterranean-like vegetation. In some localities, plants of *Dianthus acantholimonoides* Schischk. are still numerous but are rapidly disappearing due to the construction of resorts, roads, and gas and oil pipelines, and from over-collection as ornamental plants.

CHENOPODIACEAE

25. *Salsola daghestanica* (Turcz. ex Bunge) Lipsky EN

Endemic to Dagestan, this species grows on clay soil on slopes in low and mid-mountain zones. There are ca. 20 populations, each with less than 50 plants. The main threats are land transformation, overgrazing, and poor reproduction.

CONVOLVULACEAE

26. *Convolvulus ruprechtii* Boiss. EN

Endemic to Dagestan, this species occurs on dry limestone slopes. The number of individuals and its competitive ability are very low. It could become extinct due to overgrazing.

EUPHORBIACEAE

27. *Euphorbia aristata* Schmalh. EN

Endemic to the Stavropol Heights in the northern Caucasus, this species is closely related to the Central Asian *Euphorbia soongarica* Boiss. and the Balkan *Euphorbia velenovskyi* Bornm. but is quite disjunct from the distributional range of either species. It occurs in slightly saline, wet depressions on the steppes and is disappearing due to overgrazing and plowing of these steppes.

FABACEAE

28. *Astragalus daghestanicus* Grossh. CR

Endemic to Dagestan, this species occurs on dry limestone slopes and in pine forests. The number of individuals is very low and it could become extinct due to overgrazing and road construction.

29. *Astragalus fissuralis* T. Alex. VU

Endemic to Dagestan, this species occurs on limestone rocks and in dry gorges. It could disappear from some localities because of overgrazing.

30. *Genista angustifolia* Schischk. VU

Endemic to the western and central Caucasus region, this species occurs on limestone in dry steppes in an area of intensive recreation. Its seed production is very low.

31. *Hedysarum daghestanicum* Rupr. ex Boiss. EN

Endemic to Dagestan, this species occurs on dry limestone slopes, often in shrub thickets. It is disappearing due to various kinds of land transformation.

FUMARIACEAE

32. *Corydalis tarkiensis* Prokh. VU

Endemic to Dagestan, this species occurs in shrub thickets and sparse forests. It is disappearing because of lumbering, the replacement of native shrubs by cultivated trees, and its collection as an ornamental plant.

GERANIACEAE

33. *Erodium stevenii* M. Bieb. EN

Endemic to Pre-Caucasia and most common in the Stavropol Heights, this species occurs on sand or sandstone slopes with sparse vegetation. The number of individuals is very low, and it could become extinct due to road construction and other land transformations.

HYPERICACEAE

34. *Hypericum maleevii* Zernov & Elenevsky CR

Endemic to northwestern Transcaucasia, this species occurs on stony and steep marl slopes in the low mountain belt. The number of individuals is very low and most of the known localities are situated in an area of intensive land transformation.

IRIDACEAE

35. *Iris marschalliana* Bobrov EN

Endemic to the central Caucasus and Pre-Caucasia, this species occurs in montane steppes. It is disappearing due to overgrazing, plowing of steppes, afforestation of steppe slopes, and its over-collection for bouquets and transplanting into gardens.

36. *Iris notha* M. Bieb. VU

Endemic to the northern Caucasus region, this species occurs in steppes and meadow-steppes. It is disappearing due to overgrazing and the plowing of steppes, as well as its over-collection for bouquets and transplanting into gardens.

37. *Iris timofejewii* Woronow EN

Endemic to Dagestan, this species occurs on dry montane slopes and rocks. It is disappearing due to the low number of individuals and its over-collection for bouquets and transplanting into gardens.

LAMIACEAE

38. *Thymus pulchellus* C. A. Mey. EN

Endemic to the western Caucasus region, this species occurs on gypsum outcrops. It could become extinct due to intensive gypsum mining.

PAPAVERACEAE

39. *Papaver lisae* N. Busch EN

Endemic to the central Caucasus region, this species occurs in the alpine belt among stones, on moraines, and on talus slopes. It is disappearing due to its narrow ecological amplitude and low competitive ability, in combination with overgrazing.

POACEAE

40. *Hyalopoa lakia* (Woronow) Tzvelev **CR**

Endemic to Dagestan, this species occurs in wet meadows and on shale outcrops. It is disappearing due to the low number of individuals and overgrazing.

41. *Psathyrostachys dagestanica* (T. Alex.) Nevski **EN**

Endemic to Dagestan, this species occurs on clay shale outcrops. It is disappearing due to shale mining, road construction, and overgrazing.

42. *Psathyrostachys rupestris* (T. Alex.) Nevski **EN**

Endemic to Dagestan, this species occurs on limestone outcrops and rocks. It is disappearing due to limestone mining, road construction, and overgrazing.

43. *Stipa sosnowskyi* Seredin **EN**

Endemic to Dagestan, this species occurs on dry shale slopes in montane steppes and pine forests. It is disappearing due to overgrazing and land transformation.

PRIMULACEAE

44. *Primula renifolia* Volgunov **EN**

Endemic to the western Caucasus region, this species occurs in cracks in rocks near the upper margin of pine forests. The species is rare but effectively protected in the Teberda Nature Reserve.

SAXIFRAGACEAE

45. *Saxifraga columnaris* Schmalh. **VU**

Endemic to the central and eastern Caucasus region, this species occurs in the alpine and subalpine belts on wet limestone rocks. It is disappearing due to climate change, narrow ecological amplitude, and low competitive ability.

46. *Saxifraga dinnikii* Schmalh. ex Akinf. **CR**

Endemic to the central and eastern Caucasus region, this species occurs on wet limestone rocks and taluses. The number of localities and individuals is very low. It is disappearing due to its narrow ecological amplitude and low competitive ability, in combination with its collection as an ornamental plant.

SCROPHULARIACEAE

47. *Veronica bogosensis* Tumadz. **EN**

Endemic to the eastern Caucasus region, this species occurs in the alpine and subnival belts on talus, pebble slopes, moraines, and in habitats with scarce vegetation cover. It is disappearing due to climate change and low competitive ability, in combination with overgrazing in high mountain areas.

48. *Veronica filifolia* Lipsky **EN**

Endemic to northwestern Transcaucasia, this species occurs on marl and limestone outcrops and rocks. It is disappearing due to intensive land development for resorts, construction of roads and gas and oil pipelines, and its collection as an ornamental plant.

THYMELAEACEAE

49. *Daphne baksanica* Pobed. **CR**

A narrow endemic of the Baksan valley, in the central Caucasus region, this species occurs in dry rocky places. Its main threats are low seed production and low competitive ability, in combination with overgrazing.

50. *Stelleropsis caucasica* Pobed. **EN**

Endemic to the central Caucasus region, this species occurs in the alpine belt on dry stony slopes and rock outcrops. It is disappearing due to its low competitive ability and land transformation in high mountain areas.

RUSSIAN FEDERATION MAPS

1. *Allium circassicum* Kolak.

2. *Allium daghestanicum* Grossh.

3. *Allium grande* Lipsky

4. *Allium mirzojevii* Tscholok.

5. *Allium psebaicum* Mikheev

6. *Allium samurense* Tscholok.

7. *Galanthus bortkewitschianus* Koss

8. *Asphodeline tenuior* (Fisch. ex M. Bieb.) Ledeb.

9. *Asplenium daghestanicum* Christ

10. *Centaurea avarica* Tzvelev

11. *Centaurea ruprechtii* (Boiss.) Wagenitz

12. *Psephellus galushkoi* Alieva

13. *Tanacetum akinfiewii* (T. Alex.) Tzvelev

14. *Crambe cordifolia* Steven

15. *Hornungia angustilimbata* V. I. Dorof.

16. *Campanula andina* Rupr.

17. *Campanula autraniana* Albov

18. *Campanula besenginica* Fomin

19. *Campanula czerepanovii* Fed.

20. *Campanula dolomitica* E. A. Busch

21. *Campanula komarovii* Maleev

22. *Campanula kryophila* Rupr.

23. *Muehlbergella oweriniana* (Rupr.) Feer

24. *Dianthus acantholimonoides* Schischk.

25. *Salsola daghestanica* (Turcz. ex Bunge) Lipsky

26. *Convolvulus ruprechtii* Boiss.

27. *Euphorbia aristata* Schmalh.

28. *Astragalus daghestanicus* Grossh.

29. *Astragalus fissuralis* T. Alex.

30. *Genista angustifolia* Schischk.

31. *Hedysarum daghestanicum* Rupr. ex Boiss.

32. *Corydalis tarkiensis* Prokh.

33. *Erodium stevenii* M. Bieb.

34. *Hypericum maleevii* Zernov & Elenevsky

35. *Iris marschalliana* Bobrov

36. *Iris notha* M. Bieb.

37. *Iris timofejewii* Woronow

38. *Thymus pulchellus* C. A. Mey.

39. *Papaver lisae* N. Busch

40. *Hyalopoa lakia* (Woronow) Tzvelev

41. *Psathyrostachys dagestanica* (T. Alex.) Nevski

42. *Psathyrostachys rupestris* (T. Alex.) Nevski

43. *Stipa sosnowskyi* Seredin

44. *Primula renifolia* Volgunov

45. *Saxifraga columnaris* Schmalh.

46. *Saxifraga dinnikii* Schmalh. ex Akinf.

47. *Veronica bogosensis* Tumadz.

48. *Veronica filifolia* Lipsky

49. *Daphne baksanica* Pobed.

50. *Stelleropsis caucasica* Pobed.

BIBLIOGRAPHY

Alper, V. N. 1960. Kratkiy ocherk flory i rastitelnosti izvestnyakovogo massiva Fishta i Oshtena. [A brief essay on the flora and vegetation of Fisht and Oshten limestone massif.] Trudy Kavkaz. Zapov. 6: 3–56 [in Russian].

Dmitrieva, A. A. 1990. Opredelitel Rasteniy Adzharii, 2-izd. [Guide to Plants of Adjara, Second edition, Vols. 1–2.] Metzniereba, Tbilisi [in Russian].

Galushko, A. I. 1978–1981. Flora Severnogo Kavkaza: Opredelitel', 1–3. [Flora of the North Caucasus: A Guide, Vols. 1–3.] Rostov University, Rostov-on-Don [in Russian].

Grossheim, A. A. 1946. Rastitel'nye resursy Kavkaza. [Plant Resources of the Caucasus.] Akad. Nauk, Baku [in Russian].

Grossheim, A. A. 1952. Rastitel'nye Bogatstva Kavkaza, 2-izd. [Plant Wealth of the Caucasus, Second edition.] Moscow Society of Naturalists, Moscow [in Russian].

Güldenstädt, J. A. 1787–1791. Reisen durch Russland und im Kaukasischen Gebirge. Anz. Kaiserl. Akad. Wiss. Wien, Math.-Naturwiss. Vol. 1–2.

Iliashenko, V. Yu. & E. I. Iliashenko. 2000. Krasnaya kniga Rossii: pravovye akty [Red Data Book of Russia: legislative acts]. State Committee of the Russian Federation for Environmental Protection, Moscow [in Russian]. <http://2mn.org/engl/rdbrf_en.htm>, accessed 19 September 2013.

Kolakovsky, A. A. 1938–1949. Flora Abkhazii, Tt. 1–4. [Flora of Abkhazia, Vols. 1–4.] Abgiz, Sukhumi [in Russian].

Kosenko, I. S. 1970. Opredelitel' vysshykh Rasteniy Severo-Zapadnogo Kavkaza i Predkavkazya. [Guide to Higher Plants of the North-West Caucasus and Pre-Caucasian Area.] Kolos, Moscow [in Russian].

Lvov, P. L. 1960. Opredelitel' rasteniy Dagestana. [A Guide to Plants of Dagestan.] Dagestan State University, Makhachkala [in Russian].

Marschall von Bieberstein, F. A. 1808–1819. Flora taurico-caucasica, exhibens stirpes phaenogamas in Chersoneso Taurica et regionibus Caucasicis sponte crescentes, Tt. 1–3. Typis Acad., Charkouiae.

Radde, G. 1901. Die Sammlungen des Kaukasischen Museums, Band II, Botanik. Typographie der Kanzelei des Landeschefs, Tiflis.

Ruprecht, F. J. 1869. Flora Caucasi. Pars 1. Mém. Acad. Imp. Sci. St.-Pétersbourg, Sér. 7, 15(2).

Takhtajan, A. L. 1978. Floristicheskie Oblasti Zemli. [The Floristic Regions of the World.] Nauka, Leningrad (St. Petersburg) [in Russian].

Takhtajan, A. L. (editor). 2003–2012. Caucasian Flora Conspectus, Vols. 1–3. St. Petersburg University Press, St. Petersburg [in Russian].

Takhtajan, A. L. & An. A. Fedorov. 1946. Flora Erevana: Opredelitel' rastenii okrestnostei Erevana. [Erevan Flora: Guide to Plants of the Vicinity of Erevan.] Acad. Sci., Erevan [in Armenian].

Takhtajan, A. L. & An. A. Fedorov. 1972. Flora Erevana: Opredelitel' Dikorastushchikh Rastenii Araratskoi Kotloviny, 2-izd. [Erevan Flora: Guide to Wild Plants of Ararat Depression, Second edition.] Nauka, Leningrad (St. Petersburg) [in Russian].

Tanfiliev, V. G. & V. N. Kononov. 1987. Katalog Dikorastushchikh Rasteniy Stavropolskogo Kraya. [Catalogue of the Wild Plants of the Stavropol Territory.] Stavropol S/X Inst., Stavropol [in Russian].

Zernov, A. S. 2000. Rasteniya Severo-Zapadnogo Zakavkazya. [Plants of North-West Transcaucasia.] Moskovsk. Gosud. Pedagog. Univ., Moscow [in Russian].

Zernov, A. S. 2002. Opredelitel' sosudistykh rasteniy severa Rossiyskogo Prichernomorya. [Guide to vascular plants of the northern part of Russian Black Sea Area.] Nauch. Izd. KMK, Moscow [in Russian].

Zernov, A. S. 2006. Flora Severo-Zapadnogo Kavkaza. [Flora of North-West Caucasus.] Tov-vo Nauch. Izd. KMK, Moscow [in Russian].

TURKEY

Tuna Ekim, Salih Terzioğlu, Özgür Emanağaoğlu, and Kamil Coşkunçelebi

CAPITAL

Ankara, the capital and second largest city in the country behind Istanbul.

REGIONAL CITIES AND POPULATIONS

Ağrı (532,000), Ardahan (106,000), Artvin (166,000), Bayburt (108,000), Erzurum (770,000), Giresun (421,000), Gümüşhane (131,000), Kars (312,000), Iğdır (182,000), Ordu (720,000), Rize (320,000), and Trabzon (763,000).

REGIONAL POPULATION

4,530,000 (Turkish Statistical Institute, 2010 estimate)

LANGUAGE

Turkish

FLORA

About 4000 to 4500 (total plant species); ca. 210 (total national endemics in the Caucasus region); ca. 400 (Caucasian region endemics).

The *Red Book of Turkey's Plants* (Ekim et al., 2000) enumerated ca. 3000 endemic plant species. A basic source of information on the plants of Turkey is the *Flora of Turkey* (Davis, 1965–1988).

BOTANICAL INSTITUTIONS

The primary and oldest institutes belong to the Forestry Faculty and Arts and Sciences Faculty at Karadeniz Technical University in Trabzon. There are two herbaria at the university. KATO, in the Faculty of Forestry, is the largest in the region with a collection of ca. 10,000 specimens, primarily from northeastern Anatolia. Address: Karadeniz Technical University, Faculty of Forestry, Forestry Engineering Department, Forestry Botany subdepartment, Trabzon, Turkey; Tel.: +90 (462) 377 2852 or +90 (462) 377 2841; FAX: +90 (462) 325 7499; e-mail: kato@ktu.edu. tr or sterzi@ktu.edu.tr. The second herbarium, KTUB, in the Faculty of Arts and Sciences, contains ca. 10,000 specimens. Address: Karadeniz Technical University, Faculty of Arts and Sciences, Biology Department, Botany subdepartment, Trabzon, Turkey; Tel: +90 (462) 377 2581; FAX: +90 (462) 325 3195; e-mail: biology@ktu.edu.tr.

There is a newly established herbarium at the Faculty of Forestry, Çoruh University, in Artvin, with ca. 7000 specimens. Address: Çoruh University, Faculty of Forestry, Artvin, Turkey; Tel: +90 (466) 215 1035; FAX: +90 (466) 215 1034; e-mail: orman@artvin.edu. tr.

Botanical gardens are connected to the following three universities: Artvin Çoruh University (Faculty of Forestry, Forestry Engineering Department, Forestry Botany subdepartment, Artvin, Turkey); Kafkas University (Faculty of Arts and Sciences, Biology Department, Botany subdepartment, Kars, Turkey); and Atatürk University (Faculty of Arts and Sciences, Biology Department, Botany subdepartment, Erzurum, Turkey). There are no botanical gardens in the immediate territory of the Caucasus region.

PROTECTED AREAS

In the region, there are several national parks, conservation areas, nature monuments, and additional protected areas. All protected areas belong to the government of Turkey.

BIOSPHERE RESERVE AREA
Camili in Artvin Province.

NATIONAL PARKS
Karagöl-Sahara in Artvin Province, consisting of two separate areas, the Karagöl and the Sahara plateau.
Hatila Vadisi or the Hatila Valley National Park in Artvin.
Kaçkar Dağları or the Kaçkar Mountain Range in the coastal province of Rize along the Black Sea.
Altındere Vadisi or the Altindere Valley National Park, Trabzon Province.
Sarıkamış Allahuekber Mountains, a national park between the borders of Kars and Erzurum provinces.

NATURE CONSERVATION AREAS
Örümcek Ormanı Nature Conservation Area (NCA), Gümüşhane Province.
Çamburnu NCA, Trabzon Province.
Camili-Efeler NCA, Borçka District of Artvin Province.
Camili-Gorgit NCA, Borçka District of Artvin Province.

NATURE PARKS
Uzungöl Nature Park, Trabzon Province.
Borçka Karagöl Nature Park, Artvin Province.
Artabel Gölleri Nature Park, Gümüşhane Province.

RAMSAR AREA
(Ramsar Convention on Wetlands, 1971)
Kuyucuk Lake, near Arpaçay, established in 2010.

RECREATIONAL AREA
Çamburnu Forest Recreational Area, Trabzon Province.

NATURE MONUMENTS
Komilet Nature Monument, *Fagus orientalis* L. (Fagaceae), Artvin Province.
Melodere Nature Monument, *Picea orientalis* (L.) Peterm. (Pinaceae), Artvin Province.
Kirani Evliya Nature Monument, *Juniperus* L. (Cupressaceae), Gümüşhane Province.
Ali Ağanın Kavağı, the Poplars of Ali Ağa, Gümüşhane Province.
Örümcek Forest, *Picea orientalis* (L.) Link (Pinaceae), Gümüşhane Province.
Örümcek Forest, *Abies nordmanniana* (Steven) Spach subsp. *nordmanniana* (Pinaceae), Gümüşhane Province.

The Caucasian region of Turkey occupies the northeastern portion of the country, bounded on the east by Georgia and Azerbaijan and on the southeast by Iran. The region in Turkey ranges from approximately 39°21′N–41°35′N and 37°42′E–44°48′E.

Throughout most of northern Anatolia the coast plunges steeply into the Black Sea. In the east, the mountainous range is high, continuous, and close to the sea, rising at many points over 3000 m, with the highest peak being Kaçkar Dağı (Kaçkar Mountain) at 3937 m, in Rize District. Most of the rocks are basic igneous, though many of the highest jagged peaks are granitic. The rivers cleave their way through deep gorges to the sea. On the south side, the range falls steeply to the plateau. The Çoruh River follows the fault line parallel with the main range until it cuts through the mountains near Batumi (Georgia). The majority of rocks of the interior (or southern) portion of the region are again basic igneous rocks, though some limestone areas occur. The highest peak at this eastern extent is Ağrı Dağı or Mount Ağrı (ca. 5165 m, also known as Mount Ararat), which is the highest peak in Turkey. A mountainous belt known as the Anatolian Diagonal extends from the Anti-Taurus northeast to Suşehri, Sivas Province, which is close to the Caucasian region of Turkey and starts at its western edge. Drainage of the eastern Anatolian area flows into the Black Sea through the Çoruh River and into the Caspian Sea by the Aras River.

CLIMATE

In contrast to the rest of the country, the most striking climatic feature, especially in the northern part of the Caucasian region of Turkey, is the heavy rainfall that is continuous all year long. When it is not raining, the hills are frequently swathed in mist. So wet is the climate at Rize and Hopa that tea has become a successful crop. The coastal climate in the same area is mild, allowing cultivation of some species of *Citrus* L. (Rutaceae). In this area, particularly in the Çoruh valley near Artvin, luxuriant euxine forest almost excludes the development of Mediterranean vegetation with its characteristic macchie shrubs and trees such as *Pinus pinea* L. and *Pinus brutia* Ten. (both Pinaceae) and *Olea europaea* L. (Oleaceae). Although precipitation in the area is often believed to increase with altitude, anyone camping out in summer on the peaks of the mountain areas will find the skies are usually clearer above the timberline than they are below it. Indeed, the alpine pasture may be quite sunny while the valleys are filled by clouds.

The climate changes immediately in the interior parts of the region, as the precipitation drops sharply and occurs mostly during the winter period. Winter temperatures are much lower than those in the northern portion of the region. Therefore, the vegetation is mostly herbs or subshrubs and precludes the growth of evergreen trees and shrubs, with the exception of some conifers near Sarıkamış.

FLORA AND VEGETATION

From the phytogeographic point of view, the eastern Black Sea region is included in the Colchic sector of the Euxine province of the Euro-Siberian region, which is one of the three floristic regions in Turkey (Davis, 1971). However, in the interior or southern parts of the region, Irano-Turanian el-

Epimedium colchicum (Boiss.) Trautv., endemic to the western Caucasus region.

Euonymus leiophloea Steven, endemic to the western and central Caucasus region.

Hypericum xylosteifolium (Spach) N. Robson, endemic to the Caucasus region.

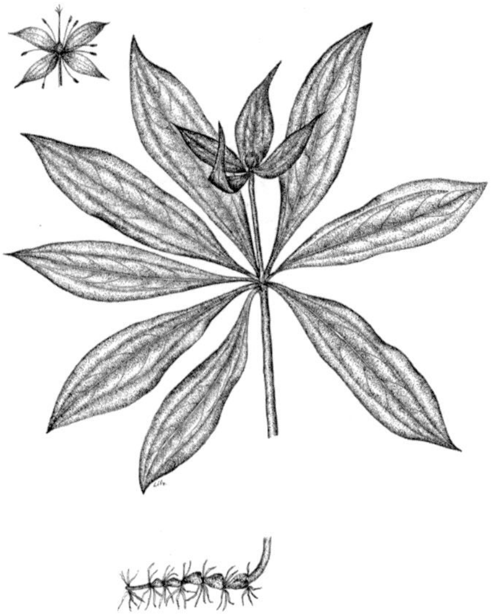

Paris incompleta M. Bieb., endemic to the western and central Caucasus region.

ements have become dominant (Zohary, 1973).

One of the important floristic investigations of the region was carried out by Anşin (1983) in the Colchic sector. He ascertained that the flora there consists of 163 families, 693 genera, and 2239 taxa. The largest families included the Asteraceae (360 taxa), Fabaceae (202 taxa), Brassicaceae (141 taxa), Rosaceae (125 taxa), Apiaceae (122 taxa), Caryophyllaceae (111 taxa), and Scrophulariaceae (111 taxa). The dominant and most widespread vegetation of the Colchic sector is a belt of broad-leaved deciduous forests, penetrated by coniferous forest at higher elevations. This vegetation belt is most closely related (especially in the east) to the Hyrcanian province of northern Iran and adjacent Azerbaijan (Talysh) (Davis, 1971). It has been pointed out that numerous endemics and relict taxa are found in the vegetation of this area, which is an indication of the influence of Caucasian elements (Anşin, 1983). The other primary vegetation types of the region are pseudomacchie, alpine, subalpine, rocky, and hydrophytic.

PSEUDOMACCHIE VEGETATION

This type of vegetation is distributed throughout the area, especially at altitudes of 0–400(–600) m in the eastern Caucasian region of Turkey, in some valleys in Trabzon and Artvin provinces. Although the main vegetation is of Euxine origin, it is also penetrated by some Mediterranean elements. The Euxine elements of this belt are *Rhododendron ponticum* L., *Rhododendron luteum* Sweet (both Ericaceae), *Corylus avellana* L. (Betulaceae), *Cornus sanguinea* L. (Cornaceae), *Staphylea pinnata* L. (Staphyleaceae), *Acer cappadocicum* Gled. (Aceraceae), *Frangula alnus* Mill. (Rhamnaceae), *Buxus sempervirens* L. (Buxaceae), *Ilex colchica* Pojark. (Aquifoliaceae), and *Lauro-cerasus officinalis* M. Roem. (Rosaceae). There are a few stands of *Pinus pinea* L. (Pinaceae) distributed sporadically at the junction of the Hatila valley and the Çoruh River near Fıstıklı village (formerly Naşviye). In the stands of *Pinus pinea* L. (Pinaceae) in Fıstıklı village, several herbaceous and woody plants specifically from the Mediterranean region are located in a transition zone between the Euro-Siberian and Mediterranean regions. Many authors believe that the stands of *Pinus pinea* L. (Pinaceae) are of natural distribution because of the many other associated Mediterranean taxa such as *Erica arborea* L. (Ericaceae); *Cotinus coggygria* Scop. (Anacardiaceae); *Ficus carica* L. (Moraceae); *Trachomitum venetum* (L.) Woodson subsp. *sarmatiense* V. E. Avet. (Apocynaceae); *Cistus creticus* L. and *Cistus salviifolius* L. (both Cistaceae); *Laurus nobilis* L. (Lauraceae); *Olea europaea* L. var. *sylvestris* (Mill.) Lehr. and *Jasminum fruticans* L. (both Oleaceae).

FOREST VEGETATION

The principal and most common vegetation type of the Caucasus region in Turkey, particularly in the northern portion, is forest. This starts from sea level or after the pseudomacchie, if present, and climbs to 1900–2000 m. Between 300 and 800 m in elevation, associations of *Alnus* Mill. and *Corylus* L. (both Betulaceae) or *Castanea* Mill. (Fagaceae) are found. Above 800 m and up to 1500 m, *Fagus orientalis* Lipsky (Fagaceae) and *Carpinus* L. (Betulaceae) forests predominate. Although most of the *Fagus* L. forests are pure stands, conifers and other deciduous trees are also mixed in some places. This dominant deciduous forest is replaced by coniferous forests up to the timberline. The main tree of this upper forest belt is *Picea orientalis* (L.) Peterm., although *Pinus sylvestris* L. and *Abies nordmanniana* (Steven) Spach subsp. *nordmanniana* (these three, Pinaceae) occur as pure or mixed conifer forests.

The forest vegetation of this region includes the following predominant taxa:

Conifers: *Abies nordmanniana* (Steven) Spach subsp. *nordmanniana*, *Picea orientalis* (L.) Peterm., and *Pinus sylvestris* L. (these three, Pinaceae); *Taxus baccata* L. (Taxaceae).

Deciduous trees: *Ulmus glabra* Huds. (Ulmaceae); *Castanea sativa* Mill. and *Fagus orientalis* Lipsky (both Fagaceae); *Alnus glutinosa* (L.) Gaertn. subsp. *barbata* (C. A. Mey.) Yalt. (Betulaceae); *Quercus petraea* L. ex Liebl. subsp. *iberica* (Steven ex M. Bieb.) Krassiln., *Quercus hartwissiana* Steven, and *Quercus pontica* K. Koch (these three, Fagaceae); *Populus tremula* L. and *Salix caucasica* Andersson (both Salicaceae); *Carpinus betulus* L., *Corylus avellana* L., and *Ostrya carpinifolia* Scop. (these three, Betulaceae); *Acer campestre* L. var. *campestre* and *Acer cappadocicum* Gled. (both Aceraceae); *Fraxinus angustifolia* Vahl subsp. *oxycarpa* (M. Bieb. ex Willd.) Franco & Rocha Afonso (Oleaceae); *Tilia rubra* DC. subsp. *caucasica* (Rupr.) V. Engl. (Tiliaceae); *Sorbus subfusca* (Ledeb.) Boiss. (Rosaceae); *Euonymus latifolius* (L.) Mill. and *Euonymus europaeus* L. (both Celastraceae); *Sambucus nigra* L. (Caprifoliaceae); *Rubus platyphyllos* K. Koch, *Crataegus microphylla* K. Koch, and *Crataegus monogyna* Jaq. (these three, Rosaceae); *Rhamnus imeretina* Booth, Petz & Kirchn. (Rhamnaceae); *Betula medwediewii* Regel (Betulaceae); *Daphne glomerata* Lam. (Thymelaeaceae).

Evergreens: *Lauro-cerasus officinalis* M. Roem. (Rosaceae); *Ilex colchica* Pojark. (Aquifoliaceae); *Hedera helix* L. and *Hedera colchica* (K. Koch) K. Koch (both Araliaceae).

Herbs: *Pachyphragma macrophyllum* (Hoffm.) N. Busch (Brassicaceae); *Hypericum bupleuroides* Griseb. (Hypericaceae); *Lathyrus roseus* Steven (Fabaceae); *Rhamphicarpa medwedewii* Albov (Scrophulariaceae); *Psoralea acaulis* Steven ex M. Bieb. (Fabaceae); species of *Lilium* L. (Liliaceae); *Galanthus woronowii* Losinsk. (Amaryllidaceae).

HYDROPHYTIC VEGETATION

Found along streams and lakes, the characteristic species are *Alnus glutinosa* (L.) Gaertn. subsp. *barbata* (C. A. Mey.) Yalt. (Betulaceae); *Salix alba* L. and *Salix rizeensis* Güner & Ziel. (both Salicaceae); *Tamarix tetrandra* Pall. ex M. Bieb. (Tamaricaceae); the fern *Thelypteris limbosperma* (All.) H. P. Fuchs (Thelypteridaceae); *Petasites hybridus* (L.) G. Gaertn., B. Mey. & Schreb. and *Petasites albus* (L.) Gaertn. (both Asteraceae); *Impatiens noli-tangere* L. (Balsaminaceae); *Mentha longifolia* (L.) Huds. subsp. *longifolia* (Lamiaceae); *Lythrum salicaria* L. (Lythraceae); *Polygonum amphibium* L. (Polygonaceae); *Myriophyllum spicatum* L. (Haloragaceae); *Alisma plantago-aquatica* L. (Alismataceae); *Primula auriculata* Lam. (Primulaceae); and the sphenopsid horsetail, *Equisetum ramosissimum* Desf. (Equisetaceae), as well as a number of fern species.

SUBALPINE VEGETATION

Representative taxa include *Betula medwediewii* Regel, *Betula recurvata* (I. V. Vassil.) V. N. Vassil., *Betula litwinowii* Doluch., and *Betula browicziana* Güner (these four, Betulaceae); *Quercus pontica* K. Koch (Fagaceae); *Rhododendron luteum* Sweet, *Rhododendron ponticum* L., and *Rhododendron caucasicum* Pall. (these three, Ericaceae); the conifer *Juniperus communis* L. subsp. *saxatilis* (Pall.) A. E. Murray (Cupressaceae); *Vaccinium myrtillus* L. and *Vaccinium arctostaphylos* L. (both Ericaceae); *Daphne glomerata* Lam. (Thymelaeaceae); *Acer trautvetteri* Medw. (Aceraceae); *Ribes biebersteinii* Berland. ex DC. (Grossulariaceae); *Rubus idaeus* L., *Sorbus aucuparia* L., and *Sorbus umbellata* (Desf.) Fritsch var. *cretica* (Lindl.) C. K. Schneid. (these three, Rosaceae); *Lonicera caucasica* Pall. var. *caucasica* (Caprifoliaceae); *Dianthus carmelitarum* Reut. ex Boiss. (Caryophyllaceae); *Viburnum lantana* L. (Caprifoliaceae); *Empetrum nigrum* L. subsp. *hermaphroditum* (Hagerup) Böcher (Empetraceae); and *Galanthus alpinus* Sosn. (Amaryllidaceae).

ALPINE VEGETATION

This type of vegetation is quite widespread above the timberline (2000–3500 m) where the most common species are *Anemone narcissiflora* L., *Anemone blanda* Schott & Kotschy, *Trollius ranunculinus* (Sm.) Stearn, and *Caltha polypetala* Hochst. (these four, Ranunculaceae); *Colchicum speciosum* Steven (Colchicaceae); *Primula elatior* (L.) Hill (Primulaceae); *Ajuga orientalis* L. (Lamiaceae); species of *Alchemilla* L., such as *Alchemilla caucasica* Buser (Rosaceae); *Gentiana septemfida* Pall. (Gentianaceae); *Scutellaria pontica* K. Koch and *Stachys macrantha* (K. Koch) Stearn (both Lamiaceae); *Helichrysum graveolens* (M. Bieb.) Sweet (Asteraceae); *Geum coccineum* Sibth. & Sm. (Rosaceae); *Trifolium polyphyllum* C. A. Mey. (Fabaceae); *Viola altaica* Ker-Gawl. subsp. *oreades* (M. Bieb.) W. Becker (Violaceae); *Erophila verna* (L.) DC. (Brassicaceae); *Geranium cinereum* Cav. (Geraniaceae); *Astragalus viciifolius* DC. (Fabaceae); *Ornithogalum balansae* Boiss. (Hyacinthaceae); *Asperula affinis* Boiss. & A. Huet (Rubiaceae); *Draba polytricha* Ledeb. (Brassicaceae); *Potentilla ruprechtii* Boiss. (Rosaceae); *Pedicularis atropurpurea* Nordm. and *Pedicularis pontica* Boiss. (both Scrophulariaceae); *Silene alba* (Mill.) E. Krause subsp. *divaricata* (Rchb.) Walters (Caryophyllaceae); *Veronica peduncularis* M. Bieb. (Scrophulariaceae); *Crocus aerius* Herb., *Crocus scharojanii* Rupr., and *Crocus vallicola* Herb. (these three, Iridaceae); and *Scilla siberica* Haw. subsp. *armena* (Grossh.) Mordak and *Scilla bifolia* L. (both Hyacinthaceae).

CHASMOPHYTES

Rhodothamnus sessilifolius P. H. Davis (Ericaceae), *Dianthus andronakii* Woronow ex Schischk. and *Silene dianthoides* Pers. (both Caryophyllaceae), *Rhamnus microcarpa* Boiss.

(Rhamnaceae), *Potentilla oweriniana* Boiss. (Rosaceae), and *Campanula troegerae* Damboldt, as well as other species of *Campanula* L. (Campanulaceae), are commonly found among rock outcrops in the region.

VEGETATION OF THE INTERIOR OR SOUTHERN PORTION OF THE REGION

In the interior (or southern) portion of the Caucasian region of Turkey the vegetation is highly complex. Although steppe vegetation is the predominant type for this region, different plant associations occur according to climate, topography, and underlying rock and soil types. Although detailed vegetation studies have not yet been carried out in this area, it can be said that the floristic composition of these vegetation types is quite different from the alpine vegetation that occurs in the northern portion of the region. In general, areas where the influence of the Irano-Turanian Floristic Region is important, the cushion forms of spiny *Astragalus* L. (Fabaceae) and other Irano-Turanian subshrub elements such as *Thymus* L. (Lamiaceae), *Onobrychis* Mill. (Fabaceae), *Artemisia* L. (Asteraceae), and *Ephedra* L. (Ephedraceae) are dominant. Hemicryptophyte and chamaephyte life forms are also dominant within the floristic composition. This tragacanthic steppe is normally distributed up to 2000 m in elevation, but if overgrazing is heavy, then it may extend higher to 2500 m. Above 2500 m in elevation, some relatively level areas also occur, and in these places wet grass steppes composed of mostly Euro-Siberian elements are found.

Steppe vegetation starts in the western Caucasus region in Turkey from Gümüşhane and Bayburt provinces and continues to the east and interior. Characteristic genera of this vegetation include *Astragalus* L. (Fabaceae), *Acantholimon* Boiss. (Plumbaginaceae), *Thymus* L. (Lamiaceae), *Eryngium* L. (Apiaceae), *Artemisia* L. (Asteraceae), *Dianthus* L. (Caryophyllaceae), *Teucrium* L. (Lamiaceae), and *Stipa* L. and *Festuca* L. (both Poaceae), with most of these being cushion-forming plants. There are many therophytes and geophytes within the floristic composition of this zone as well (Uslu, 1962; Anşin, 1980).

TAHIR DAĞI (TAHIR MOUNTAIN)

Tahir Mountain lies between Ağrı and Erzurum provinces in eastern Turkey and is included in the Irano-Turanian Floristic Region. Overgrazing is heavier here than in other parts of the region and therefore some tragacanth species of *Astragalus* L., such as *Astragalus microcephalus* Willd., and also *Onobrychis cornuta* (L.) Desv. (both Fabaceae), occur in the alpine belt on the mountain. In the wet pastures of this area some mesophytic grasses have become dominant, such as *Alopecurus arundinaceus* Poir. [*Alopecurus ventricosus* Pers., nom. illeg. superfl.], *Deschampsia caespitosa* (L.) P. Beauv., and *Festuca woronowii* Hack. (these three, Poaceae).

GAVUR DAĞI (GAVUR MOUNTAINS)

There are pastures in the subalpine belt (2500–3100 m) of the Gavur Mountains where mesophytic pontic species are dominant such as *Hedysarum hedysaroides* (L.) Schinz & Thell. (Fabaceae) and *Festuca chalcophaea* V. I. Krecz. & Bobrov, *Poa bulbosa* L., and *Festuca varia* Haenke (these three, Poaceae). In the parts of these mountains where overgrazing occurs some Irano-Turanian elements have become dominant and constant. Lower elevations and the southern portion of these mountains are under the influence of the Irano-Turanian Floristic Region, whereas the flora of the northern portion and higher elevations are Euro-Siberian. Apart from the forest vegetation, in the steppe portions, cushion forms of spiny species of *Astragalus* L. and *Onobrychis cornuta* (L.) Desv. (both Fabaceae) have become dominant because of overgrazing. On rocky scree slopes, *Centranthus longiflorus* Steven (Caprifoliaceae) is the dominant species.

ALLAHUEKBER DAĞLARI (ALLAHUEKBER MOUNTAIN RANGE)

In the high-elevation steppes of these mountains in northeastern Turkey, cushion-forming plants such as *Astragalus microcephalus* Willd. (Fabaceae), *Artemisia spicigera* K. Koch (Asteraceae), and *Thymus transcaucasicus* Ronniger (Lamiaceae) predominate. *Deschampsia caespitosa* (L.) P. Beauv., *Festuca chalcophaea* V. I. Krecz. & Bobrov, and *Festuca pratensis* Huds. (these three, Poaceae) and *Trifolium repens* L. (Fabaceae) become dominant at higher elevations depending on increasing humidity.

KARS PROVINCE, ARPAÇAY DISTRICT

The pastures of this area in northeastern Turkey are represented by grass and tragacanthic steppes with some wet pastures. *Festuca valesiaca* Schleich. ex Gaudin and *Festuca cratericola* Markgr.-Dann. (both Poaceae) are the dominant taxa of the grass-dominated steppes. *Astragalus microcephalus* Willd. (Fabaceae) and *Spiraea hypericifolia* L. (Rosaceae) are the dominant species consistently appearing in tragacanthic steppes. Wet pastures are dominated by the grasses *Deschampsia caespitosa* (L.) P. Beauv. and *Elymus repens* (L.) Gould (both Poaceae) as well as the sedges *Carex kurdica* Kük. ex Hand.-Mazz. and *Carex disticha* Huds. (both Cyperaceae), with *Iris spuria* L. (Iridaceae) occurring in some limited places.

ÇILDIR GÖLÜ (LAKE ÇILDIR) AND KISIR DAĞI (KISIR MOUNTAIN)

Mesophytic pastures and steppes occur between 2000 and 3100 m in elevation and are the principal vegetation types of this area in Ardahan Province, northeastern Turkey. Some common steppe plants are found in the wet pastures. This is an ecotone where Euro-Siberian and Irano-Turanian plants grow together. The common and consistently present taxa are mostly grasses; however, the subshrub *Thymus transcaucasicus* Ronniger (Lamiaceae) is another important component. The common taxa of this mesophytic vegetation include *Bromus variegatus* M. Bieb., *Festuca brunnescens* (Tzvelev) Galushko, *Nardus stricta* L., *Festuca valesiaca* Schleich. ex Gaudin, *Stipa tirsa* Steven, and *Koeleria eriostachya* Pančić (these six, Poaceae); *Artemisia splendens* Willd. and *Cirsium ciliatum* (Murr.) Moench (both Asteraceae); *Campanula sibirica* L. (Campanulaceae); *Heracleum trachyloma* Fisch. & C. A. Mey. (Apiaceae); *Scabiosa caucasica* M. Bieb. (Dipsacaceae); and *Alchemilla retinervis* Buser (Rosaceae). *Rosa canina* L. and *Rosa iberica* Steven ex M. Bieb. (both Rosaceae) have become dominant in some places. The wet pastures start from the lake level at 2000 m and extend to 2500 m in flat areas. These sites are dominated by the sedges *Carex acutiformis* Ehrh., *Carex cilicica* Boiss., *Carex melanostachya* M. Bieb. ex Willd., and *Carex orbicularis* Boott (these four, Cyperaceae), as well as by *Oxytropis albana* Steven and *Onobrychis viciifolia* Scop. (both Fabaceae), *Menyanthes trifoliata* L. (Menyanthaceae), *Agrostis stolonifera* L. (Poaceae), and *Caltha polypetala* Hochst. (Ranunculaceae).

Other widely distributed and characteristic taxa of the Caucasus region in Turkey include *Bromus tomentellus* Boiss., *Koeleria pyramidata* (Lam.) P. Beauv. [= *Koeleria cristata* (L.) Pers., nom. illeg.], *Poa bulbosa* L., and *Stipa pontica* P. A. Smirn. (these four, Poaceae); *Centaurea virgata* Lam. (Asteraceae); *Globularia trichosantha* Fisch. & C. A. Mey., *Plantago lanceolata* L., and *Plantago atrata* Hoppe (these three, Plantaginaceae); *Potentilla recta* L. (Rosaceae); *Scabiosa argentea* L. (Dipsacaceae); *Asperula glomerata* (M. Bieb.) Griseb., *Cruciata taurica* (Pall. ex Willd.) Ehrend., and *Galium incanum* Sm. (these three, Rubiaceae); *Marrubium globosum* Montbret & Aucher ex Benth., *Phlomis armeniaca* Willd., *Salvia multicaulis* Vahl, *Scutellaria orientalis* L., *Teucrium chamedrys* L., and *Teucrium polium* L. (these six, Lamiaceae); *Veronica multifida* L.

(Scrophulariaceae); *Linum mucronatum* Bertol. (Linaceae); *Hypericum scabrum* L. (Hypericaceae); *Acantholimon acerosum* (Willd.) Boiss. (Plumbaginaceae); *Onosma armenum* DC. (Boraginaceae); and *Astragalus condensatus* Ledeb. and *Astragalus plumosus* Willd. (both Fabaceae).

Characteristic taxa found in wet pastures of the Caucasus region in Turkey include *Lotus corniculatus* L. and *Trifolium pratense* L. (both Fabaceae); *Alopecurus arundinaceus* Poir., *Phleum pratense* L., and *Molinia caerulea* (L.) Moench (these three, Poaceae); *Eleocharis palustris* (L.) M. Roem. & Schult. (Cyperaceae); *Juncus gerardii* Loisel. (Juncaceae); *Ranunculus repens* L. (Ranunculaceae); *Erigeron acer* L., *Scorzonera parviflora* Jacq., and *Tragopogon pratensis* L. (these three, Asteraceae); and *Pedicularis comosa* L. (Scrophulariaceae).

IGDIR PLAIN

This broad plain is located in Igdır Province in eastern Turkey, to the north of Mount Ağrı and is the only halophytic area in this region, with a diverse halophytic or partly psammophytic vegetation. The plateau is ca. 850 m in elevation and is one of the driest parts of the country. Its distinctive flora penetrates up the Aras valley from southern Armenia and also from adjacent northwestern Iran. This adds several genera to the Turkish flora, such as *Allochrusa* Bunge ex Boiss. (Caryophyllaceae), *Seidlitzia* Bunge ex Boiss. and *Aellenia* Ulbr. (both Amaranthaceae), *Calligonum* L. (Polygonaceae), and *Halimodendron* Fisch. ex DC. (Fabaceae) (Davis, 1971). The dominant and consistently observed taxa here include *Ceratocarpus arenarius* L., *Halanthium rarifolium* K. Koch, *Kochia prostrata* (L.) Schrad., *Panderia pilosa* Fisch. & C. A. Mey., *Salsola macera* Litv., *Salsola dendroides* Pall., and *Seidlitzia florida* (M. Bieb.) Bunge ex Boiss. (these seven, Amaranthaceae); *Alhagi pseudalhagi* (M. Bieb.) Desv. ex B. Keller & Shap. (Fabaceae); and *Juncus gerardii* Loisel. (Juncaceae). In the driest part of the plain, a shrubby vegetation is observed, dominated by *Ephedra major* Host and *Ephedra distachya* L. (both Ephedraceae).

MAIN THREATS

The vegetation throughout the Caucasus region in Turkey is experiencing human-mediated impacts, the most important of which are agriculture, tourism, and road and dam construction. In particular, the pseudomacchie vegetation around urban areas has lost its Mediterranean elements, as these areas have been converted into tea plantation fields or invaded by *Corylus-Alnus* stands (Anşin, 1980; Terzioğlu, 1998).

Anşin indicated that there were 220 endemic plants in his research area that covered the eastern Black Sea region (Anşin, 1982). The number of known endemic species has since increased as other floristic studies have been carried out (Güner et al., 1987; Terzioğlu, 1998, 1999; Palabaş, 2002; Uzun, 2002; Eminağaoğlu & Anşin, 2003, 2004; Eminağaoğlu et al., 2006, 2007, 2008). The endemic plants of this region (except the interior portion) and their habitats are shown in Table 1.

Table 1. Distribution of endemic species within different types of vegetation of the east Black Sea region of Turkey (Ordu, Giresun, Gümüşhane, Trabzon, Rize).

HABITAT	SPECIES
Alpine pastures and steppes	*Aethionema caespitosum* (Boiss.) Boiss., *Alchemilla oriturcica* Pawł., *Alchemilla rizensis* Pawł., *Alchemilla sintenisii* Rothm., *Alchemilla tiryalensis* Pawł., *Allium balansae* Boiss., *Allium djimilense* Boiss. ex Regel, *Alopecurus laguroides* Balansa, *Alyssum aizoides* Boiss., *Alyssum artwinense* N. Busch, *Anthemis anthemiformis* (Freyn & Sint.) Grierson, *Anthemis melanoloma* Trautv. subsp. *melanoloma*, *Anthemis melanoloma* Trautv. subsp. *trapezuntica* Grierson, *Arenaria acutisepala* Hausskn. ex F. N. Williams, *Astragalus acmophyllus* Bunge, *Astragalus baibutensis* Bunge, *Astragalus coadunatus* Hub.-Mor. & D. F. Chamb., *Astragalus czorochensis* Charadze, *Astragalus lineatus* Lam. var. *jildisianus* (Bornm.) V. A. Matthews, *Astragalus trachytrichus* Bunge, *Barbarea integrifolia* DC., *Carduus lanuginosus* Willd., *Carex melanorrhyncha* Nelmes, *Centaurea appendicigera* K. Koch, *Centaurea armena* Boiss., *Cerastium gnaphalodes* Fenzl, *Cerastium lazicum* Boiss., *Cochlearia aucheri* Boiss., *Cochlearia sintenisii* Hausskn. ex Bornm., *Crepis bupleurifolia* (Boiss. & Kotschy ex Boiss.) Freyn & Sint., *Doronicum balansae* Cavill., *Doronicum macrolepis* Freyn & Sint., *Doronicum tobeyi* J. R. Edm., *Draba rigida* Willd. var. *bryoides* (DC.) Boiss., *Draba rosularis* Boiss., *Echinops sintenisii* Freyn, *Elymus sosnowskyi* (Hack.) Melderis, *Erodium absinthoides* Willd. subsp. *latifolium* P. H. Davis, *Euphorbia djimilensis* Boiss., *Euphrasia minima* Jacq. ex DC. subsp. *davisii* Yeo, *Festuca amethystina* L. var. *turcica* Markgr.-Dann., *Festuca anatolica* Markgr.-Dann. subsp. *anatolica*, *Festuca anatolica* Markgr.-Dann. subsp. *borealis* Markgr.-Dann., *Festuca cyllenica* Boiss. & Heldr. subsp. *uluana* Markgr.-Dann., *Festuca lazistanica* E. B. Alexeev subsp. *lazistanica*, *Festuca paphlagonica* (St.-Yves) Markgr.-Dann. subsp. *villosula* (St.-Yves) Markgr.-Dann., *Festuca pontica* E. B. Alexeev ex Markgr.-Dann., *Festuca woronowii* Hack. subsp. *woronowii*, *Geranium cinereum* Cav. var. *ponticum* P. H. Davis & J. Roberts, *Geranium ibericum* Cav. subsp. *jubatum* (Hand.-Mazz.) P. H. Davis, *Heracleum sphondylium* L. subsp. *artvinense* (Manden.) P. H. Davis, *Hieracium lasiochaetum* (Bornm. & Zahn) P. D. Sell & C. West, *Hieracium tamderense* Hub.-Mor., *Jasione supina* Sieber ex Spreng. subsp. *pontica* (Boiss.) Damboldt, *Lamium sulfureum* Hausskn. & Sint. ex R. R. Mill, *Minuartia umbellulifera* (Boiss.) McNeill var. *umbellulifera*, *Muscari aucheri* (Boiss.) Baker, *Muscari bourgaei* Baker, *Muscari coeleste* Fomin, *Nonea pulmonarioides* Boiss. & Balansa, *Onobrychis araxina* Schischk., *Onobrychis armena* Boiss. & A. Huet, *Onobrychis lasistanica* Širj., *Pastinaca armena* Fisch. & C. A. Mey. subsp. *dentata* (Freyn & Sint.) Chamb., *Potentilla cappadocica* Boiss., *Potentilla doddsii* P. H. Davis, *Potentilla savvalensis* Pawł., *Primula longipes* Freyn & Sint., *Ptilostemon afer* (Jacq.) Greuter subsp. *eburneus* Greuter, *Ranunculus dissectus* M. Bieb. subsp. *rigidulus* (Boiss.) P. H. Davis, *Ranunculus dissectus* M. Bieb. var. *huetii* (Boiss.) P. H. Davis, *Ranunculus tempskyanus*

HABITAT	SPECIES
Alpine pastures and steppes (continued)	Freyn & Sint., *Saxifraga artvinensis* V. A. Matthews, *Scorzonera pygmaea* Sibth. & Sm. subsp. *nutans* (Czeczott) D. T. Chamb., *Scorzonera sericea* Aucher ex DC., *Sempervivum furseorum* Muirhead, *Senecio cilicius* Boiss., *Senecio inops* Boiss. & Balansa, *Senecio ovatifolius* Boiss., *Stachys rizeensis* R. Bhattacharjee, *Thlaspi lilacinum* Boiss. & A. Huet, *Thlaspi sintenisii* Hausskn. ex Bornm., *Tragopogon aureus* Boiss., *Tragopogon fibrosus* Freyn & Sint., *Trinia scabra* Boiss. & Noë, *Tripleurospermum monticolum* Bornm., *Verbascum songaricum* Schrenk ex Fisch. & C. A. Mey., *Vicia alpestris* Steven subsp. *hypoleuca* (Boiss. & Noë) P. H. Davis
Forest clearings	*Arenaria armeniaca* Boiss., *Centaurea sessilis* Willd., *Galium fissurense* Ehrend. & Schönb.-Tem., *Hieracium djimilense* Boiss. & Balansa, *Hieracium lazistanum* Avr.-Touv., *Hieracium radiatellum* (Woronow & Zahn) Üksip, *Hieracium virosiforme* Woronow & Zahn, *Papaver lateritium* K. Koch, *Senecio integrifolius* (L.) Clairv. subsp. *karsianus* V. A. Matthews, *Silene armena* Boiss.
Forests	*Acer cappadocicum* Gled. var. *stenocarpum* Yalt., *Acer divergens* K. Koch & Pax var. *divergens*, *Adonis cyllenea* Boiss., Heldr. & Orph., *Aethionema sintenisii* Hausskn. & Bornm., *Alcea calvertii* (Boiss.) Boiss., *Alyssum callichroum* Boiss., *Alyssum paphlagonicum* (Hausskn.) T. R. Dudley, *Angelica sylvestris* L. var. *stenoptera* Boiss., *Arum orientale* M. Bieb. subsp. *amoenum* (Engl.) R. R. Mill, *Astragalus acmophylloides* Grossh., *Astragalus aucheri* Boiss., *Astragalus campylosema* Boiss. subsp. *nigripilis* Hub.-Mor. & D. F. Chamb., *Astragalus fumosus* Boriss., *Astragalus taochius* Woronow, *Astragalus vexillaris* Boiss., *Astragalus viridissimus* Freyn & Sint., *Astragalus voronovianus* (Boriss.) Boriss., *Barbarea trichopoda* Hausskn. ex Bornm., *Betula browicziana* Güner, *Bupleurum brachiatum* K. Koch ex Boiss., *Campanula latiloba* A. DC., *Carduus onopordioides* Fisch. ex M. Bieb. subsp. *turcicus* (Kazmi) P. H. Davis, *Centaurea drabifolioides* Hub.-Mor., *Centaurea pecho* Albov, *Centaurea woronowii* Bornm. ex Sosn., *Cirsium trachylepis* Boiss., *Clypeola raddeana* Albov, *Digitalis cariensis* Boiss. ex Jaub. & Spach, *Echium orientale* L., *Epipactis pontica* Taubenheim, *Euonymus latifolius* (L.) Mill. subsp. *cauconis* Coode & Cullen, *Geranium davisianum* Peşmen & Güner, *Glycyrrhiza flavescens* (Boiss.) Boiss., *Gypsophila glandulosa* (Boiss.) Walp., *Gypsophila simulatrix* Bornm. & Woronow, *Haplophyllum armenum* Spach, *Helichrysum artvinense* P. H. Davis & Kupicha, *Heracleum crenatifolium* Boiss., *Heracleum pastinacifolium* K. Koch subsp. *incanum* P. H. Davis, *Heracleum platytaenium* Boiss., *Herniaria argaea* Boiss., *Hesperis buschiana* Tzvelev, *Hesperis cappadocica* E. Fourn., *Hieracium artvinense* (Woronow & Zahn) Üksip, *Hieracium asterodermum* (Woronow & Zahn) Üksip, *Hieracium bornmuelleri* Freyn, *Hieracium debilescens* (Woronow & Zahn) Üksip, *Hieracium diaphanoidiceps* (Woronow & Zahn) Üksip, *Hieracium floccicomatum* (Woronow & Zahn) Üksip, *Hieracium foliosissimum* (Woronow & Zahn) Üksip, *Hieracium gentiliforme* (Zahn) P. D. Sell & C. West, *Hieracium hypopityforme* Üksip, *Hieracium* [illegible] (Zahn) P. D. Sell & C. West, *Hieracium lazicum* Boiss. & Balansa, *Hieracium*

HABITAT	SPECIES
Forests (continued)	*leptogrammoides* Üksip, *Hieracium microtum* Boiss., *Hieracium odontophyllum* Freyn & Sint., *Hieracium onosmopsis* (Zahn) P. D. Sell & C. West, *Hieracium ovalifrons* (Woronow & Zahn) Üksip, *Hieracium reducatum* (Norrl.) Üksip, *Hypericum fissurale* Woronow, *Knautia degenii* Borbás ex Formánek, *Lamium gundelsheimeri* K. Koch, *Lamium ponticum* Boiss. & Balansa, *Lilium carniolicum* Bernh. ex K. Koch subsp. *ponticum* (K. Koch) P. H. Davis & D. M. Hend., *Lilium carniolicum* Bernh. ex K. Koch var. *artvinense* (Miscz.) P. H. Davis & D. M. Hend., *Lilium ciliatum* P. H. Davis, *Lilium monadelphum* M. Bieb. subsp. *armenum* (Miscz. ex Grossh.) Kudrjasch., *Lonicera caucasica* Pall. subsp. *orientalis* (Lam.) D. F. Chamb. & D. G. Long, *Melampyrum arvense* L. var. *elatius* Boiss., *Onosma ambigens* Lacaita, *Onosma circinnata* Riedl, *Onosma trapezuntea* Boiss. & A. Huet ex Hand.-Mazz., *Papaver triniifolium* Boiss., *Ranunculus bulbosus* L. subsp. *aleae* (Wilk.) Rouy & Foucaud, *Reseda armena* Boiss., *Scaligera lazica* Boiss., *Scorzonera inaequiscapa* Boiss., *Sempervivum armenum* Boiss. & A. Huet var. *armenum*, *Sempervivum armenum* Boiss. & A. Huet var. *insigne* Muirhead, *Sempervivum glabrifolium* Boriss., *Sempervivum minus* Turrill var. *glabrum* Wale, *Sempervivum minus* Turrill var. *minus*, *Sempervivum staintonii* Muirhead, *Silene scythicina* Coode & Cullen, *Stenotaenia macrocarpa* Freyn & Sint., *Symphyandra lazica* Boiss. & Balansa, *Taraxacum turcicum* Soest, *Tripleurospermum fissurale* (Sosn.) E. Hossain, *Verbascum artvinense* E. Wulff, *Verbascum eriorrhabdon* Boiss., *Vicia quadrijuga* P. H. Davis
Forests and rocky places	*Hieracium insolitum* (Zahn) Üksip
Forests and alpine communitites	*Alkanna cordifolia* K. Koch, *Astragalus pinetorum* Boiss., *Centaurea simplicicaulis* Boiss. & A. Huet, *Cerastium armeniacum* Gren., *Cirsium pseudopersonata* Boiss. & Balansa subsp. *pseudopersonata*, *Cirsium sommieri* Petr., *Cyclamen parviflorum* Pobed., *Dianthus carmelitarum* Reut. ex Boiss., *Dianthus zederbaueri* Vierh., *Ferulago platycarpa* Boiss. & Balansa, *Geranium asphodeloides* Burm. subsp. *sintenisii* (Freyn) P. H. Davis, *Hedysarum nitidum* Willd., *Helictotrichon argaeum* (Boiss.) Parsa, *Lathyrus czeczottianus* Bassler, *Lathyrus tukhtensis* Czeczott, *Linaria genistifolia* (L.) Mill. subsp. *confertiflora* (Boiss.) P. H. Davis, *Myosotis platyphylla* Boiss., *Senecio lazicus* Boiss. & Balansa, *Symphytum longipetiolatum* Wickens, *Taraxacum revertens* G.E. Haglund, *Veronica hispidula* Boiss. & A. Huet subsp. *ixodes* (Boiss. & Balansa) M. A. Fisch., *Vicia freyniana* Bornm.
Forests and steppes	*Astragalus squalidus* Boiss. & Noë, *Iris histrioides* (G. F. Wilson) S. Arn., *Saponaria picta* Boiss., *Trifolium pannonicum* Jacq. subsp. *elongatum* (Willd.) Zohary
Forests and subalpine communities	*Alchemilla ziganadagensis* Pawł., *Arenaria scariosa* Boiss., *Astragalus albifolius* (M.E. Jones) Abrams, *Astragalus densifolius* Lam., *Astragalus hirsutus* Vahl, *Bupleurum schistosum* Woronow, *Campanula betulifolia* K. Koch, *Campanula saxonorum* Gand., *Centaurea rhizocalathium* (K. Koch) Tchich., *Crocus aerius*

HABITAT	SPECIES
Forests and subalpine communities (continued)	Herb., *Erysimum kotschyanum* J. Gay, *Erysimum uncinatifolium* Boiss. & A. Huet, *Hieracium managettae* Freyn, *Phlomis russeliana* (Sims) Lag. ex Benth., *Swertia hemshinica* Güner
Pseudomacchie	*Genista involucrata* Spach, *Senecio trapezuntinus* Boiss.
Rocks	*Hieracium giresunense* Hub.-Mor., *Hieracium patentissimum* Freyn & Sint.
Steppes	*Alyssum blepharocarpum* T. R. Dudley & Hub.-Mor., *Alyssum pseudomouradicum* Hausskn. & Bornm. ex Baumg., *Alyssum stylare* (Boiss. & Balansa) Boiss., *Anchusa leptophylla* Roem. & Schult., *Astragalus asciocalyx* Bunge, *Astragalus brachypterus* Fisch., *Astragalus cinereus* Willd., *Astragalus compactus* Lam., *Astragalus condensatus* Ledeb., *Astragalus crinitus* Boiss., *Astragalus cylindraceus* C. A. Mey., *Astragalus noaeanus* Boiss., *Astragalus ovatus* DC., *Astragalus strigillosus* Bunge, *Camelina hispida* Boiss. var. *grandiflora* (Boiss.) Hedge, *Cousinia bicolor* Freyn & Sint., *Cousinia woronowii* Bornm., *Ferulago latiloba* Schischk., *Frangula alnus* Mill. subsp. *pontica* (Boiss.) P. H. Davis & Yalt., *Genista aucheri* Boiss., *Linum flavum* L. subsp. *scabrinerve* (P. H. Davis) P. H. Davis, *Pilosella hoppeana* (Schult.) F. W. Schultz & Sch. Bip. subsp. *isaurica* Hub.-Mor., *Rosa pisiformis* Sosn., *Scorzonera tomentosa* L., *Scrophularia capillaris* Boiss. & Balansa, *Tanacetum cadmeum* (Boiss.) Heywood, *Tchihatchewia isatidea* Boiss., *Verbascum oreophilum* K. Koch
Subalpine communities	*Centaurea helenioides* Boiss. & Hausskn., *Centaurea mucronifera* DC., *Centaurea pulcherrima* Willd. var. *freynii* (Sint.) Wagenitz, *Centaurea pyrrhoblephara* Boiss., *Cicerbita boissieri* (Rouy) C. Jeffrey, *Dactylorhiza osmanica* P. F. Hunt & Summerh., *Delphinium formosum* Boiss. & A. Huet, *Draba bruniifolia* Steven subsp. *armeniaca* Coode & Cullen, *Elymus lazicus* (Boiss.) Melderis subsp. *lazicus*, *Helichrysum chionophilum* Boiss. & Balansa, *Hieracium subhastulatum* (Zahn) Üksip, *Verbascum songaricum* Schrenk ex Fisch. & C. A. Mey. subsp. *subdecurrens* Hub.-Mor.

TOP 50 NATIONAL CONSERVATION PRIORITIES FOR TURKEY

ACERACEAE

1. *Acer divergens* K. Koch & Pax var. ***divergens*** **EN**

The typical variety of *Acer divergens* is one of the rarer endemics of northeastern Anatolia. There are several populations, but these are restricted to the Çoruh River valley, through the Mescit Mountains. The main threats are from dam and road construction.

2. *Acer divergens* K. Koch & Pax var. ***trilobum*** Yalt. **CR**

This variety of *Acer divergens* is a very rare local endemic of northeastern Anatolia. A few populations are known from the area corresponding to the older Ottoman vilayet known as the Erzurum Vilayet, approximately similar to Erzurum Province in northeastern Turkey today. This variety grows in dry habitats, and its main threats are erosion and mining activities.

ALLIACEAE

3. *Allium baytopiorum* Kollmann & Özhatay **CR**

This is a rare species, endemic to the eastern portion of the Caucasus region in Turkey. It is only known from two quite close localities in a single region of northeastern Anatolia, where it grows on mountain steppe in Kars Province. The main threats are harvesting to feed livestock, overgrazing, and over-collecting in the wild as medicinal or ornamental plants.

4. *Allium czelghauricum* Bordz. **CR**

This is a rare species, endemic to northeastern Anatolia. Its status was considered to be Data Deficient (DD) but was relocated again recently. *Allium czelghauricum* Bordz. was found in the mountain steppes of Kars and Ardahan provinces. These populations are stable but under threat of harvesting for livestock fodder, overgrazing, and over-collecting in the wild as medicinal or ornamental plants.

AMARYLLIDACEAE

5. *Galanthus koenenianus* Lobin, C. D. Brickell & A. P. Davis **CR**

This is a recently described endemic species of northeastern Anatolia, where the taxon is known only from two or three subpopulations. *Galanthus koenenianus* Lobin, C. D. Brickell & A. P. Davis is a potential ornamental species but is not represented in any living collection so far. There is no possibility to make mixed collections with *Galanthus woronowii* Losinsk., as both species are widely separated geographically.

APIACEAE

6. *Chaerophyllum posofianum* Erik & Demirkuş **CR**

This very rare endemic plant species grows in Ardahan Province in fields between 1500 and 1700 m in elevation and is known only from the type locality in northeastern Turkey. Its main threats are from harvesting for livestock fodder and overgrazing.

7. *Heracleum sphondylium* L. subsp. ***artvinense*** (Manden.) P. H. Davis **EN**

This subspecies is endemic to the Caucasian portion of Turkey but very sparsely distributed. It grows on rocky, igneous slopes in gullies, between 1500 and 2950 m in elevation. Threats are ecotourism and overharvesting for livestock.

ASTERACEAE

8. *Centaurea appendicigera* K. Koch **EN**

This species is endemic to northeastern Turkey, where it grows in alpine meadows. There is only one known locality close to the southern frontier between the Caucasian and non-Caucasian portions of northeastern Anatolia. The population is stable but under the threat of ecotourism and other human activities, as well as from overgrazing by chamois (*Rupicapra rupicapra* L., Bovidae).

9. *Centaurea gracillima* Wagenitz **CR**

This is one of the rarest endemic species known from the Caucasian region of eastern Turkey. Only one population from Ağrı Province is known, where it is under threat from heavy overgrazing and harvesting for livestock fodder.

10. *Centaurea helenioides* Boiss. & Hausskn. **EN**

This species is endemic to the Turkish Caucasus. There are several populations in northeastern Anatolia that are known to be decreasing. It is under threat from overgrazing, by cutting for ornamental purposes, and by over-collecting for local use in folk medicine.

11. *Doronicum tobeyi* J. R. Edm. **CR**

This is a rare species restricted to only one population from the Karagöl Mountains in Giresun Province, in the eastern Black Sea region. It grows along streamsides and in wet places in alpine habitats. The one population is known to be decreasing. The main threats are ecotourism and road construction.

12. *Echinops sintenisii* Freyn EN

This species is one of the rarer endemics known from northeastern Anatolia. The main threats are grazing and harvesting for livestock fodder.

13. *Helichrysum artvinense* P. H. Davis & Kupicha CR

This species is one of the very rare endemics of northeastern Turkey and is known only from Artvin Province. It grows on dry, calcareous soils in forests. The main threats are from ecotourism and overgrazing.

14. *Senecio integrifolius* (L.) Clairv. subsp. *karsianus* V. A. Matthews EN

This species is endemic to the northeastern portion of the Caucasus region in Turkey, where it grows in wet meadows, alpine pastures, and along streamsides, at elevations between 1050 and 2500 m. The main threats are from harvesting for livestock fodder, grazing, agricultural activities, and trampling.

15. *Senecio ovatifolius* Boiss. EN

This Turkish Caucasian endemic is quite widely distributed, particularly in the Rize valley, where it grows in alpine meadows of the Kaçkar Mountain Range at elevations from 2000 to 3000 m. Some localities are included with the boundaries of one of the national parks. The main threats are from over-grazing, harvesting for livestock fodder, and ecotourism activities.

16. *Senecio trapezuntinus* Boiss. CR

This is a species endemic to northeast ern Anatolia, where it grows in scrubby and stony habitat in Trabzon Province at very low elevations (ca. 100–150 m). The population is known to be decreasing and is particularly under threat from urbanization and recreational activities.

17. *Tripleurospermum fissurale* (Sosn.) E. Hossain EN

Endemic to Kars Province, this species is found growing on rocky slopes and crevices between 380 and 1400 m in elevation. The primary threats are from dam and road construction. Some subpopulations will be flooded by the man-made dam lake within the next five years.

BETULACEAE

18. *Betula browicziana* Güner EN

This species is endemic to northeastern Anatolia, where several populations are known. It grows in *Rhododendron* L. scrub on igneous slopes and in humid forests. The main threats are from collection for fuel wood and by clear-cutting and mining activities.

BORAGINACEAE

19. *Nonea pulmonarioides* Boiss. & Balansa EN

This species is endemic to the northeastern portion of the Caucasus region in Turkey, where it grows in alpine meadows, grassy screes, and on rocky slopes between 1600 and 3400 m in elevation. There are some localities known within the boundaries of one of the national parks. The major threats to *Nonea pulmonarioides* Boiss. & Balansa are from harvesting for livestock fodder, grazing, and ecotourism.

20. *Onosma nigricaulis* Riedl CR

This rare species is endemic to Kars Province, where there are several populations that are currently decreasing. *Onosma nigricaulis* Riedl grows on rocky, stony, igneous slopes. The main threats are from pipeline construction and by road expansion and new construction.

21. *Onosma obtusifolia* Hausskn. & Sint. ex Riedl CR

This is a very rare endemic to northeastern Anatolia and known to be distributed among three vilayets, but the populations are not thriving. *Onosma obtusifolia* Hausskn. & Sint. ex Riedl grows in wet meadows, where it is threatened by grazing and from a variety of different agricultural activities.

BRASSICACEAE

22. *Alyssum artwinense* N. Busch EN

This species is endemic to northeastern Turkey. There are several stable populations in Erzurum and Artvin provinces. It grows on rocky, stony, igneous slopes and in forest clearings. The main threats are from erosion and from dam and road construction.

23. *Draba narmanensis* Yıld. CR

This is a recently described and very rare endemic that is known only from the type locality in Erzurum Province, where it grows on rocky slopes between 1700 and 2200 m in elevation. It is under threat from grazing and by harvesting for livestock fodder.

CAMPANULACEAE

24. *Campanula choruhensis* Kit Tan & Sorger EN

Endemic to northeastern Turkey, this is a beautiful species with showy flowers, growing in rocky places. The main threats are from dam construction and microclimate changes caused by the dams.

25. *Campanula latiloba* A. DC. subsp. *rizeensis* Güner EN

This subspecies grows on metamorphic bedrock that forms very steep, rocky slopes. The main threats to *Campanula latiloba* subsp. *rizeensis* Güner are from urbanization and road construction.

26. *Campanula troegerae* Damboldt CR

Endemic to northeastern Turkey, *Campanula troegerae* Damboldt grows at relatively low elevations, from 350 to 850 m. This is a very handsome species with beautiful, showy white flowers and has possible horticultural potential. Its main threats are dam and road construction as well as microclimatic changes because of the dam.

CARYOPHYLLACEAE

27. *Cerastium lazicum* Boiss. EN

This is endemic to Trabzon and Rize provinces in the Caucasian region of Turkey. Some populations are protected because they are within the boundaries of one of the national parks. Its main threats are from grazing and ecotourism.

CRASSULACEAE

28. *Sempervivum furseorum* Muirhead EN

This species is endemic to Kaçkar Dağları or the Kaçkar Mountain Range that forms the coastal range along the eastern Black Sea coast. *Sempervivum furseorum* grows on rocky slopes at elevations of 1800–2500 m. Its only threat is from grazing.

29. *Sempervivum glabrifolium* Boriss. EN

This species is endemic to Artvin Province in northeastern Turkey, where it occurs on igneous, conglomerate bedrock, at elevations of 250–730 m. The main and most serious threat is construction of several dams along the Çoruh River.

ERICACEAE

30. *Rhodothamnus sessilifolius* P. H. Davis CR

This is a very rare local endemic restricted to Artvin Province. It is known from only two localities in northeastern Anatolia, and both are very close to each other. The current populations appear to be decreasing. *Rhodothamnus sessilifolius* P. H. Davis grows on moist banks and igneous cliff ledges. Its main threats are from human settlement, overgrazing and harvesting, trampling by goats, and overcollecting for scientific purposes.

FABACEAE

31. *Astragalus owirensis* Ali ex Podlech CR

This is a very rare local endemic of northeastern Anatolia. The species is known only from the type locality between Olur and Yusufeli. The current population of *Astragalus owirensis* Ali ex Podlech is decreasing. It grows on scree and roadsides, and the main threats are from erosion and from dam and road construction.

32. *Astragalus tatlii* Peşmen **CR**

This is a very rare local endemic in northeastern Anatolia and has been recorded only from the type locality in the Erzurum region, where it grows in steppe. The main threats are from over-grazing and harvesting for livestock fodder.

33. *Astragalus trabzonicus* Ekici, Aytaç & Akan **CR**

This recently described species is a very rare, local endemic with only one population known from Trabzon Province. *Astragalus trabzonicus* Ekici, Aytaç & Akan grows in mountain pastures, where the main threats are from over-grazing and harvesting for livestock fodder.

34. *Chesneya elegans* Fomin **EN**

Endemic to northeastern Turkey, *Chesneya elegans* Fomin grows on eroded hillsides on shale, at elevations from 650 to 1650 m. The main threats are from continuing erosion, dam construction, agricultural activities, and microclimatic changes caused by the dam.

35. *Lathyrus karsianus* P. H. Davis **EN**

This is the typical endemic plant for Kars Province. Recorded by many botanists from different localities across the vilayet of Kars, *Lathyrus karsianus* P. H. Davis is known to be scarce but common in subpopulations. The species grows on meadow banks and at the edge of *Pinus sylvestris* L. forest, where the main threats are from overgrazing and harvesting for livestock fodder.

36. *Lathyrus woronowii* Bornm. **CR**

Endemic to Artvin Province, *Lathyrus woronowii* grows on schistose scree at elevations from 320 to 1000 m. Many botanists have collected this species from the Yusufeli area, particularly from the Çoruh River valley. Because of its specialized habitat requirements, this species is in greater danger of extinction than *Lathyrus karsianus* P. H. Davis. In the Artvin region where this species occurs, there are various human activities, such as several dam and road construction projects, agricultural activities, urbanization, and mining, that potentially threaten it.

37. *Vicia erzurumica* Demirkuş & Erik **CR**

This recently described species grows only in the vilayet of Erzurum, where it is found mostly on stony slopes in *Pinus sylvestris* L. forest at elevations between 1900 and 1950 m. The main threats to this species are from overharvesting for livestock fodder and from other agricultural activities.

GERANIACEAE

38. *Geranium davisianum* Peşmen & Güner **CR**

Known only from Rize Province along the eastern Black Sea coast, *Geranium davisianum* Peşmen & Güner grows in a very restricted area of evergreen forest and shrubs between 2000 and 2400 m in elevation. The species is under threat from grazing, habitat disturbance, and climate change.

IRIDACEAE

39. *Crocus aerius* Herb. **EN**

Endemic to northeastern Turkey, there are several stable populations of *Crocus aerius* Herb. in Trabzon, Rize, and Artvin provinces. The species grows on grassy hillsides and in montane Scots pine (*Pinus sylvestris* L.) forests. The main threats are overgrazing and global climate change.

40. *Iris taochia* Woronow ex Grossh. **EN**

This very handsome iris grows mostly in Erzurum Province and has been recorded from different localities of the province by many botanists. Populations of this species are severely fragmented in distribution but generally grow in open places and on somewhat dry rocky slopes, at elevations between 980 and 2000 m. Although the wild collection of this species, as well as the genus, is prohibited, according to the national flower bulb regulation of Turkey, *Iris taochia* Woronow ex Grossh. may still be collected for scientific purposes. This handsome, potentially ornamental plant is used locally in gardens but is not presently in the horticultural trade.

LAMIACEAE

41. *Lamium sulfureum* Hausskn. & Sint. ex R. R. Mill **EN**

Endemic to northeastern Anatolia, *Lamium sulfureum* Hausskn. & Sint. ex R. R. Mill has one locality that is found close to the southern frontier between the Caucasian and non-Caucasian portions of northeastern Anatolia. The species grows on mostly non-calcareous unstable scree at elevations from 2750 to 3100 m. Some populations are known from one of the national parks in Turkey. The species is under threat from overharvesting for livestock fodder, grazing, and collecting for medicinal purposes.

42. *Lamium tschorochense* A. P. Khokhr. **CR**

This is one of the very rare and local endemics known from northeastern Anatolia. Known from only a few localities in the Çoruh valley of Artvin Province, the species grows on stony slopes and under scrub. The main threat is a continuing decline in the extent and quality of habitat in the area due to dam construction.

43. *Stachys rizeensis* R. Bhattacharjee **EN**

Known only from Rize Province, this species grows on scree and slopes at elevations between 2700 and 3200 m. Most of the known localities for *Stachys rizeensis* are in one of the national parks. The main threats are from harvesting for livestock fodder, overgrazing, and ecotourism activities.

LILIACEAE

44. *Lilium ciliatum* P. H. Davis **EN**

This species is endemic to northeastern Anatolia, where there are several populations that are decreasing. *Lilium ciliatum* P. H. Davis grows along forest margins in clearings, scrub, and rich meadows above the tree line and on acid soils. It is under threat from human disturbances such as ecotourism, recreational activities, and road construction.

45. *Tulipa gumusanica* Terzioğlu **CR**

This is an extremely rare endemic species from Gümüşhane Province, where it is known from only one locality with a decreasing population. *Tulipa gumusanica* Terzioğlu grows on rocky slopes and scree, where the main threats to it are from overgrazing, harvesting for livestock fodder, ecotourism, clear-cutting, and road construction.

RANUNCULACEAE

46. *Delphinium munzianum* P. H. Davis & Kit Tan **CR**

Known only from Erzurum Province, this species grows on rocky limestone slopes at elevations between 2000 and 2200 m, where it is under threat from overgrazing and road construction.

ROSACEAE

47. *Crataegus turcica* Dönmez **CR**

This species is a very rare, local, and recently described woody plant that is endemic to northeastern Turkey. It is known only from the type locality in Artvin Province, and the current population is decreasing. *Crataegus turcica* Dönmez grows in the transition zone between forest and alpine meadow scree where the main threats are from cutting and the effects of insects and fungal pathogens.

48. *Potentilla savvalensis* Pawł. **VU**

Endemic to the northeastern corner of the Turkish Caucasian region, this species grows at high elevations from 2000 to 2700 m. It is a typical chasmophyte found in crevices of sunny sites, on vertical igneous rocks. The small population size of *Potentilla savvalensis* Pawł. is the only known potential threat to this species.

SALICACEAE

49. ***Salix rizeensis*** Güner & Ziel. **EN**

Endemic to northeastern Turkey, there are several stable populations of this species in Rize and Trabzon provinces, where it grows in open thickets, pastures, and along streamsides. The main threats are from overgrazing, ecotourism, stagnant water, avalanches, fishing, and other human activities.

SCROPHULARIACEAE

50. ***Verbascum varians*** Freyn & Sint. var. ***trapezunticum*** Murb. **CR**

Endemic to northeastern Anatolia, this variety grows on slopes and in fallow fields at low elevations, ca. 150 m. The population is decreasing very seriously as it is under threat from urbanization and hazelnut- and tobacco-farming.

TURKEY MAPS

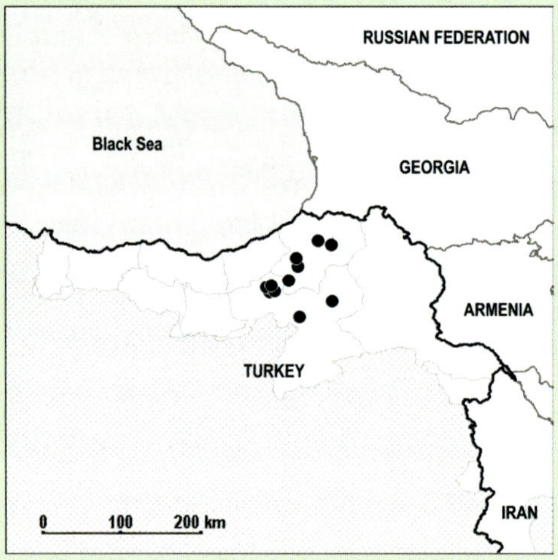

1. *Acer divergens* K. Koch & Pax var. *divergens*

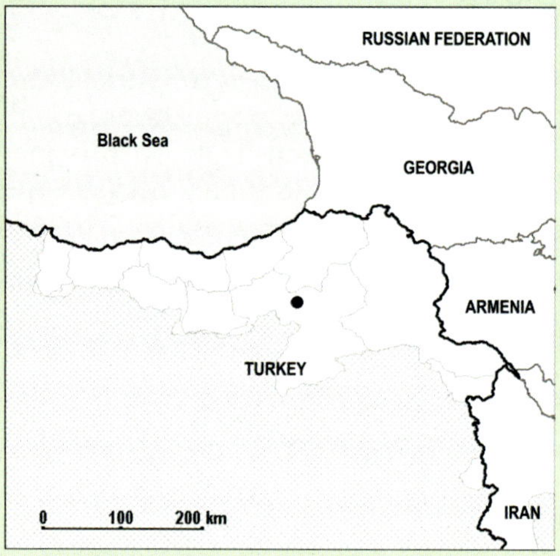

2. *Acer divergens* K. Koch & Pax var. *trilobum* Yalt.

3. *Allium baytopiorum* Kollmann & Özhatay

4. *Allium czelghauricum* Bordz.

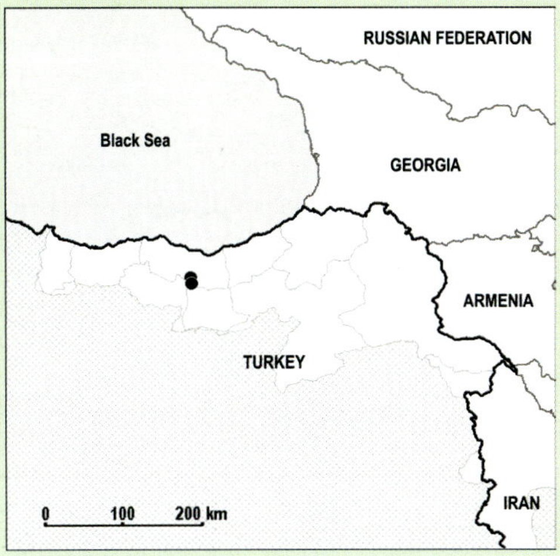

5. *Galanthus koenenianus* Lobin, C. D. Brickell & A. P. Davis

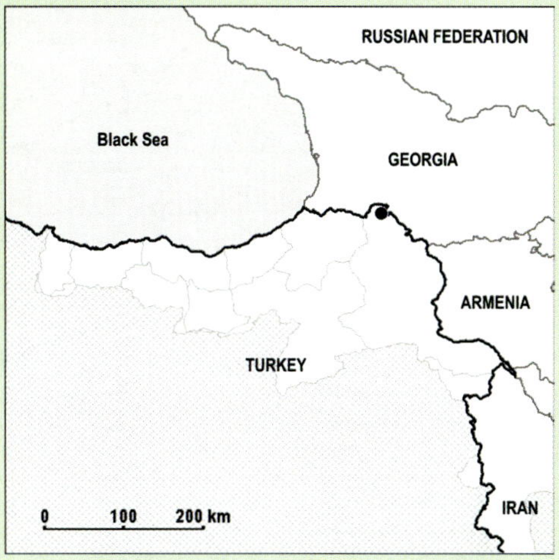

6. *Chaerophyllum posofianum* Erik & Demirkuş

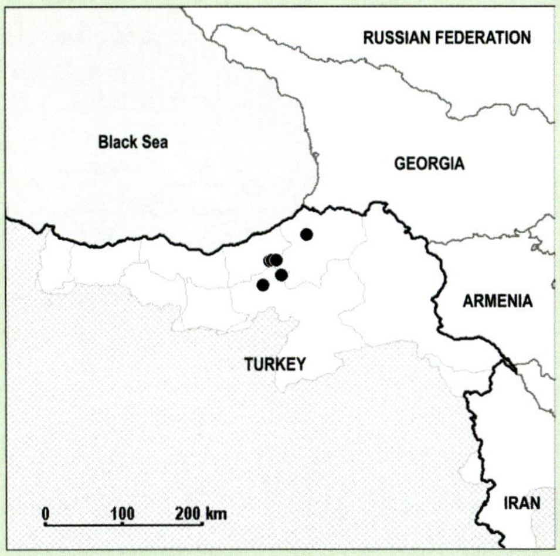

7. *Heracleum sphondylium* L. subsp. *artvinense* (Manden.) P. H. Davis

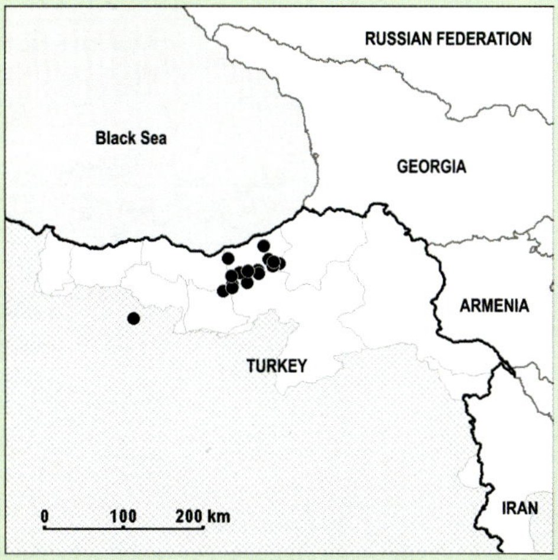

8. *Centaurea appendicigera* K. Koch

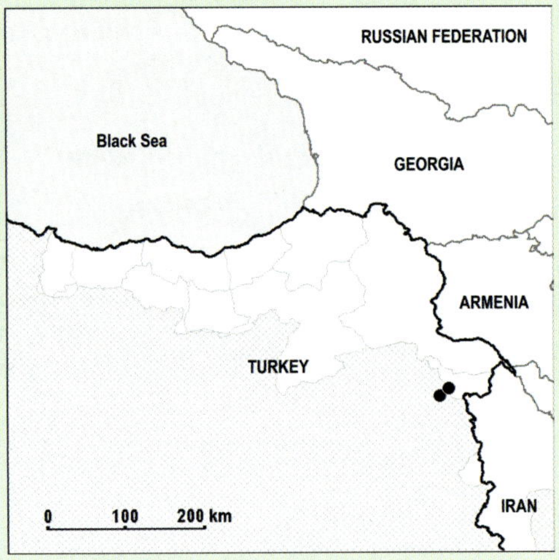

9. *Centaurea gracillima* Wagenitz

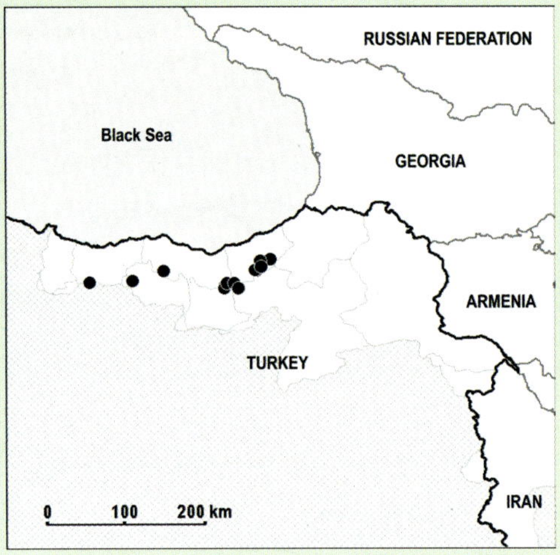

10. *Centaurea helenioides* Boiss. & Hausskn.

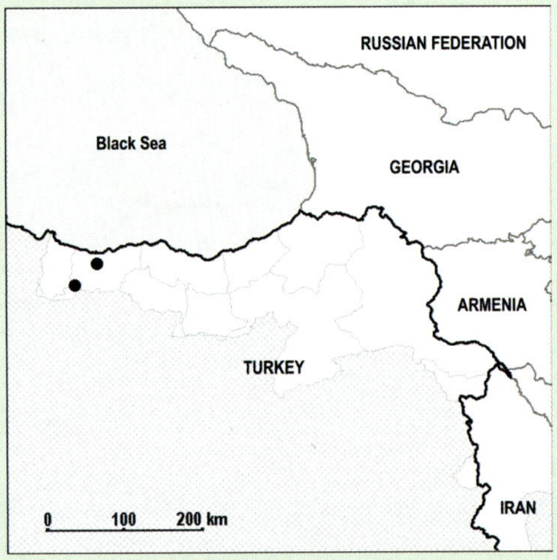

11. *Doronicum tobeyi* J. R. Edm.

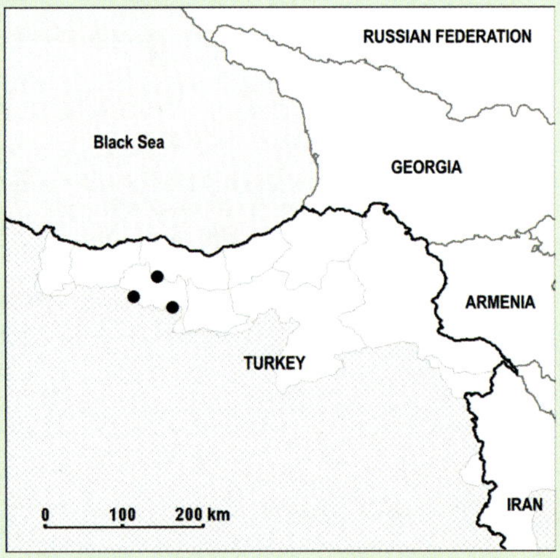

12. *Echinops sintenisii* Freyn

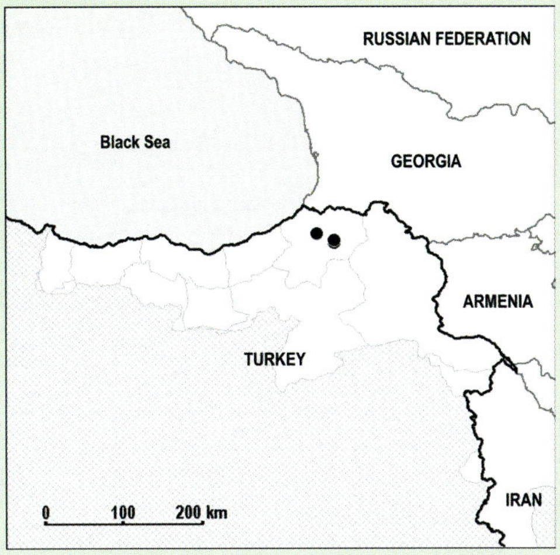

13. *Helichrysum artvinense* P. H. Davis & Kupicha

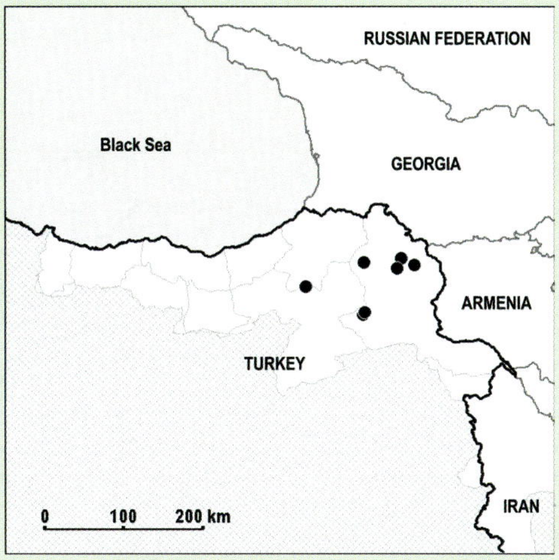

14. *Senecio integrifolius* (L.) Clairv. subsp. *karsianus* V. A. Matthews

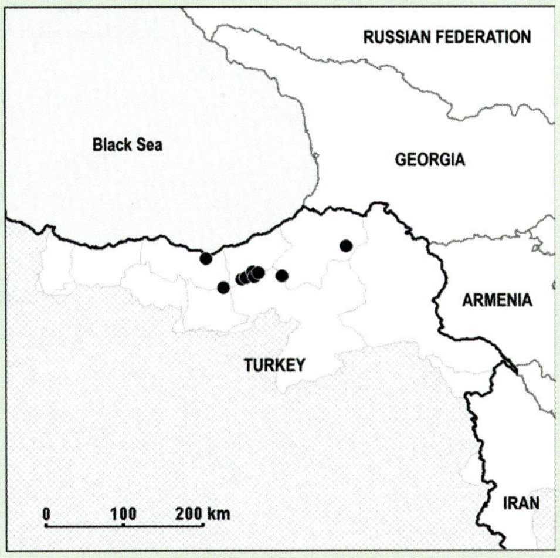

15. *Senecio ovatifolius* Boiss.

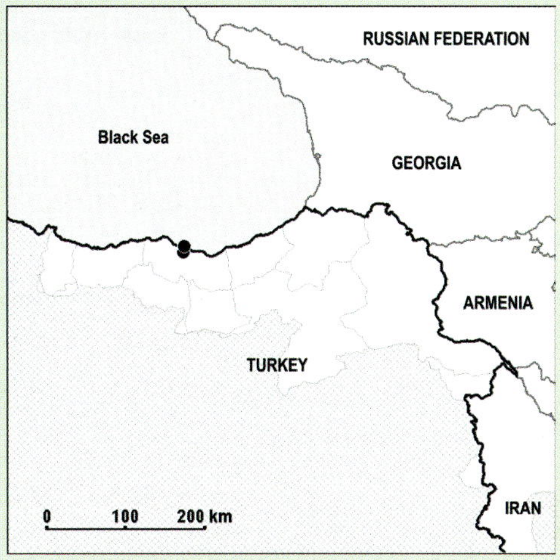

16. *Senecio trapezuntinus* Boiss.

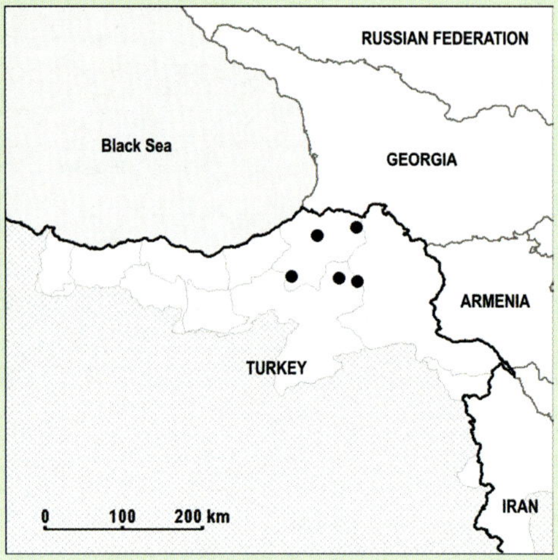

17. *Tripleurospermum fissurale* (Sosn.) E. Hossain

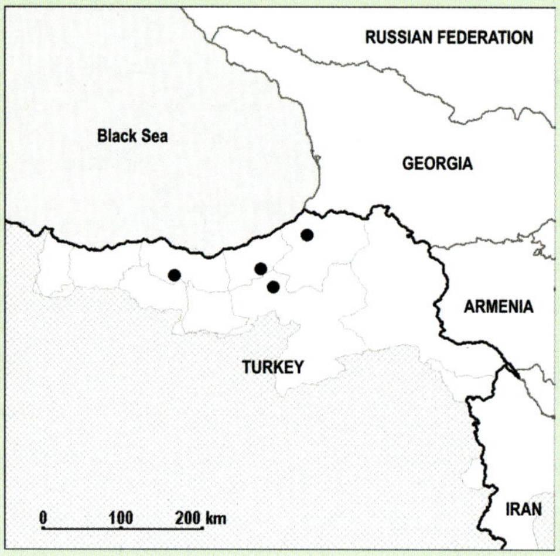

18. *Betula browicziana* Güner

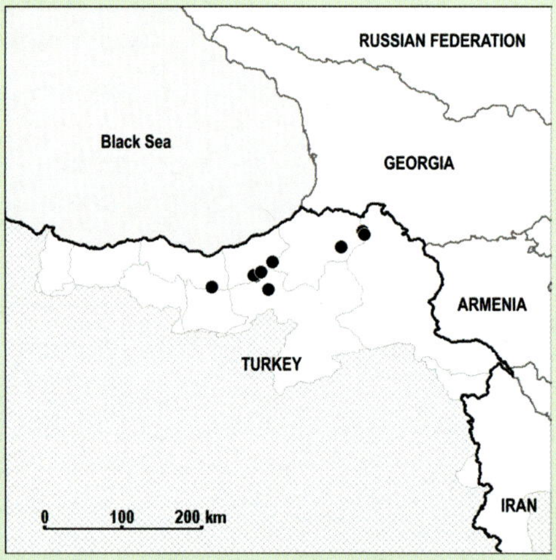

19. *Nonea pulmonarioides* Boiss. & Balansa

20. *Onosma nigricaulis* Riedl

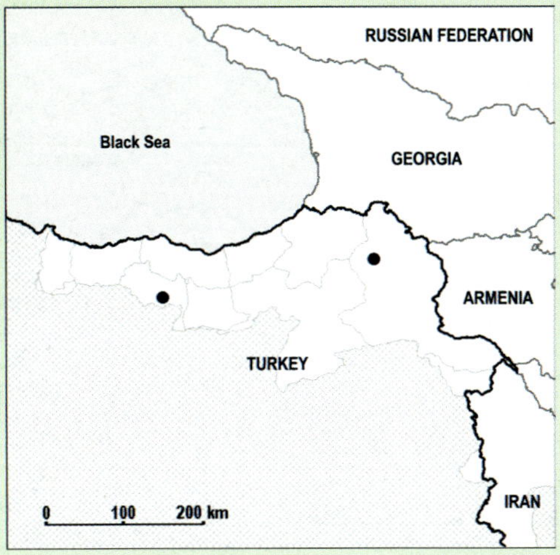

21. *Onosma obtusifolia* Hausskn. & Sint. ex Riedl

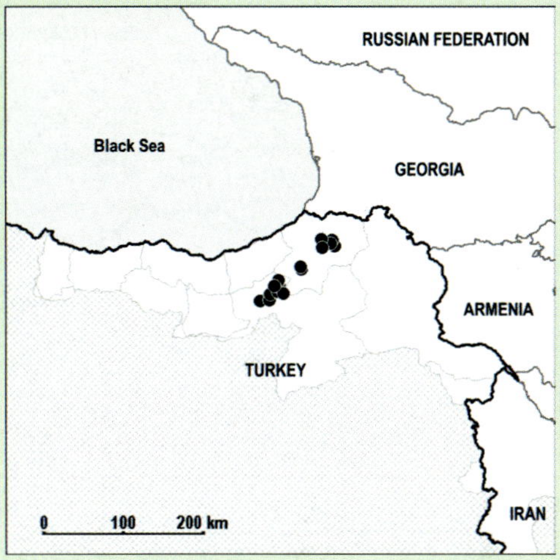

22. *Alyssum artwinense* N. Busch

23. *Draba narmanensis* Yıld.

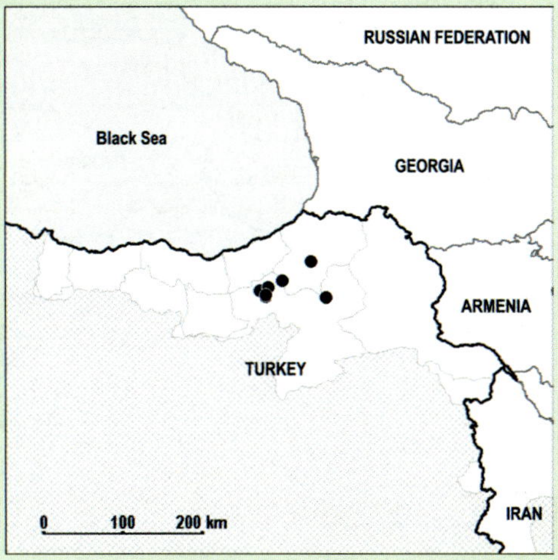

24. *Campanula choruhensis* Kit Tan & Sorger

25. *Campanula latiloba* A. DC. subsp. *rizeensis* Güner

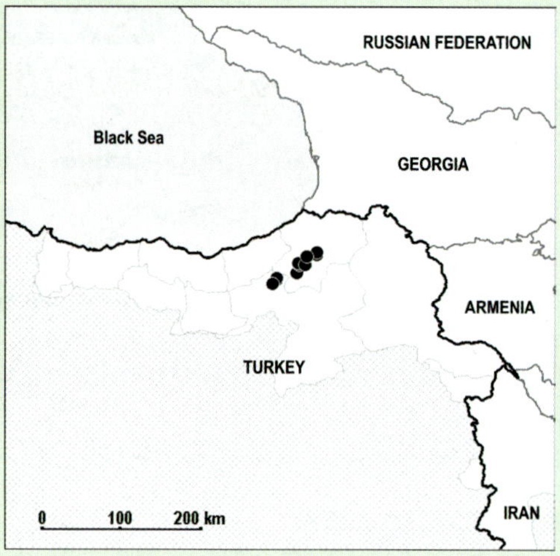

26. *Campanula troegerae* Damboldt

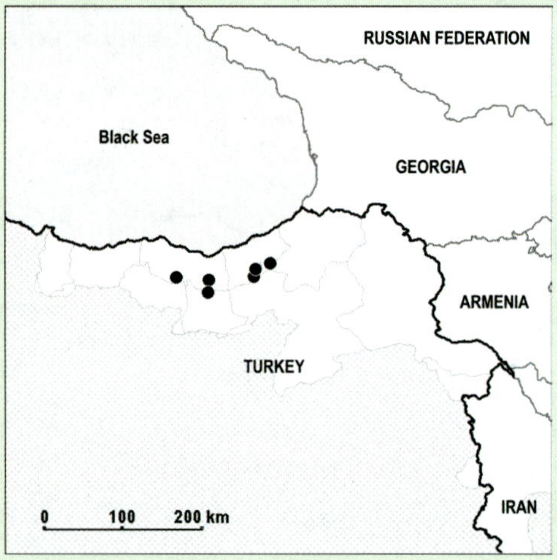

27. *Cerastium lazicum* Boiss.

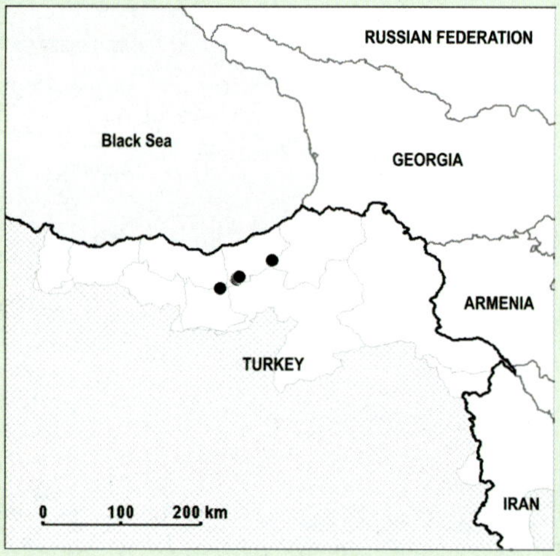

28. *Sempervivum furseorum* Muirhead

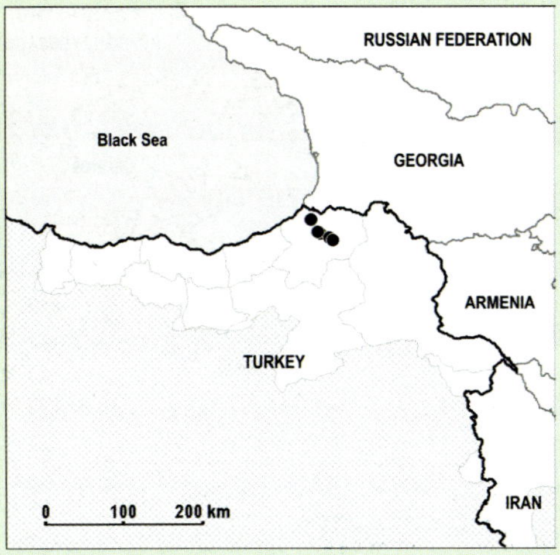

29. *Sempervivum glabrifolium* Boriss.

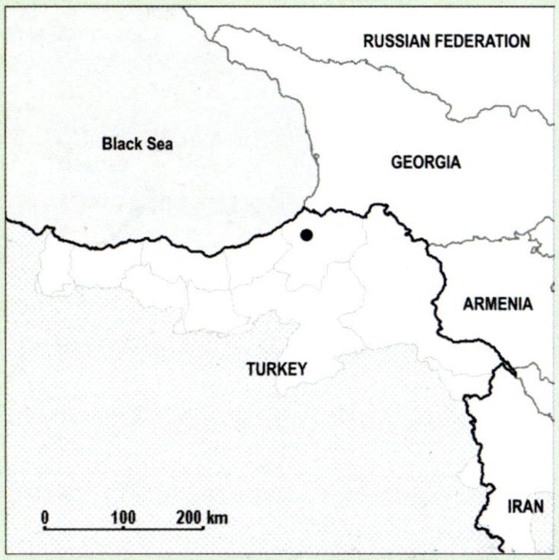

30. *Rhodothamnus sessilifolius* P. H. Davis

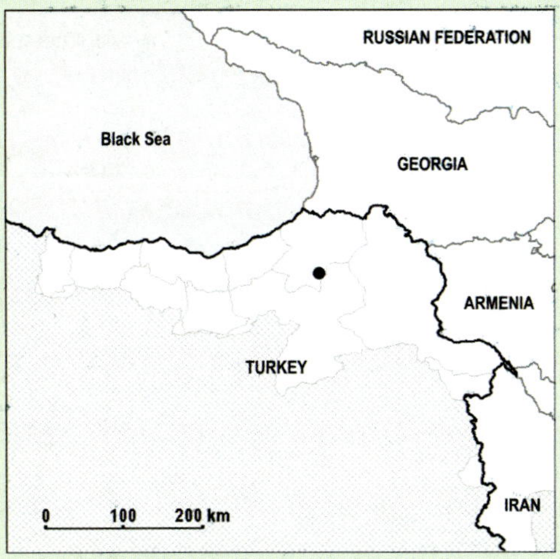

31. *Astragalus owirensis* Ali ex Podlech

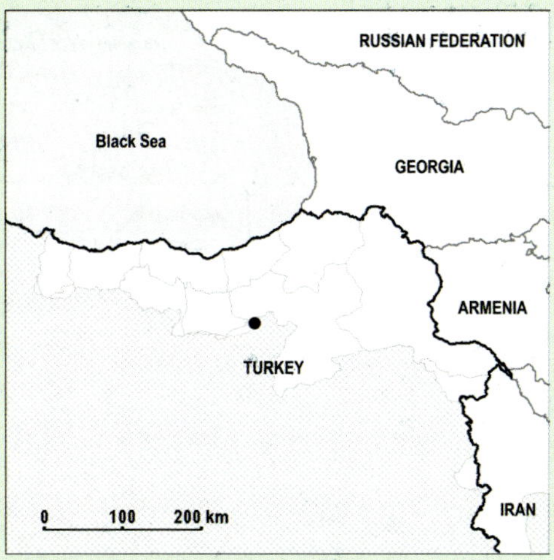

32. *Astragalus tatlii* Peşmen

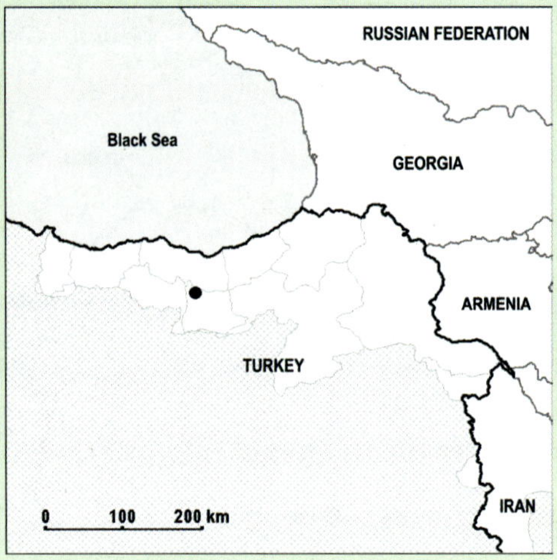

33. *Astragalus trabzonicus* Ekici, Aytaç & Akan

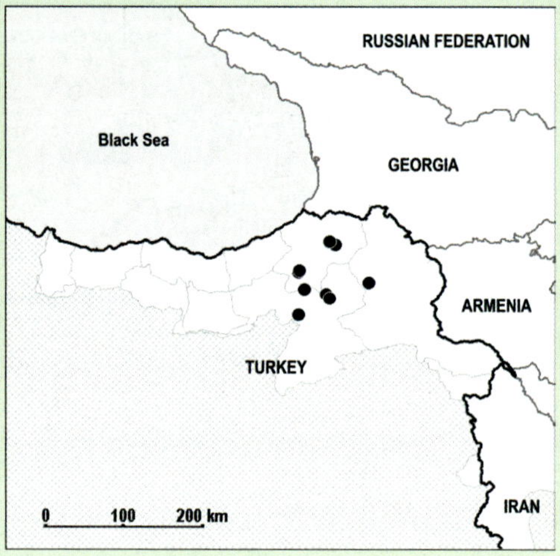

34. *Chesneya elegans* Fomin

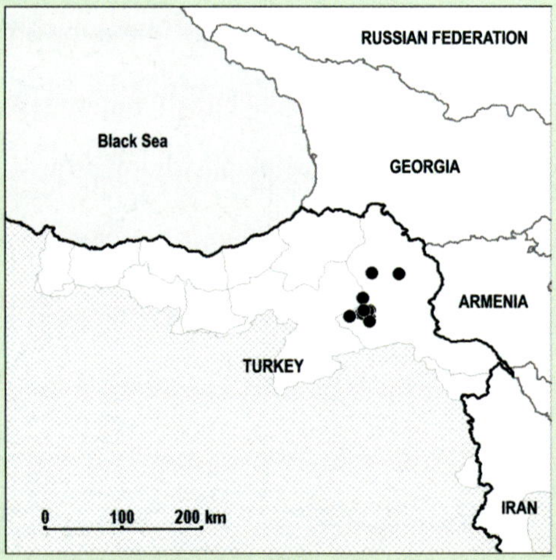

35. *Lathyrus karsianus* P. H. Davis

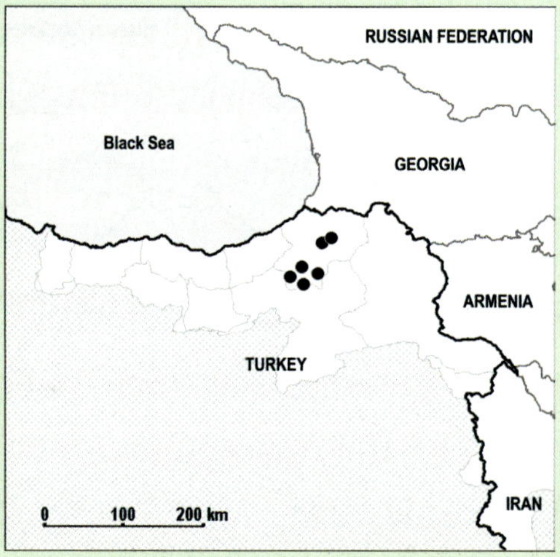

36. *Lathyrus woronowii* Bornm.

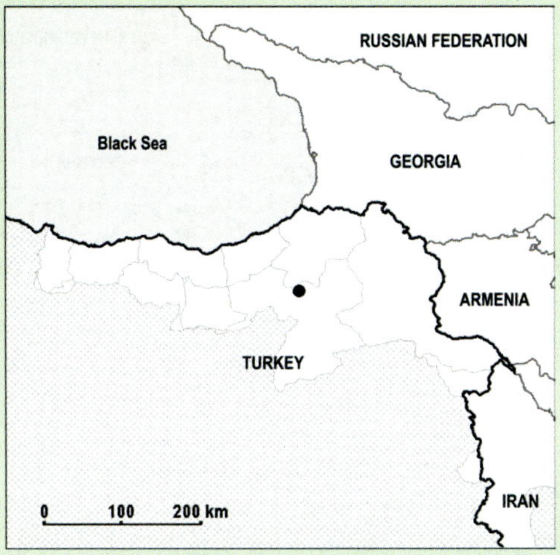

37. *Vicia erzurumica* Demirkuş & Erik

38. *Geranium davisianum* Peşmen & Güner

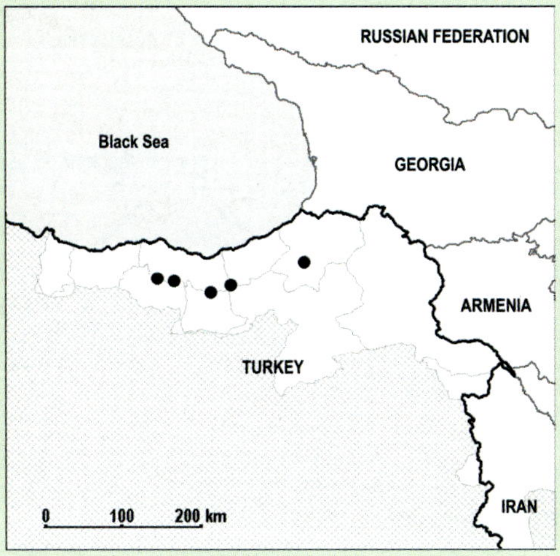

39. *Crocus aerius* Herb.

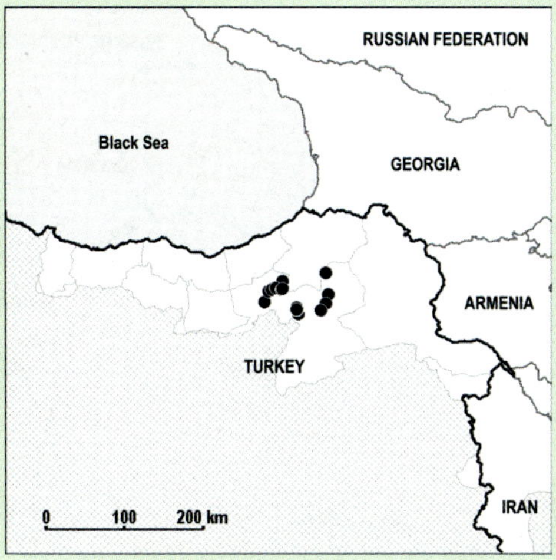

40. *Iris taochia* Woronow ex Grossh.

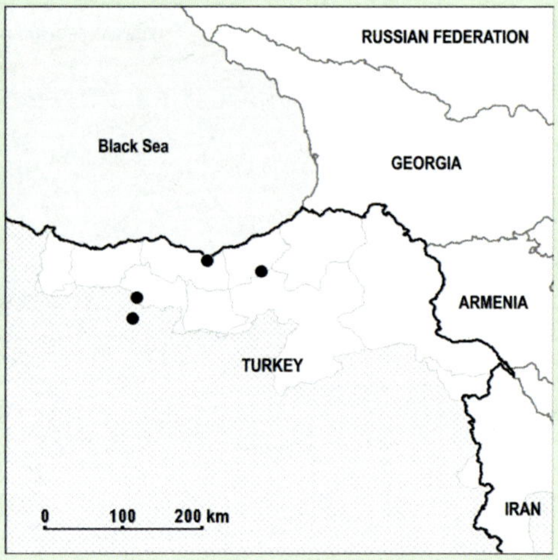

41. *Lamium sulfureum* Hausskn. & Sint. ex R. R. Mill

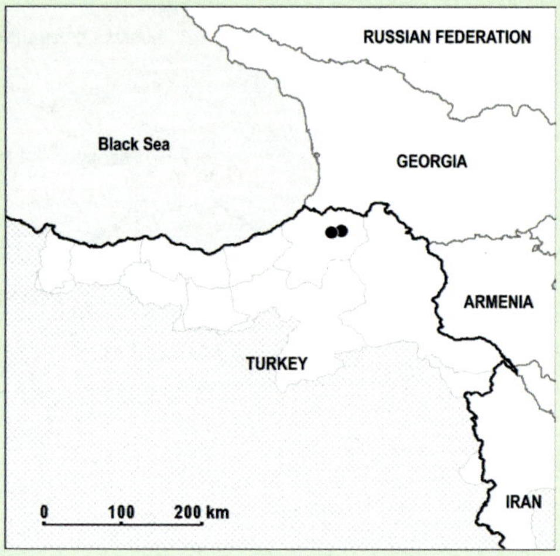

42. *Lamium tschorochense* A. P. Khokhr.

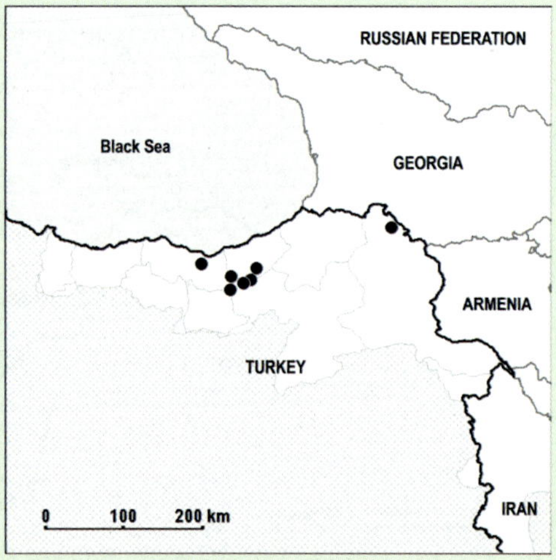

43. *Stachys rizeensis* R. Bhattacharjee

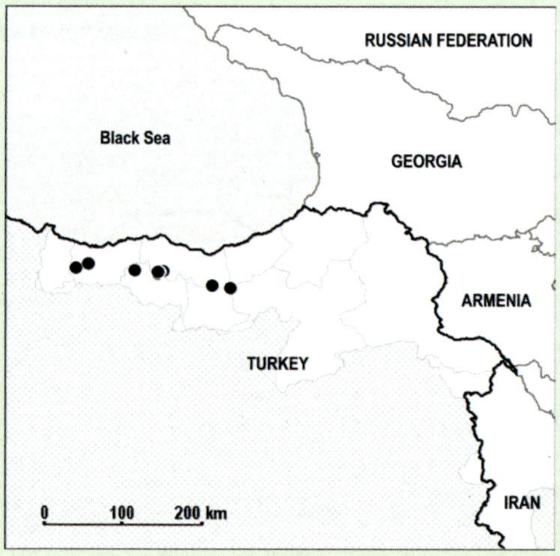

44. *Lilium ciliatum* P. H. Davis

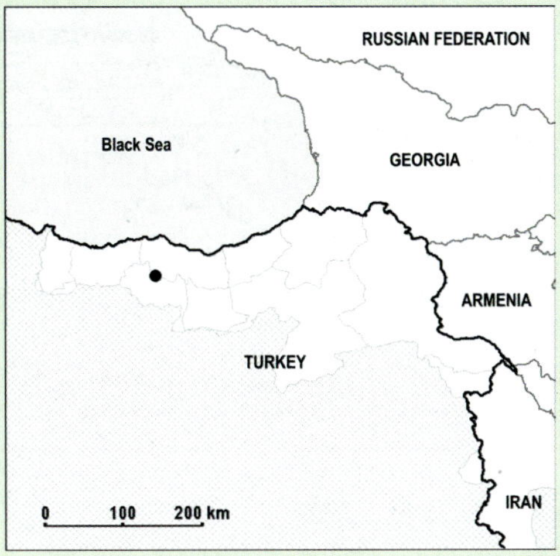

45. *Tulipa gumusanica* Terzioğlu

46. *Delphinium munzianum* P. H. Davis & Kit Tan

47. *Crataegus turcica* Dönmez

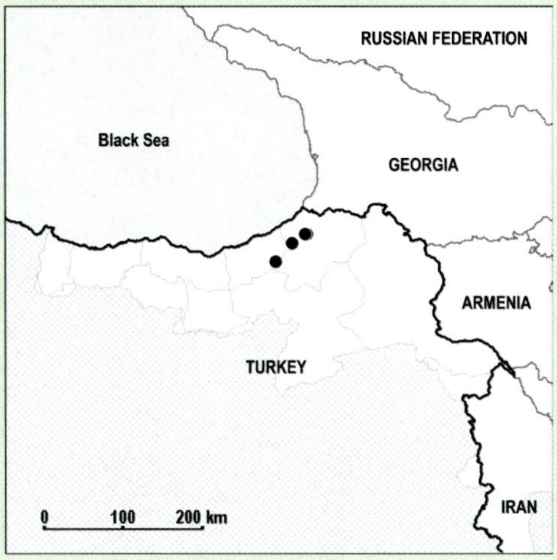

48. *Potentilla savvalensis* Pawł.

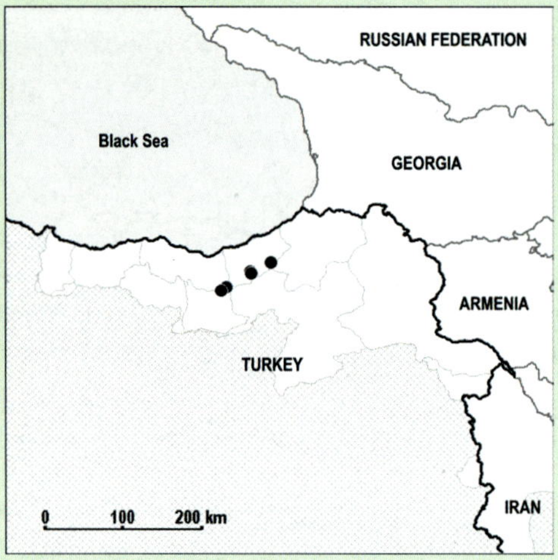

49. *Salix rizeensis* Güner & Ziel.

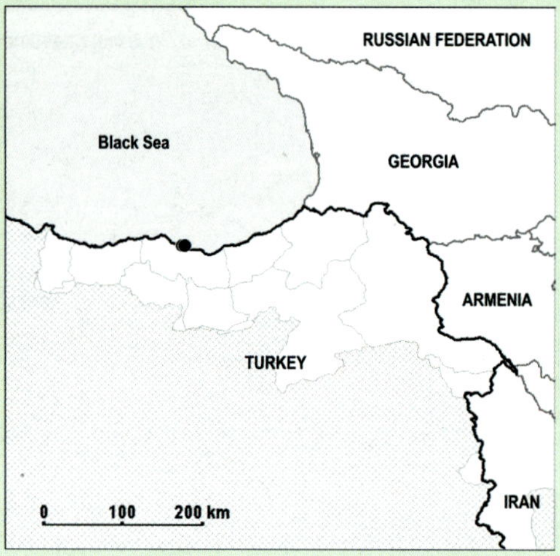

50. *Verbascum varians* Freyn & Sint. var. *trapezunticum* Murb.

BIBLIOGRAPHY

Anşin, R. 1980. Doğu Karadeniz Bölgesi Florası ve Asal Vejetasyon Tiplerinin Floristik İçerikleri. [Flora of the East Black Sea Region and the Floristic Composition of Its Main Vegetation Types.] Doçentlik Tezi, KTÜ Orman Fakültesi, Trabzon [in Turkish].

Anşin, R. 1981. Doğu Karadeniz Bölgesi Sahil ve İç Kesimlerinde Yayılan Ana Vejetasyon Tipleri. [The Main Vegetation Types of the Coast and the Inner Parts of Black Sea Region.] KTÜ Orman Fakültesi Dergisi 4(4): 239–252 [in Turkish].

Anşin, R. 1982. Endemizm ve Doğu Karadeniz Bölgesi'nde Yetişen Endemik Bitki Taksonları. [Endemism and Endemic Plants of the East Black Sea Region.] KTÜ Orman Fakültesi Dergisi 5(2): 311–326 [in Turkish].

Anşin, R. 1983. Türkiye'nin Flora Bölgeleri ve Bu Bölgelerde Yayılan Asal Vejetasyon Tipleri. [The Floristic Regions and the Main Vegetation Types of Turkey.] KTÜ Orman Fakültesi Dergisi 6(2): 318–339 [in Turkish].

Davis, P. H. 1965–1988. Flora of Turkey and the East Aegean Islands, Vols. 1–10. The University Press, Edinburgh.

Davis, P. H. 1971. Distribution patterns in Anatolia with particular reference to endemism. Pp. 15–27 *in* P. H. Davis, P. C. Harper & J. C. Hedge (editors), Plant Life of South-West Asia. The Botanical Society of Edinburgh, Edinburgh.

Ekim, T., M. Koyuncu, M. Vural, H. Duman, Z. Aytaç & N. Adıgüzel. 2000. Türkiye Bitkileri Kırmızı Kitabı. [The Red Book of Turkey's Plants.] TTKD and Van Yüzüncü Yıl University Publications, Ankara [in Turkish].

Eminağaoğlu, Ö. & R. Anşin. 2003. The flora of Hatila Valley National Park and its close environs (Artvin). Turkish J. Bot. 27: 1–27.

Eminağaoğlu, Ö. & R. Anşin. 2004. Flora of the Karagöl-Sahara National Park (Artvin) and its environs. Turkish J. Bot. 28: 557–590.

Eminağaoğlu, Ö., H. G. Kutbay, A. Bilgin & E. Yalçın. 2006. Contribution to the phytosociology and conservation of tertiary relict species in northeastern Anatolia (Turkey). Belg. J. Bot. 139(1): 124–130.

Eminağaoğlu, Ö., R. Anşin & H. G. Kutbay. 2007. Forest vegetation of Karagöl-Sahara National Park (Artvin-Turkey). Turkish J. Bot. 31: 421–449.

Eminağaoğlu, Ö., H. G. Kutbay, Z. C. Özkan & A. Ergül. 2008. Flora of the Camili Biosphere Reserve Area (Borçka, Artvin, Turkey). Turkish J. Bot. 32: 43–90.

Güner, A., M. Vural & K. Sorkun. 1987. Rize Florası, Vejetasyonu ve Yöre Ballarının Polen Analizi. [Flora and Vegetation of Rize District and Pollen Analysis of its Honeys.] TÜBİTAK-TBAG, Proje No: TBAG–650 [in Turkish].

Palabaş, S. 2002. Altındere Vadisi (Maçka-Trabzon) Subalpin ve Alpin Florası. Yüksek Lisans Tezi. [Alpine and Subalpine Flora of Altındere Valley (Maçka-Trabzon). Master's Thesis.] KTÜ Fen Bilimleri Enstitüsü, Trabzon [in Turkish].

Radde, G. 1899. Grundzüge der Pflanzenverbreitung in dem Kaukasuslandern. Wilhelm Engelmann, Leipzig [in German].

Ramsar Convention on Wetlands (RCW). 1971 [2 Feb.]. Convention on Wetlands of International Importance especially as Waterfowl Habitat. Ramsar (Iran). UN Treaty Series No. 14583. As amended by the Paris Protocol, 3 December 1982, and Regina Amendments, 28 May 1987. <http:// www.ramsar.org/cda/en/ramsar-about-about-ramsar/main/ramsar/1-36%5E7687_4000_0>, accessed 30 September 2013.

Terzioğlu, S. 1998. Uzungöl (Trabzon-Çaykara) ve Çevresinin Flora ve Vejetasyonu. Doktora Tezi. [Flora and Vegetation of Uzungol (Trabzon-Çaykara) and Its Environs. Ph.D. Thesis.] KTÜ Fen Bilimleri Enstitüsü, Trabzon [in Turkish].
Terzioğlu, S. 1999. Trabzon İlinin Endemik Bitkileri. [Endemic Plants of Trabzon District.] T. C. Trabzon Valiliği İl Çevre Müdürlüğü, Çevre Durum Raporu [in Turkish].
Uslu, S. 1962. İç Anadolu Step Problemi (H. Walter'den Çeviri). [Problem of the Steppe of Inner Anatolia (Translated from H. Walter.)] İÜ Yayın No: 943, Orman Fakültesi yayın No: 79, Ankara Matbaası, İstanbul [in Turkish].
Uzun, A. 2002. Altındere Vadisi (Maçka-Trabzon) Orman Vejetasyonu Florası. Yüksek Lisans Tezi. [Flora of Forest Vegetation of Altındere (Maçka-Trabzon) Valley. Master's Thesis.] KTÜ Fen Bilimleri Enstitüsü, Trabzon [in Turkish].
Zohary, M. 1973. Geobotanical Foundations of the Middle East, Vols. I–II, Gustav Fischer Verlag, Stuttgart.

Plate 1. The natural landscape of the Caucasus is remarkably luxuriant and rich, ranging from high mountains with peaks over 4000 m to semi-deserts. —A. The Greater Caucasus Range. —B. Kazbeg peak in Georgia. —C. Semi-desert. —D. Relict forests with Persian ironwood. —E. Evergreen forest in the southwestern Caucasus. —F. Subalpine and alpine meadows. (A, B, E photos by Otar Abdaladze; C photo by Vugar Kerimov; D photo by Jalil Noroozi; F photo by Kamilla Tamanian.)

Plate 2. —A. Aragats Mountain is the highest elevation in Armenia (4090 m), home to rare species such as *Pseudovesicaria digitata* (C. A. Mey.) Rupr. and *Cerastium pseudokasbek* Vysok. —B. Beech forest in northern Armenia, in a region with adequate precipitation. (A photo by George Fayvush; B photo by Kamilla Tamanian.)

Plate 3. —A. Sheki region of Azerbaijan. Steppe meadows in the alpine zone, home to many endemic plants. —B. The Greater Caucasus Range in Azerbaijan supports alpine vegetation with cushion-like plants and those that grow on cliffs (Lezhe area). (Photos by Vugar Kerimov.)

Plate 4. —A. The immense Sarykum Sand Dune rises above the surrounding steppes of Dagestansky National Reserve, Russian Federation. Its vegetation consists of unusual plant species, such as *Eremosparton aphyllum* Fisch. & C. A. Mey., and many endemics. —B. Chokh Village in the Gunib region, the most botanically rich region in Dagestan. (Photos by Ramazan Murtazaliev.)

Plate 5. —A. The medieval fortress of Hambert in Armenia, surrounded by meadows and steppe-meadows with *Campanula armena* Steven growing on cliffs. —B. Mount Eranos, part of the Eranos Range, in Armenia, characterized by phrygana vegetation. (Photos by Kamilla Tamanian.)

Plate 6. —A. Sevan Lake, at 1900 m in elevation, is the largest freshwater lake in Armenia. This high-elevation lake is surrounded by meadows, with fragments of forest now found only along the Gyuneisk shoreline. —B. Red cliffs near Naravank, Armenia. The soil, with a characteristic reddish tint from iron, supports a number of endemic plants. (Photos by George Fayvush.)

Plate 7. —A. The Caucasian State Nature Biosphere Reserve (Russian Federation), situated in the northwestern part of the Caucasus, is a World Heritage Site, with different types of forest. —B. Novalishinsky canyon (Russian Federation). Rivers run through gorges in the western Caucasus that are often composed of limestone or dolomite, offering spectacular scenic views. (Photos by Svetlana Litvinskaya.)

Plate 8. —A. The Teberda Nature Reserve is situated in the central Caucasus region in the Russian Karachay-Cherkess territory. This is an area with high mountains, over 3000 m, and steep slopes that are populated by many rare and endemic plants. —B. Fantastic dolomite cliffs in the Gerpegem Mountain Range in the northern Caucasus region (Russian Federation). (Photos by Svetlana Litvinskaya.)

Plate 9. —A. In the central Caucasus, the Nakhchivan region is very rich in plant endemics based on the great habitat diversity, ranging from mountain meadows to dry phrygana vegetation. —B. Significant parts of Azerbaijan are covered by forest in which beech, *Fagus orientalis* Lipsky, and species of oak, *Quercus* L., predominate. This photo is of the Ismayilly area in central Azerbaijan. (Photos by Vugar Kerimov.)

Plate 10. —A. Forests in Trabzon, Turkey, consist of many conifers, with the columnar spires of Caucasian spruce, *Picea orientalis* (L.) Peterm., seen here. —B. Slope in the northeastern Iranian Caucasus covered by the bright red poppy *Papaver lasiothrix* Fedde, a species endemic to Turkey and Iran. (A photo by Salih Terzioğlu; B photo by Jalil Noroozi.)

Plate 11. —A. *Rhododendron caucasicum* Pall. is a common shrub in the alpine zone of the Caucasus region. —B. Spruce woodland, *Picea orientalis* (L.) Peterm., in the western Caucasus, Turkey. —C. *Pyrus zangezura* Maleev occurs in the southern Caucasus. —D. The Eldar pine, *Pinus eldarica* Medw., occurs in the steppe plateau, Azerbaijan. —E. *Rhodothamnus sessilifolius* P. H. Davis (Ericaceae), endemic to Turkey, where it is critically endangered. —F. Also endangered, *Sambucus tigranii* Troitzky (Caprifoliaceae) from the southern Caucasus, a species endemic to Armenia and Georgia. (A, C photo by George Fayvush; B photo by Salih Terzioğlu; D photo by Vahid Farzaliev; E photo by Özgür Eminagaoglu; F photo by Kamilla Tamanian.)

Plate 12. —A. The autumn-flowering *Crocus vallicola* Herb. occurs in the western and central Caucasus. —B. *Iris paradoxa* Steven, endemic to the southeastern Caucasus. —C. *Iris iberica* Hoffm., endemic to southern Caucasus. —D. *Iris medwedewii* Fomin, from the southern and eastern Caucasus. —E. Only recently described in 2004, *Gladiolus murgusicus* Mikheev occurs in the northeastern part of Armenia. —F. *Iris camillae* Grossh., a vulnerable and endemic iris of Azerbaijan. (A, B, D, E photos by Anatoly Mikheev; C photo by Otar Abdaladze; F photo by Vugar Kerimov.)

Plate 13. —A. *Galanthus platyphyllus* Traub & Moldenke, from subalpine and alpine zones in the western Greater Caucasus. —B. *Tulipa eichleri* Regel occurs in the eastern Caucasus, endemic to the southern Caucasus. —C. *Iris lazica* Albov, endemic to the southwestern Caucasus. —D, E. *Iris winogradowii* Fomin was once considered extinct in the wild, but after its reintroduction in western Georgia, the status of this endemic iris has now improved to vulnerable. —F. *Aethionema edentulum* N. Busch, a semishrubby endemic in the Brassicaceae, growing on cliffs and in rocky soil. (A–E photos by Otar Abdaladze; F photo by Lamara Asieshvili.)

Plate 14. —A. *Lilium ledebourii* (Baker) Boiss., one of the floral gems of the genus, endemic to southeastern Caucasus. —B. *Fritillaria orientalis* Adams, from alpine meadows and endemic to the central Caucasus. —C. *Lilium monadelphum* M. Bieb. subsp. *georgicum* (Manden.) Gagnidze, an endemic to Georgia, occurs in meadows. —D. The dark-flowered *Fritillaria caucasica* Adams is found in the Caucasus and northeastern Turkey, in open forests and on dry slopes and scrub. —E. *Erythronium caucasicum* Woronow, one of the earliest flowering species in the western Caucasus and northern Iran, found in lower montane forests and tall grass communities. —F. *Lilium ciliatum* P. H. Davis, a rare and endangered endemic from Turkey. (A photo by Jalil Noroozi; B, D photos by Anatoly Mikheev; C photo by Lamara Asieshvili; E photo by Svetlana Litvinskaya; F photo by Nursel Ikinci.)

Plate 15. —A. *Aquilegia caucasica* (Ledeb.) Rupr., an endemic that occurs in subalpine meadows and montane forests in the Caucasus region. —B. The white-flowered *Paeonia steveniana* Kem.-Nath., an endangered endemic in Georgia. —C. The endemic *Cerastium kasbek* Parrot grows on rocky cliffs in the central Caucasus. —D. Pink-flowered *Paeonia caucasica* (Schipcz.) Schipcz. is widely distributed in the Caucasus. —E. *Delphinium caucasicum* C. A. Mey. occurs in the subnival zone in the Greater Caucasus. —F. The golden peony, *Paeonia mlokosewitschii* Lomakin, is a vulnerable Caucasian endemic from Azerbaijan, Georgia, and Dagestan. B, D, F. Note the distinctly different flower colors among the peony species. (A, B, D–F photos by Otar Abdaladze; C photo by Zaliko Kikodze.)

Plate 16. —A. The willow, *Salix rizeensis* Güner & Ziel., is an endangered endemic in northeastern Turkey where it grows along streams. —B. *Calligonum bakuense* Litv., an endangered endemic from Azerbaijan. —C. *Lathyrus woronowii* Bornm. is a critically endangered endemic from the Artvin region, northeastern Turkey and Georgia, bordering the Black Sea. —D. *Astragalus igniarius* Popov, another endangered endemic from Azerbaijan. Both species in photos B and D are protected in the Absheron National Park, Azerbaijan. —E. *Helleborus abchasicus* A. Braun, with variable flower color, is endemic to the western Caucasus, from Georgia and adjacent Russia. —F. *Gymnospermium smirnovii* (Trautv.) Takht. is a very rare, endangered endemic species in the Berberidaceae from Georgia. (A photo by Salih Terzioğlu; B, D photos by Vugar Kerimov; C photo by Özgür Eminagaoglu; E photo by Otar Abdaladze; F photo by Lamara Asieshvili.)

Plate 17. —A. *Galanthus artjuschenkoae* Gabrieljan occurs in the southern Caucasus, endemic to Armenia and Azerbaijan. —B. *Papaver bracteatum* Lindl., a highly ornamental poppy, is endemic to subalpine zones in the Alborz mountains. —C. *Dianthus seidlitzii* Boiss. is a critically endangered endemic in Iran and among the country's Top 50 listing for conservation. —D. *Ranunculus dissectus* M. Bieb. subsp. *aragazi* (Grossh.) Bulany & Derv.-Sok., from Armenia, blooming very early in spring. —E. *Astragalus daghestanicus* Grossh. is critically endangered and endemic to Dagestan. —F. *Euphorbia sahendi* Bornm. is also critically endangered and endemic to Iran. (A photo by David Kikodze; B, C, F photos by Jalil Noroozi; D photo by Kamilla Tamanian; E photo by Ramazan Murtazaliev.)

Plate 18. —A. *Scorzonera ketzkhowelii* Sosn. ex Grossh., a critically endangered endemic in Georgia, but potentially horticultural. —B. *Psephellus barbeyi* Albov, a promising ornamental plant, is a vulnerable endemic in the low mountains of the western Caucasus. —C. *Grossheimia macrocephala* (Muss.-Puschk. ex Willd.) Sosn. & Takht., an endemic species within tall herb communities across the Caucasus. —D. *Campanula sarmatica* Ker Gawl., endemic to the Caucasus region, with ornamental potential. —E. *Muehlbergella oweriniana* (Rupr.) Feer, a generic endemic in Dagestan and critically endangered. —F. *Campanula pendula* M. Bieb. occurs on cliffs in the middle mountain zone. (A, C, D photos by Otar Abdaladze; B photo by Lamara Asieshvili; E photo by Ramazan Murtazaliev; F photo by Svetlana Litvinskaya.)

Plate 19. —A. *Cerastium argenteum* M. Bieb., a Caucasian endemic found on rocky soil in Georgia. —B. *Draba ossetica* (Rupr.) Sommier & Levier, a vulnerable endemic from the western and central Greater Caucasus. —C. *Paeonia lagodechiana* Kem.-Nath., a rare peony from Georgia and a Caucasian endemic. —D. *Hibiscus ponticus* Rupr., restricted to a small area in Georgia and critically endangered. —E. Wittmann's peony, *Paeonia wittmanniana* Hartwiss. ex Lindl., with large flowers, is native to the western Caucasus. —F. *Galanthus woronowii* Losinsk., a Caucasian endemic distributed in Georgia, Russia, and Turkey, is threatened from over-collection. (A, B, D–F photos by Otar Abdaladze; C photo by Shalva Sikharulidze.)

Plate 20. —A. *Dianthus imereticus* (Rupr.) Schischk., with pink flowers ca. 1 cm across, is endemic to the southwestern Caucasus. —B. *Muscari alpanicum* Schchian grows in forests in the western Caucasus, a vulnerable endemic from Georgia. —C. *Iris caucasica* Steven, also placed in the genus *Juno* Tratt., is an endemic iris distributed across the southern Caucasus. —D. *Paeonia macrophylla* (Albov) Lomakin, with the largest flowers in the genus, is a vulnerable Caucasian endemic. —E. Recognized by its many-flowered inflorescence, *Ornithogalum magnum* Krasch. & Schischk. is a vulnerable endemic from Georgia and adjacent Russia. —F. *Iris carthaliniae* Fomin, with either light blue or white flowers, is endemic across the Caucasus region. (A, B, D–F photos by Lamara Asieshvili; C photo by Shalva Sikharulidze.)

Plate 21. —A. *Sempervivum sosnowskyi* Ter-Chatsch., endemic to the central Caucasus and considered vulnerable in Georgia and Turkey, is a succulent plant with rosettes 12–15 cm in diameter. —B. *Anthemis saguramica* Sosn. grows in the central Caucasus where it is found in Georgian forests. —C. *Iris timofejewii* Woronow is an endangered endemic, restricted to dry montane slopes in the northeastern Caucasus of Dagestan. —D. The endemic *Primula ruprechtii* Kusn. ex Lipsky is distributed across the central and southern Caucasus. —E. Used medicinally, *Dioscorea caucasica* Lipsky is an endangered endemic from the western Caucasus of Georgia and Russia. —F. The endemic *Anthemis marschalliana* Willd. grows in the alpine zone across the northern Caucasus. (A, B, F photos by Lamara Asieshvili; C photo by Anatoly Mikheev; D, E photos by Shalva Sikharulidze.)

Plate 22. —A. *Minuartia biebersteinii* (Rupr.) Schischk. grows on cliffs and rocky soil, endemic to Georgia and Russia. —B. Introduced into horticulture from Georgia, *Primula juliae* Kusn. is an endangered endemic recorded from Azerbaijan, Georgia, and Russia. —C. A promising ornamental plant, *Salvia garedji* Troitzky is a vulnerable endemic, known from Georgia and Azerbaijan. —D. Known in horticulture, *Scabiosa olgae* Albov originated from the western Caucasus where it is a vulnerable endemic from Georgia and Russia. —E. Recognized by its succulent rosettes, *Sempervivum caucasicum* Rupr. ex Boiss. is a Caucasian endemic. —F. *Campanula trautvetteri* Grossh. ex Fed., with its remarkable deep blue flowers, is a Caucasian endemic from Azerbaijan and Georgia. (A, C photos by Shalva Sikharulidze; B, D–F photos by Lamara Asieshvili.)

Plate 23. —A. The Caucasian endemic *Convallaria transcaucasica* Utkin ex Grossh. has narrow acute leaves. —B. The Persian ironwood, *Parrotia persica* (DC.) C. A. Mey., is a relictual Hyrcanian species in the witchhazel family, seen here in flower. —C. *Scilla rosenii* K. Koch is endemic to the alpine zone of the western Caucasus. —D. *Papaver lateritium* K. Koch subsp. *lateritium* is a vulnerable endemic from Turkey. —E. *Thymus pulchellus* C. A. Mey. occurs in the northwestern Caucasus. —F. The fruits of *Sorbus caucasigena* Kom. ex Gatsch. are used in drinks. The tree is a Caucasian endemic known from Azerbaijan and Georgia. (A, C photos by Lamara Asieshvili; B photo by Vahid Farzaliev; D photo by Salih Terzioğlu; E photo by Svetlana Litvinskaya; F photo by Otar Abdaladze.)

Plate 24. —A. *Campanula ochroleuca* Kem.-Nath. is endemic to the Greater Caucasus Range in the subalpine zone. —B. An endangered endemic in the Caucasus, *Campanula armazica* Charadze is recorded from only two localities in Georgia. —C. Found in Caucasian forests, *Aristolochia steupii* Woronow is a vulnerable endemic in Georgia and Russia. —D. Endemic to the Caucasus, *Anemone speciosa* Adams ex Pritz. is an early spring-flowering herb. —E. *Campanula kemulariae* Fomin, a vulnerable endemic with hanging stems, can be found on cliffs in Georgia. —F. *Leptopus colchicus* (Fisch. & C. A. Mey. ex Boiss.) Pojark. is the only representative in the Caucasus region of this Himalayan genus. (A, B, D, E photos by Lamara Asieshvili; C photo by Svetlana Litvinskaya; F photo by Shalva Sikharulidze.)

Plate 25. —A. Meryemana Village seen behind evergreen forest in Trabzon, northeastern Turkey, along the Black Sea coast. —B. *Gadellia lactiflora* (M. Bieb.) Shulkina, a Caucasian endemic, is widely distributed in subalpine meadows. —C. *Amygdalus georgica* Desf., a vulnerable endemic dwarf almond, is from Georgia. —D. *Sternbergia colchiciflora* Waldst. & Kit. var. *alexandrae* (Sosn.) Artjush. is a critically endangered narrow endemic from Azerbaijan. —E. *Dryopteris raddeana* (Fomin) Fomin is a rare, endangered endemic fern from Azerbaijan. —F. Participants at the workshop in Bakuriani, 2008. (A photo by Salih Terzioğlu; B photo by Lamara Asieshvili; C photo by Shalva Sikharulidze; D, E photos by Vugar Kerimov; F photo by George Schatz.)

Plate 26. —A. The Caucasian endemic *Tulipa sosnowskyi* Achv. & Mirzoeva, photographed in Armenia. —B. *Hypericum formosissimum* Takht. is a medicinally useful Caucasian endemic from Armenia and Azerbaijan. —C. A critically endangered endemic from Armenia, *Centaurea vavilovii* Takht. & Gabrieljan was named after the great Russian biologist Nikolai Vavilov. —D. *Corydalis tarkiensis* Prokh. is a rare and vulnerable endemic species in the Fumariaceae from Dagestan. —E. *Thymus markhotensis* Maleev, a low ground cover, is a vulnerable species endemic to the central Caucasus of Russia. —F. *Centaurea woronowii* Bornm. ex Sosn. is endemic to the southwestern Caucasus where it grows on dry, rocky soil and talus. (A–C photos by Kamilla Tamanian; D photo by Ramazan Murtazaliev; E photo by Svetlana Litvinskaya; F photo by Özgür Eminagaoglu.)

Plate 27. —A. *Seseli rupicola* Woronow is endemic to the western Caucasus where it grows on limestone cliffs. —B. *Ulmus georgica* Schchian is an endemic elm from Georgia. —C. *Veronica telephiifolia* Vahl, found in dry meadows in the western Caucasus of Georgia, Russia, and Turkey. —D. *Quercus pontica* K. Koch, from the southwestern Caucasus of Georgia and Turkey, is a vulnerable endemic oak with conspicuous, large leaves. —E. *Ziziphora serpyllacea* M. Bieb., broadly distributed in the Caucasus region, is an endemic mint found on dry hills. —F. *Salvia canescens* C. A. Mey. var. *daghestanica* (Sosn.) Menitsky is another endemic mint with narrow distribution in Dagestan. (A–C photos by Shalva Sikharulidze; D–F photos by Lamara Asieshvili.)

Plate 28. —A. *Centaurea appendicigera* K. Koch is an endangered endemic species from Turkey. —B. The endemic *Galanthus alpinus* Sosn. occurs in the western Caucasus, with three varieties under threat from over-collection. —C. *Paeonia carthalinica* Ketsk. is an endangered endemic from Georgia. —D. *Iris marschalliana* Bobrov is an endangered endemic from montane steppes in the central Caucasus. —E. *Digitalis schischkinii* Ivanina is endemic to Georgia, Russia, and Turkey and has both medicinal and ornamental uses. —F. *Iris notha* M. Bieb. is a vulnerable endemic from the northern Caucasus of Russia where it occurs in steppes and meadow steppes. (A photo by Salih Terzioğlu; B photo by David Kikodze; C photo by Shalva Sikharulidze; D–F photos by Anatoly Mikheev.)

A

B

Plate 29. —A. Field trip to the Bakuriani region. The white flowers of the Caucasian endemic *Anemone fasciculata* L. are visible on the hillside. —B. The first workshop held in Georgia, Tbilisi, in 2003. Turkish and Iranian botanists would join this group one year later. Front row from left to right: K. Batsatsashvili (Georgia), D. Cao (CEPF, USA), K. Tamanian† (Armenia), D. Atha (NY, USA), R. Gagnidze† (Georgia). Back row from left to right: L. Asieshvili (Georgia), T. Shulkina (MO, USA), W. Strahm (IUCN), E. Gabrielyan (Armenia), V. Ali-zade (Azerbaijan), O. Abdaladze (Georgia), G. Fayvush (Armenia), J. Solomon (MO, USA), P. Jørgensen (MO, USA), V. Hajiev† (Azerbaijan), G. Nakhutsrishvili (Georgia), A. Mikheev† (Russia), S. Musayev† (Azerbaijan), S. Shetekauri (Georgia). (Photos by George Schatz.)

Plate 30. A major threat for Caucasian wild ornamental plants is their over-collection. —A. Habitat photo of *Tulipa eichleri* Regel, endemic to the eastern Caucasus region. Inset photo of boy holding a bouquet of tulip blossoms collected for sale. When leaves and flowers are gathered, the bulbs die. —B. The flowers and inflorescences of *Achillea* L., *Helichrysum* Mill., and *Trifolium* L. offered for sale at a farmers market in Armenia. (A photo by Marina Eristavi; B photo by Kamilla Tamanian.)

Plate 31. —A. Local use of native plants: *Pterocarya pterocarpa* (Michx.) Kunth ex Iljinsk. and species of *Delphinium* L., as well as *Juglans regia* L., among others, are used for natural dyes for textiles and needlework. —B. A local wine festival in Georgia. Wine has been produced in the Caucasus region for over 4000 years. —C. The Caucasus region is the historic area where cereals such as the bread wheats, rye, and barley are believed to have been first domesticated. Traditional methods of baking bread are still followed in Georgia today. (Photos by Otar Abdaladze.)

Plate 32. —A. Participants and guests at the St. Louis Workshop in October 2009. Front row from left to right: Libing Zhang (MO, USA), K. Tamanian† (Armenia), M. Oganesyan (Armenia), P. Raven (MO, USA), S. Litvinskaya (Russia), S. Terzioğlu (Turkey). Second row: Hai He (China), V. Yarmishko (Russia), T. Shulkina (MO, USA), G. Nakhutsrishvili (Georgia), V. Shvanova (Russia), V. Ali-zade (Azerbaijan), N. Morin (FNA, USA), V. Farzaliyev (Azerbaijan), G. Fayvush (Armenia), R. Murtazaliev (Dagestan, Russia), K. Coşkunçelebi (Turkey), E. Alirzaeva (Azerbaijan), K. Batsatsashvili (Georgia). Third row: N. Turland (MO, USA), D. Geltman (Russia), S. Sikharulidze (Georgia), N. Zazanashvili (WWF, Georgia), R. Bussmann (MO, USA), M. Merello (MO, USA), J. Cocos (MO, USA), J. Delaney (MO, USA), B. Tankersley (CBG, USA), G. Schatz (MO, USA), J. Solomon (MO, USA). —B. While in St. Louis participants enjoyed a field trip to the Missouri Botanical Garden's Shaw Nature Reserve. (A photo by Burgund Bassüner; B photo by George Schatz.)

RED LIST OF THE ENDEMIC PLANTS OF THE CAUCASUS REGION

There are 2791 plant taxa endemic to the Caucasus biodiversity hotspot, of which 1752 (62%) are assessed in the Red List. This has been assembled from accounts initially compiled independently by botanists in each of the six countries within the region: Armenia, Azerbaijan, Georgia, Iran, Russia, and Turkey. Based on both regional and national floras, as well as more recent literature and recent field trips, each country enumerated all plant species and infra-specific taxa present within their borders that were considered to be endemic to the Caucasus region, noting in particular those thought to be national endemics. These national lists were then distributed among all countries for comparison and confirmation of the distributions of the taxa within the region as a whole.

In some cases, it was determined that a taxon also occurred outside of the Caucasus biodiversity hotspot and so was removed from the list. In other cases, it was discovered that a taxon that was considered to be a national endemic by one country, in fact, also occurred in one or more of the other countries within the region. In addition, sometimes the same taxon was known by different names in different countries, either through homotypic or heterotypic synonymy. In such cases, the question of which name should be applied, and how narrowly or broadly the taxon should be defined, requires additional taxonomic study and consensus beyond the scope of the current project.

In terms of conservation policy and practice, and especially as natural areas become increasingly fragmented within human-dominated landscapes, narrowly defined geographical variants will become more and more isolated, and therefore of increasing focus for conservation concern in the face of local threats. An example of this can be found in *Paeonia* L. Recent studies of chromosome numbers led D. Y. Hong (2010) to the conclusion to combine many species together. Local Caucasian botanists, however, distinguish these species by flower color, ranging from white to yellow and red, and by leaf forms and structure. Because species of *Paeonia* L. are highly desirable ornamental plants and are under serious threat, the total range of variation requires conservation attention. This distinction is lost if narrowly defined species are synonymized. Since the publication of the *Flora of the USSR*, botanists from Russia and other republics have continued to use the Komarov species concept, which was the basis for species delimitation in this volume. The narrow species concept used in the *Flora of the USSR* was considered preferable to large conglomerates, as this approach better serves the purposes of conservation. The Caucasus region was a refuge during the Pleistocene (see Foreword) and has been a source of speciation resulting in the evolution of additional genera (e.g., *Kemulariella* Tamamsch., *Rhodothamnus* P. H. Davis, *Vavilovia* Fed., *Woronowia* Juz.) and species (e.g., *Campanula albovii* Kolak., *Campanula kluchorica* Kolak., *Campanula leskovii* Fed.) with restricted distribution. A more inclusive species concept leads to a loss of information and misunderstanding about the levels of diversity actually represented in a species. The situation might be illustrated with the new treatment of *Campanula* L. in the *Conspectus Flora Caucasus* (Oganesian, 2006), in which 49 Caucasian species accepted in the *Flora of the USSR* and local floras were synonymized into a relatively few more broadly circumscribed species. Most of these eliminated species are Caucasian endemics and are included in regional Red Books. The compilers of this current work faced the same problem and have resolved it by including more narrowly defined species as found in the *Flora of the USSR*. In advanced floristics (e.g., natural protection frameworks, including Red Data Lists) there is a need for consistency in taxonomy to form an appropriate classification and recognition of taxa that otherwise might be neglected and therefore receive no protection.

There are some apomicts among the listed species, especially in the genera *Alchemilla* L. and *Hieracium* L. Although these taxa may be very difficult to characterize, the authors believe it is important to include them in the list.

Western botanists will find some differences between the Russian and western taxonomic literature. This is primarily a result of isolation of the Soviet botanical community for many years. Russia and the other republics of the former Soviet Union (FSU) were essentially closed to foreigners for more than 70 years. Following the 1917 Russian Revolution, this enormous area became progressively isolated from the rest of the world. Although a trickle of scientific communication was maintained throughout the years, the ability of Soviet botanists to freely interact with foreign colleagues was severely limited. The vast majority of botanical publications from the former Soviet Union countries are written in Russian. This linguistic barrier is a major reason that few taxonomists in the West are familiar with the botanical literature on the Caucasus. As a result, some taxonomic publications were not included by international indexing services, such as IPNI or The Plant List (e.g., species of *Pyrus* L.). On the other hand, current Russian botanical literature still maintains traditional taxonomic circumscriptions for many families, genera, and species. Current western taxonomic treatments of large families, such as Asteraceae and Rosaceae, are not completely accepted by the majority of Russian botanists, and this explains why the number of species in large genera, such as *Anthemis* L., *Centaurea* L., and *Prunus* L., differs significantly from that found in western literature.

Traditionally, the way of abbreviating some authors' names in older Russian botanical literature has also differed from western publications. Thus, Bunge is given in Russian publications as Bge., Regel as Rgl., Bieberstein as M.B., etc. In this publication all author names follow the standards established in R. K. Brummitt and C. E. Powell (1992) with the exception of two names: Charadze and Charkevich. Charadze is often spelled Kharadze, based on a standard transliteration of this name from Russian into English. However, Anna Charadze, who was a Georgian botanist, wrote her name in Latin transcription as Charadze for all new taxa that she described. This spelling is maintained in this publication. The other name is Sigismund Charkevich, a Ukrainian botanist who consistently put his name as Charkevich in Latin diagnoses. Therefore, it was considered more appropriate to leave his name as Charkevich instead of Kharkevich, as it is in transliteration. A commonly used author abbreviation in old and current Russian botanical literature is C. Koch. However, in Brummitt and Powell his name is given as K. Koch, which is followed here.

The Red List is arranged by major taxonomic groupings—Pteridophytes, Gymnosperms, and Angiosperms (Dicots and Monocots)—to make it easy to use by any reader. Within these divisions the families, genera, and species names are listed in alphabetical order. Sometimes what might be considered one taxon may be represented by different names in different countries. In this case all of the names used are present in the table with the one in brackets serving as a cross reference to the other name(s). It is beyond the scope of this publication to resolve the circumscription and application of these names. If the name of a Caucasian endemic is accepted by one country but not accepted by another, both names are included in the list. Thus, *Arnebia echioides* (L.) DC. and *Macrotomia echioides* (L.) Boiss. are both recorded. For a few names a decision was made by a consensus between the countries or primarily based on the preference given to the opinion of the country where the species occurs.

The IUCN categories and criteria follow IUCN standards (2001). Red List categories are abbreviated as follows: Extinct = EX; Extinct in the Wild = EW; Critically Endangered = CR; Endangered = EN; Vulnerable = VU; Near Threatened = NT; Least Concern = LC; Data Deficient = DD; and Not Evaluated = NE. Criteria and subcriteria are also given for Critically Endangered, Endangered, and Vulnerable. There is a range of quantitative criteria that are appropriate for assessing threat levels for any taxon. The criteria show reduction of population (A), decline of geographical range (B), number of population size (C, D), and probability of extinction (E). Details and tendency are shown by using the abbreviations a, b, c and i, ii, iii, and other letters following the IUCN standards.

Distribution ranges are based on information provided by each country. The six countries covering the Caucasus region are presented in the list in alphabetical order. An "x" mark in the country column indicates that a given taxon is present in its territory according to country information or *Caucasian Flora Conspectus* (Takhtajan, 2003–2012). If only one country is indicated by "x," then, in most cases, it means the species is a national endemic, most of which are mentioned in the introductory chapter for each country. The absence of an "x" indicates that a particular country does not have evidence of the presence of this species, under this name.

BIBLIOGRAPHY

Brummitt, R. K. & C. E. Powell. 1992. Authors of Plant Names. Royal Botanic Gardens, Kew.

Hong, D. Y. 2010. Peonies of the World: Taxonomy and Phytogeography. Royal Botanic Gardens, Kew.

International Union for Conservation of Nature and Natural Resources (IUCN). 2001. IUCN Red List Categories and Criteria, Version 3.1. Prepared by the IUCN Species Survival Commission. IUCN, Gland, Switzerland, and Cambridge, United Kingdom.

Oganesian, M. E. 2006. Campanulaceae. Pp. 139–162 *in* A. Takhtajan (editor), Caucasian Flora Conspectus, Vol. 3(1). St. Petersburg University Press, St. Petersburg (in Russian).

Takhtajan, A. L. (editor). 2003–2012. Caucasian Flora Conspectus, Vols. 1–3. St. Petersburg University Press, St. Petersburg (in Russian).

TAXON	RED LIST ASSESSMENT	AR	AZ	GE	IR	RU	TU
F E R N S							
ASPLENIACEAE							
Asplenium daghestanicum Christ	CR B1ab(iii)+2ab(iii)					x	
Asplenium hermanni-christii Fomin	VU D2			x			
Asplenium woronowii H. Christ	NE	x		x		x	x
DRYOPTERIDACEAE							
Dryopteris liliana Golitsin	VU D2			x			x
Dryopteris oreades Fomin	LC		x	x			
Dryopteris raddeana (Fomin) Fomin	EN B2ab(iii)		x				
Polystichum kadyrovii Askerov & A. E. Bobrov	VU D2		x				
Polystichum woronowii Fomin	NE		x	x	x	x	x
POLYPODIACEAE							
Polypodium iranicum Mazooji	CR				x		
WOODSIACEAE							
Woodsia fragilis (Trevir.) T. Moore	NE		x	x		x	
C O N I F E R S							
PINACEAE							
Abies nordmanniana (Steven) Spach	LC			x		x	x
Picea orientalis (L.) Peterm.	LC			x		x	x
Pinus eldarica Medw.	NT		x				
Pinus pityusa Steven	EN B1ab(ii,iii,v)+ 2ab(ii,iii,v)			x		x	
D I C O T S							
ACERACEAE							
Acer cappadocicum Gled. var. *stenocarpum* Yalt.	EN B1ab(i,ii,iii)						x
Acer divergens K. Koch & Pax var. *divergens*	EN B1ab(i,ii,iii,iv)+ 2ab(i,ii,iii,iv)						x
Acer divergens K. Koch & Pax var. *trilobum* Yalt.	CR B2b(i,ii,iii,iv)						x
Acer ibericum M. Bieb. [*Acer monspessulanum* L. subsp. *ibericum* (M. Bieb.) Yalt.]	NE	x	x	x	x	x	x

TAXON	RED LIST ASSESSMENT	AR	AZ	GE	IR	RU	TU
Acer monspessulanum L. subsp. *ibericum* (M. Bieb.) Yalt. [*Acer ibericum* M. Bieb.]	NE	x	x	x	x	x	x
Acer sosnowskyi Doluch.	VU D2			x			
APIACEAE							
Agasyllis latifolia (M. Bieb.) Boiss.	LC	x	x	x		x	
Alococarpum erianthum (DC.) Riedl & Kuber	NE				x		
Angelica adzharica Pimenov	EN B1ab(i,ii,iii,v)+ 2ab(i,ii,iii,v)			x			
Angelica sachokiana (Karjagin) Pimenov & V. N. Tikhom.	DD		x			x	
Angelica sylvestris L. var. *stenoptera* Boiss.	NE			x		x	x
Angelica tatianae Bordz.	NE	x	x	x		x	
Anthriscus ruprechtii Boiss.	NE		x	x		x	x
Anthriscus schmalhausenii (Albov) Koso-Pol.	LC			x		x	
Anthriscus velutina Sommier & Levier	DD					x	
Aphanopleura trachysperma Boiss.	EN B1ab(ii,iii)+2ab(ii,iii)	x	x		x		
Aphanopleura zangelanica Gogina & Matsenko	DD		x				
Arafoe aromatica Pimenov & Lavrova [*Ligusticum arafoe* Albov]	VU B1ab(ii)+2ab(ii)			x		x	
Astrantia biebersteinii Trautv. [*Astrantia trifida* Hoffm.]	NE	x	x	x		x	
Astrantia colchica Albov	EN B1ab(iii)+2b(iii)			x			
Astrantia maxima Pall.	NT	x		x	x	x	x
Astrantia pontica Albov	NT			x		x	
Astrantia trifida Hoffm. [*Astrantia biebersteinii* Trautv.]	NE	x	x	x		x	
Bunium scabrellum Korovin	NE		x		x		
Bupleurum abchasicum Manden.	DD			x			
Bupleurum brachiatum K. Koch ex Boiss.	VU B1ab(ii,iii)+2ab(ii,iii)						x
Bupleurum ghahremanii Mozaff.	NE				x		

TAXON	RED LIST ASSESSMENT	AR	AZ	GE	IR	RU	TU
Bupleurum kosopoljanskyi Grossh.	VU B1ab(i,ii,iii)+ 2ab(i,ii,iii)	x					
Bupleurum rischawii Albov	VU B1ab(ii,iv)c(ii)+ 2ab(ii,iv)c(ii)			x		x	
Bupleurum schistosum Woronow	EN B1ab(i,ii,iii)+ 2ab(i,ii,iii)						x
Bupleurum sosnowskyi Manden.	NE	x		x			
Bupleurum subnivale (Galushko) Galushko	DD					x	
Bupleurum wittmannii Steven	VU D2		x	x			
Bupleurum wolffianum Bornm. ex H. Wolff	EN				x		
Bupleurum woronowii Manden.	NT					x	
Carum grossheimii Schischk.	VU B1ab(iii)			x			
Carum komarovii Karjagin	EN B1ab(iii)+2ab(iii)	x	x				
Cervaria caucasica (M. Bieb.) Pimenov [*Peucedanum caucasicum* (M. Bieb.) K. Koch]	NE	x	x	x	x	x	x
Chaerophyllum astrantiae Boiss. & Balansa	NT			x			x
Chaerophyllum borodinii Albov	NT			x		x	
Chaerophyllum confusum Woronow ex Grossh.	LC			x		x	
Chaerophyllum humile Steven	NE	x	x	x		x	
Chaerophyllum karsianum Kit Tan & Ocakv.	CR B2ab(ii,iii)						x
Chaerophyllum posofianum Erik & Demirkuş	CR B2ab(ii,iii)						x
Chaerophyllum roseum M. Bieb.	NE	x	x	x		x	
Chaerophyllum rubellum Albov	LC	x	x	x		x	x
Chymsydia agasylloides (Albov) Albov	DD			x		x	
Chymsydia colchica (Albov) Woronow ex Grossh.	DD			x			
Cnidiocarpa physospermifolia (Albov) Pimenov [*Ligusticum physospermifolium* Albov]	LC			x		x	

TAXON	RED LIST ASSESSMENT	AR	AZ	GE	IR	RU	TU
Cnidium grossheimii Manden.	DD			x			
Cnidium mandenovae Gagnidze	DD			x			
Cnidium pauciradiatum Sommier & Levier	DD			x			
Cryptotaenia flahaultii (Woronow) Koso-Pol.	VU D2			x			
Cymbocarpum anethoides DC.	DD	x	x	x	x		
Cymbocarpum erythraeum (DC.) Boiss.	NE	x	x		x		x
Dichoropetalum pschawicum (Boiss.) Pimenov & Kljuykov [*Peucedanum pschawicum* Boiss.]	LC			x		x	
Dorema glabrum Fisch. & C. A. Mey.	VU B1ab(i,ii,iii,iv,v)+ 2ab(i,ii,iii,iv,v)	x	x		x		
Eleutherospermum lazicum Boiss. & Balansa	NE			x			x
Eleutherospermum rubellum E. A. Busch [*Tamamschjanella rubella* (E. A. Busch) Pimenov & Kljuykov]	NE	x			x	x	
Ferula calcarea Pimenov	VU D2		x			x	
Ferula caucasica Korovin	VU D2		x				
Ferula mervynii Sagiroğlu & H. Duman	CR B1ab(i,ii,iii)+ 2ab(i,ii,iii); C2a(i)						x
Ferulago latiloba Schischk.	CR B2ab(iii)						x
Ferulago trifida Boiss.	NE				x		
Heracleum aconitifolium Woronow	LC			x		x	
Heracleum apiifolium Boiss.	NE			x		x	x
Heracleum asperum (Hoffm.) M. Bieb.	NE		x	x		x	
Heracleum calcareum Albov	EN B1ab(iii)+2ab(iii)			x		x	
Heracleum chorodanum (Hoffm.) DC.	NE	x	x	x		x	
Heracleum cyclocarpum K. Koch [*Heracleum sphondylium* L. subsp. *cyclocarpum* (K. Koch) P. H. Davis]	LC			x			x
Heracleum egrissicum Gagnidze	EN B1ab(ii,iii)+2ab(ii,iii)			x			
Heracleum freynianum Sommier & Levier	LC			x		x	

TAXON	RED LIST ASSESSMENT	AR	AZ	GE	IR	RU	TU
Heracleum grandiflorum Steven ex M. Bieb.	LC		x			x	
Heracleum grossheimii Manden.	DD			x			
Heracleum leskovii Grossh.	LC			x		x	
Heracleum mandenovae Satzyperova	DD			x			
Heracleum mantegazzianum Sommier & Levier	LC			x		x	
Heracleum ossethicum Manden.	NT			x			
Heracleum pastinacifolium K. Koch subsp. *transcaucasicum* (Manden.) P. H. Davis [*Heracleum transcaucasicum* Manden.]	NE	x		x	x		x
Heracleum ponticum (Lipsky) Schischk. ex Grossh.	LC			x		x	
Heracleum rechingeri Manden.	NE				x		
Heracleum roseum Steven	NE		x	x		x	
Heracleum scabrum Albov	DD			x		x	
Heracleum schelkovnikovii Woronow	LC	x	x				
Heracleum sommieri Manden.	LC			x			
Heracleum sosnowskyi Manden.	NE	x	x	x		x	x
Heracleum sphondylium L. subsp. *artvinense* (Manden.) P. H. Davis	EN B1ab(iii)+2ab(iii)						x
Heracleum sphondylium L. subsp. *cyclocarpum* (K. Koch) P. H. Davis [*Heracleum cyclocarpum* K. Koch]	LC			x			x
Heracleum trachyloma Fisch. & C. A. Mey.	LC		x				x
Heracleum transcaucasicum Manden. [*Heracleum pastinacifolium* K. Koch subsp. *transcaucasicum* (Manden.) P. H. Davis]	NE	x		x	x		x
Heracleum wilhelmsii Fisch. & C. A. Mey.	NE		x	x	x		
Hippomarathrum caspicum Grossh.	EN B1ab(iii,v)+2ab(iii,v)		x				
Johrenia ramosissima Mozaff.	EN				x		
Laserpitium affine Ledeb.	EN B1ab(iii)+2ab(iii)			x			x
Laserpitium stevenii Fisch. & Trautv.	VU B2ab(iii)			x		x	

TAXON	RED LIST ASSESSMENT	AR	AZ	GE	IR	RU	TU
Ligusticum arafoe Albov [*Arafoe aromatica* Pimenov & Lavrova]	VU B1ab(ii)+2ab(ii)			x		x	
Ligusticum caucasicum Sommier & Levier [*Mutellina caucasica* (Sommier & Levier) Lavrova]	LC			x		x	
Ligusticum physospermifolium Albov [*Cnidiocarpa physospermifolia* (Albov) Pimenov]	LC			x		x	
Mandenovia komarovii (Manden.) Alava	VU B1ac(i,ii,iii,iv)			x		x	
Mutellina caucasica (Sommier & Levier) Lavrova [*Ligusticum caucasicum* Sommier & Levier]	LC			x		x	
Oenanthe abchasica Schischk.	VU B1ab(ii,iv)+2ab(ii,iv); C2a(i)			x		x	
Oenanthe longifoliolata Schischk.	DD		x				
Pastinaca armena Fisch. & C. A. Mey.	NE	x	x	x		x	x
Pastinaca aurantiaca (Albov) Kolak.	LC			x		x	
Pastinaca pimpinellifolia M. Bieb.	NE	x	x	x		x	x
Peucedanum adae Woronow	LC			x		x	
Peucedanum caucasicum (M. Bieb.) K. Koch [*Cervaria caucasica* (M. Bieb.) Pimenov]	NE	x	x	x	x	x	x
Peucedanum knappii Bornm.	NE				x		
Peucedanum luxurians Tamamsch.	DD		x				
Peucedanum pschawicum Boiss. [*Dichoropetalum pschawicum* (Boiss.) Pimenov & Kljuykov]	LC			x		x	
Peucedanum zedelmeyerianum Manden.	EN B2ab(i,ii,iv)	x					x
Pimpinella aromatica M. Bieb.	NE		x	x		x	x
Pimpinella idae Takht.	DD		x	x		x	
Pimpinella schatilensis Otsch.	VU D2			x			
Polylophium panjutinii Manden. & Schischk.	EN B1ab(iii)+2ab(iii)			x			
Prangos arcis-romanae Boiss. & A. Huet [*Prangos meliocarpoides* Boiss. var. *arcis-romanae* (Boiss. & A. Huet) Herrnst. & Heyn]	NE		x	x			x

TAXON	RED LIST ASSESSMENT	AR	AZ	GE	IR	RU	TU
Prangos meliocarpoides Boiss. var. *arcis-romanae* (Boiss. & A. Huet) Herrnst. & Heyn [*Prangos arcis-romanae* Boiss. & A. Huet]	NE		x	x			x
Scaligeria lazica Boiss.	VU B1ab(ii,iii)+2ab(ii,iii)						x
Seseli alexeenkoi Lipsky	EN B1ab(iii)+2ab(iii)					x	
Seseli alpinum M. Bieb.	LC			x		x	
Seseli andronakii Woronow ex Schischk.	VU B1ab(i,ii)						x
Seseli cuneifolium M. Bieb.	VU B1ab(v)+2ab(v)		x				
Seseli grandivittatum (Sommier & Levier) Schischk.	NE	x	x	x			x
Seseli leptocladum Woronow	VU D2	x					
Seseli petraeum M. Bieb.	LC			x		x	x
Seseli ponticum Lipsky	LC			x		x	
Seseli rupicola Woronow	NT			x		x	
Seseli saxicolum (Albov) Pimenov	NT			x		x	
Smyrniopsis armena Schischk.	EN B1ab(ii,v)+2ab(ii,v); D	x					
Stenotaenia daralaghezica (Takht.) Schischk.	EN B1ab(i,ii,iii)+ 2ab(i,ii,iii)	x	x				
Stenotaenia nudicaulis Boiss.	NE				x		
Symphyoloma graveolens C. A. Mey.	NE		x	x		x	
Szovitsia callicarpa Fisch. & C. A. Mey.	NE	x	x		x		x
Tamamschjanella rubella (E. A. Busch) Pimenov & Kljuykov [*Eleutherospermum rubellum* E. A. Busch]	NE	x			x	x	
Zeravschania pauciradiata (Tamamsch.) Pimenov	NE		x		x		
ARACEAE							
Arum consobrimum Schott	NE	x	x				
ARALIACEAE							
Hedera caucasigena Pojark. [*Hedera helix* L. subsp. *caucasigena* (Pojark.) Takht. & Mulk.]	NE		x			x	

TAXON	RED LIST ASSESSMENT	AR	AZ	GE	IR	RU	TU
Hedera helix L. subsp. *caucasigena* (Pojark.) Takht. & Mulk. [*Hedera caucasigena* Pojark.]	NE		x			x	
Hedera pastuchovii Woronow ex Grossh.	NE		x	x	x	x	
ARISTOLOCHIACEAE							
Aristolochia hyrcana P. H. Davis & M. S. Khan	NE				x		
Aristolochia iberica Fisch. & C. A. Mey. ex Boiss.	NE	x	x	x		x	x
Aristolochia steupii Woronow	VU B1ab(ii,v); C2a(i)			x		x	
ASCLEPIADACEAE							
Vincetoxicum albovianum (Kusn.) Pobed.	DD			x		x	
Vincetoxicum rehmanni Boiss.	NE	x	x	x	x	x	
ASTERACEAE							
Achillea griseo-virens Albov	DD			x			
Achillea latiloba Ledeb. ex Nordm.	LC			x			x
Achillea moghanica Iranshahr	NE				x		
Achillea ptarmicifolia (Willd.) Rupr. [*Ptarmica ptarmicifolia* (Willd.) Galushko]	LC			x		x	
Achillea sedelmeyeriana Sosn.	VU D2			x			
Adenostyles macrophylla (M. Bieb.) Czerep. [*Senecio rhombifolius* (Adams ex Willd.) Sch. Bip.]	NE	x	x	x		x	
Aetheopappus caucasicus Sosn.	NE		x	x		x	
Aetheopappus vvedenskii (Sosn.) Sosn.	LC			x		x	
Alboviodoxa elegans (Albov) Woronow ex Grossh. [*Amphoricarpos elegans* Albov]	DD			x		x	
Amberboa glauca (Willd.) Grossh. [*Amberboa moschata* (L.) DC.]	NE	x	x	x		x	x
Amberboa moschata (L.) DC. [*Amberboa glauca* (Willd.) Grossh.]	NE	x	x	x		x	x
Amberboa sosnovskyi Iljin	NE	x	x		x		

TAXON	RED LIST ASSESSMENT	AR	AZ	GE	IR	RU	TU
Amphoricarpos elegans Albov [*Alboviodoxa elegans* (Albov) Woronow ex Grossh.]	DD			x		x	
Anacyclus ciliatus Trautv.	LC		x				
Anthemis atropatana Iranshahr	NE				x		
Anthemis calcarea Sosn. var. *calcarea*	EN B1ab(i,ii,iii,iv,v)+ 2ab(i,ii,iii,iv,v)						x
Anthemis calcarea Sosn. var. *discoidea* Grierson	EN B1ab(i,ii,iii)+ 2ab(i,ii,iii)						x
Anthemis caucasica Chandjian [*Anthemis marschalliana* Willd. subsp. *pectinata* (Boiss.) Grierson]	NE	x		x		x	x
Anthemis cretica L. subsp. *iberica* (M. Bieb.) Grierson [*Anthemis iberica* M. Bieb.]	DD	x	x	x		x	x
Anthemis cretica L. subsp. *saportana* (Albov) Chandjian [*Anthemis saportana* Albov]	NE			x		x	
Anthemis emiliae Sosn.	DD			x			
Anthemis fruticulosa M. Bieb.	NE	x	x			x	
Anthemis iberica M. Bieb. [*Anthemis cretica* L. subsp. *iberica* (M. Bieb.) Grierson]	NE	x	x	x		x	x
Anthemis macroglossa Sommier & Levier	DD			x			
Anthemis maris-nigri Fed.	DD			x			
Anthemis marschalliana Willd. [*Anthemis marschalliana* Willd. subsp. *marschalliana*]	LC			x		x	
Anthemis marschalliana Willd. subsp. *marschalliana* [*Anthemis marschalliana* Willd.]	LC			x		x	
Anthemis marschalliana Willd. subsp. *pectinata* (Boiss.) Grierson [*Anthemis caucasica* Chandjian]	NE	x		x		x	x
Anthemis melanoloma Trautv. subsp. *trapezuntica* Grierson	VU B1ab(i,ii,iii)						x
Anthemis moghanica Iranshahr	NE				x		

TAXON	RED LIST ASSESSMENT	AR	AZ	GE	IR	RU	TU
Anthemis sachokiana Sosn. ex Grossh.	NT		x				
Anthemis saguramica Sosn.	DD			x			
Anthemis saportana Albov [*Anthemis cretica* L. subsp. *saportana* (Albov) Chandjian]	DD			x		x	
Anthemis schischkiniana Fed.	DD			x			
Anthemis sosnovskyana Fed.	NE	x	x	x		x	
Anthemis talyschensis Fed.	NE				x		
Anthemis triumfettii (L.) DC. subsp. *decumbens* Iranshahr	NE				x		
Anthemis woronowii Sosn.	DD		x	x			
Anthemis zyghia Woronow	DD			x			
Arctium palladinii (Marcow.) Grossh. [*Arctium transcaucasicum* Sosn. ex Grossh.]	NE		x	x		x	
Arctium platylepis (Boiss. & Balansa) Sosn. ex Grossh.	NE		x	x		x	
Arctium transcaucasicum Sosn. ex Grossh. [*Arctium palladinii* (Marcow.) Grossh.]	NE		x	x		x	
Artemisia araxina Takht.	NE	x					
Artemisia daghestanica Krasch. & Poretzky	LC					x	
Artemisia fedorovii Rzazade	NT		x				
Artemisia szowitziana (Besser) Grossh.	NE	x	x			x	
Aster roseus Steven [*Kemulariella rosea* (Steven ex M. Bieb.) Tamamsch.]	NE		x	x		x	
Bellis hyrcanica Woronow	VU D2		x				
Calycocorsus tuberosus (Fisch. & C. A. Mey. ex DC.) Rauschert	NE				x	x	
Carduus acanthocephalus C. A. Mey.	NE		x	x		x	
Carduus hohenackeri Kazmi	DD		x	x			
Carduus nervosus K. Koch	NT	x	x				x
Carduus novorossicus Porten.	EN B1ab(ii,iii,v)+ 2ab(ii,iii,v); C2a(i)					x	

TAXON	RED LIST ASSESSMENT	AR	AZ	GE	IR	RU	TU
Carduus onopordioides Fisch. ex M. Bieb.	NE	x	x	x	x	x	
Carduus onopordioides Fisch. ex M. Bieb. subsp. *turcicus* (Kazmi) P. H. Davis	DD						x
Carduus poliochrus Trautv.	DD			x		x	
Carthamus tamamschjaniae Gabrieljan	DD	x					
Centaurea aggregata Fisch. & C. A. Mey. subsp. *albida* (K. Koch) Bornm.	DD						x
Centaurea alexandri Bordz.	EN B1ab(i,ii,iii)+2ab(i,ii,iii)	x					
Centaurea appendicigera K. Koch	EN B1ab(iii)+2ab(iii)						x
Centaurea arpensis (Czerep.) Wagenitz	EN B1ab(ii,iii)+2ab(ii,iii)	x					
Centaurea avarica Tzvelev	CR B2ab(ii,iii,v)					x	
Centaurea aziziana Rech. f.	NE	x			x		
Centaurea bagadensis Woronow	DD			x			
Centaurea bella Trautv. [*Psephellus bellis* (Trautv.) Wagenitz]	NE	x		x			
Centaurea bella Trautv. subsp. *nathadzeae* (Sosn.) Djindjolia	DD			x			
Centaurea caspia Grossh.	NE	x	x			x	
Centaurea cheiranthifolia Willd. subsp. *cheiranthifolia*	NE			x	x	x	x
Centaurea cheiranthifolia Willd. subsp. *purpurascens* (DC.) Wagenitz [*Centaurea fischeri* Willd.]	DD		x				x
Centaurea cheiranthifolia Willd. subsp. *willdenowii* (Czerep.) Mikheev	LC		x	x		x	
Centaurea daghestanica (Lipsky) Wagenitz	VU B1ab(iii)		x			x	
Centaurea demirizii Wagenitz	CR B2ab(iii)						x
Centaurea drabifolioides Hub.-Mor.	CR B1ab(i,ii,iii)+2ab(i,ii,iii)						x
Centaurea elbrusensis Boiss. & Buhse	NE	x			x		
Centaurea erivanensis (Lipsky) Bordz.	VU B1ab(ii,iii)+2ab(ii,iii)	x	x				x
Centaurea erivanensis (Lipsky) Bordz. subsp. *holargyrea* (Bornm. & Woronow) Gabrieljan	NE	x					x

TAXON	RED LIST ASSESSMENT	AR	AZ	GE	IR	RU	TU
Centaurea eugenii Sosn.	DD						x
Centaurea fajvuschii Gabrieljan	DD	x					
Centaurea fischeri Willd. [*Centaurea cheiranthifolia* Willd. subsp. *purpurascens* (DC.) Wagenitz]	NE		x				x
Centaurea gracillima Wagenitz	CR B2ab(iii)						x
Centaurea gulissashvilii Dumbadze	NE	x		x			
Centaurea helenioides Boiss. & Hausskn.	EN B1ab(i,ii,iii)+ 2ab(i,ii,iii)						x
Centaurea incanescens (Fisch. & C. A. Mey. ex DC.) Sch. Bip.	CR				x		
Centaurea kobstanica Tzvelev	NT		x				
Centaurea leptophylla (K. Koch) Tchich.	CR B1ab(i,ii,iii,iv,v); C2a(i)						x
Centaurea macrocephala Muss.-Puschk. ex Willd. [*Grossheimia macrocephala* (Muss.-Puschk. ex Willd.) Sosn. & Takht.]	NE	x	x	x	x		x
Centaurea meyeriana Tzvelev	NE		x		x		
Centaurea micrantha S. G. Gmel. subsp. *kemulariae* (Dumbadze) Mikheev	DD					x	
Centaurea nigrofimbria (K. Koch) Sosn.	LC			x		x	x
Centaurea ovina Pall. ex Willd., s. str.	NE	x		x	x	x	
Centaurea pecho Albov	EN B1ab(i,ii,iii,iv)+ 2ab(i,ii,iii,iv)						x
Centaurea phaeopappoides Bordz.	NE	x	x		x		
Centaurea polyphylla Ledeb. ex Nordm. [*Grossheimia polyphylla* (Ledeb.) Holub]	NT			x		x	
Centaurea pseudophrygia C. A. Mey. subsp. *abnormis* (Czerep.) Mikheev	NT					x	
Centaurea pseudoscabiosa Boiss. & Buhse subsp. *ossetica* (Sosn.) Gabrieljan	DD	x	x	x		x	
Centaurea pterocaula (Trautv.) Czerep.	NE	x			x		
Centaurea pulcherrima Willd. var. *freynii* (Sint.) Wagenitz	VU B1ab(iii)+2ab(iii)						x
Centaurea rhizocalathium (K. Koch) Tchich.	EN B2ab(i,ii,iii,v)	x					x

TAXON	RED LIST ASSESSMENT	AR	AZ	GE	IR	RU	TU
Centaurea ruprechtii (Boiss.) Wagenitz	EN B1ab(ii,iii,v)+ 2ab(ii,iii,v)					x	
Centaurea schelkovnikovii Sosn.	CR B1ab(i,ii,iii)+ 2ab(i,ii,iii)	x	x				
Centaurea scripczinskyi Mikheev	VU B1ab(i); C2a(i)					x	
Centaurea simplicicaulis Boiss. & A. Huet	LC			x			x
Centaurea stevenii M. Bieb.	NE	x	x	x			x
Centaurea stevenii M. Bieb. subsp. *armeniaca* Gabrieljan	NE	x	x				
Centaurea straminicephala Hub.-Mor.	EN B1ab(i,ii,iii,v)+ 2ab(i,ii,iii,v); C2a(i)						x
Centaurea takhtajanii Gabrieljan & Tonjan	CR B2ab(ii,iii)	x					
Centaurea taochia (Sosn.) Sosn. [*Psephellus taochius* Sosn.]	EN B2ab(iii,v)	x					x
Centaurea transcaucasica Sosn. ex Grossh.	NE	x	x	x		x	
Centaurea transcaucasica Sosn. ex Grossh. subsp. *georgica* (Klokov) Mikheev	DD			x			
Centaurea transcaucasica Sosn. ex Grossh. subsp. *transcaucasica*	NE	x		x			
Centaurea vavilovii Takht. & Gabrieljan	CR B1ab(i,ii,iii,v)+ 2ab(i,ii,iii,v)	x					
Centaurea woronowii Bornm. ex Sosn.	EN B1ab(iii)+2ab(iii)			x			x
Centaurea xanthocephala (DC.) Sch. Bip. subsp. *xanthocephala*	NE	x					x
Centaurea xanthocephala (DC.) Sch. Bip. subsp. *xanthocephaloides* (Tzvelev) Gabrieljan	NE	x	x		x		
Cephalorrhynchus kirpicznikovii Grossh.	DD		x				
Cephalorrhynchus takhtadzhianii (Sosn.) Kirp.	NE	x			x		
Chamaemelum silaifolium (Steven) Trautv. [*Tanacetum silaifolium* (DC.) Sch. Bip.]	DD		x			x	
Cicerbita boissieri (Rouy) C. Jeffrey	EN B1ab(iii)+2ab(iii)						x

TAXON	RED LIST ASSESSMENT	AR	AZ	GE	IR	RU	TU
Cicerbita deltoidea (M. Bieb.) Beauverd [*Mycelis deltoidea* (M. Bieb.) Sennikov]	NE		x	x		x	
Cicerbita macrophylla (Willd.) Wallr. [*Prenanthes cacaliifolia* (M. Bieb.) Beauverd]	NE	x		x		x	x
Cicerbita madatapensis Gagnidze	DD			x			
Cicerbita olgae Leskov	DD		x	x			
Cicerbita petiolata (K. Koch) Gagnidze [*Prenanthes petiolata* (K. Koch) Sennikov]	DD			x		x	
Cicerbita prenanthoides (M. Bieb.) Beauverd	NE			x		x	x
Cicerbita racemosa (Willd.) Beauverd	NE	x	x	x		x	x
Cirsium abkhasicum (Petrak) Grossh.	LC			x		x	
Cirsium aggregatum Ledeb.	LC			x		x	x
Cirsium albowianum Sommier & Levier	EN B1ab(iii)			x			
Cirsium arachnoideum (M. Bieb.) M. Bieb.	LC			x		x	
Cirsium argillosum Petrov ex Charadze	NT		x			x	
Cirsium balkharicum Charadze	NT					x	
Cirsium buschianum Charadze	LC			x		x	
Cirsium caput-medusae Sommier & Levier	DD			x		x	
Cirsium caucasicum (Adams) Petr.	NE			x			x
Cirsium charadzeae (Kimer.) Czerep. [*Lamyropsis charadzeae* Kimer.]	DD			x		x	
Cirsium chlorocomos Sommier & Levier	LC			x		x	
Cirsium ciliatum (Murr.) Moench	NE	x	x	x		x	
Cirsium cosmelii (Adams) Fisch. ex Hohen. [*Cirsium kosmelii* (Adams) Fisch. ex Hohen.]	NE	x	x	x	x	x	x
Cirsium czerkessicum Charadze	EN B1ab(iii)+2ab(iii)					x	
Cirsium daghestanicum Charadze [*Cirsium ketzkhovelii* Charadze]	DD		x	x		x	
Cirsium davisianum Kit Tan & Sorger	CR B2ab(iii); D						x

TAXON	RED LIST ASSESSMENT	AR	AZ	GE	IR	RU	TU
Cirsium dealbatum M. Bieb.	LC					x	
Cirsium depilatum Boiss. & Balansa	DD	x	x	x		x	
Cirsium elbrusense Sommier & Levier	NT					x	
Cirsium eliasianum Kit Tan & Sorger	CR B2ab(iii); D						x
Cirsium erythrolepis K. Koch	DD			x		x	
Cirsium euxinum Charadze	VU B1ab(ii,iii)+2ab(ii,iii)			x		x	
Cirsium fominii Petr.	LC			x			
Cirsium frickii Fisch. & C. A. Mey.	DD			x			x
Cirsium gagnidzei Charadze	NT			x		x	
Cirsium hydrophiloides Charadze	DD			x			
Cirsium imereticum Boiss.	LC			x			
Cirsium iranicum Petr.	EN				x		
Cirsium kemulariae Charadze	DD			x			
Cirsium ketzkhovelii Charadze [*Cirsium daghestanicum* Charadze]	DD		x	x		x	
Cirsium kosmelii (Adams) Fisch. ex Hohen. [*Cirsium cosmelii* (Adams) Fisch. ex Hohen.]	NE	x	x	x	x	x	x
Cirsium kusnezowianum Sommier & Levier [*Cirsium pseudopersonata* Boiss. & Balansa subsp. *kuznezowianum* (Sommier & Levier) Petr.]	NE			x		x	x
Cirsium macrocephalum C. A. Mey.	NE		x	x		x	
Cirsium munitum (M. Bieb.) Fisch. [*Cirsium pugnax* Sommier & Levier]	LC		x	x		x	x
Cirsium oblongifolium K. Koch	VU B1ab(iii)+2ab(iii)			x			
Cirsium obvallatum (M. Bieb.) Fisch.	NE	x		x	x	x	x
Cirsium pseudopersonata Boiss. & Balansa subsp. *pseudopersonata*	VU B1ab(iii)+2ab(iii)						x
Cirsium pseudopersonata Boiss. & Balansa subsp. *kuznezowianum* (Sommier & Levier) Petr. [*Cirsium kusnezowianum* Sommier & Levier]	NE			x		x	x
Cirsium pugnax Sommier & Levier [*Cirsium munitum* (M. Bieb.) Fisch.]	LC		x	x		x	x

TAXON	RED LIST ASSESSMENT	AR	AZ	GE	IR	RU	TU
Cirsium rigidum DC.	NE	x		x			x
Cirsium simplex C. A. Mey., s. str.	NE		x	x		x	x
Cirsium sinuatum (Trautv.) Boiss. & Kotschy [*Lamyropsis sinuata* (Trautv.) Dittrich]	NE		x	x		x	
Cirsium sosnowskyi Charadze	LC			x			
Cirsium strigosum (M. Bieb.) Fisch.	NE		x	x			
Cirsium svaneticum Sommier & Levier	NT			x		x	
Cirsium sychnosanthum Petrak	NT			x		x	
Cirsium trachylepis Boiss.	VU B1ab(iii)+2ab(iii)						x
Cirsium uliginosum (M. Bieb.) Fisch.	NT			x		x	
Cladochaeta candidissima DC.	NE		x	x		x	
Cousinia alfredii Bornm. & Gauba	NE				x		
Cousinia araxena Takht.	EN B1ab(i,ii,iii)+ 2ab(i,ii,iii)	x	x				
Cousinia armena Takht.	NE	x	x				
Cousinia atropatana Bunge	CR				x		
Cousinia azerbaijanica Djavadi, Attar & Najafi	NE				x		
Cousinia brachyptera DC.	LC	x		x			x
Cousinia calolepis Boiss.	NE				x		
Cousinia chlorocephala C. A. Mey. ex DC.	NE	x	x		x		
Cousinia daralaghezica Takht.	LC	x	x				
Cousinia erinacea Jaub. & Spach	NE				x		
Cousinia erivanensis Bornm.	EN B1ab(i,ii,iii)+ 2ab(i,ii,iii)	x	x				
Cousinia fedorovii Takht.	EN B1ab(i,ii,v)+ 2ab(i,ii,v)	x					
Cousinia gabrieljaniae Takht. & Tamanian	EN B1ab(iii)+2ab(iii)	x	x				
Cousinia gigantolepis Rech. f.	NE	x	x		x		
Cousinia gilanica Bornm.	NE				x		
Cousinia gilliatii Rech. f.	EN				x		

TAXON	RED LIST ASSESSMENT	AR	AZ	GE	IR	RU	TU
Cousinia gmelinii C. Winkl.	NE				x		
Cousinia grandis C. A. Mey. ex DC.	NE				x		
Cousinia hablitzii C. A. Mey. ex DC.	VU				x		
Cousinia hypochionea Bornm.	NE				x		
Cousinia iljinii Takht.	EN B1ab(iii)+2ab(iii)	x	x				
Cousinia intermedia C. A. Mey. ex DC.	EN B1ab(iii)+2ab(iii)	x	x				
Cousinia lomakinii C. Winkl.	EN B1ab(iii)+2ab(iii)	x	x				
Cousinia macrocephala C. A. Mey.	LC	x	x				
Cousinia meghrica Takht.	EN B1ab(i,ii,iii,v)+ 2ab(i,ii,iii,v)	x	x				
Cousinia microcephala C. A. Mey. ex DC.	NE	x	x		x		
Cousinia orientalis (Adams) K. Koch	NE	x	x	x		x	
Cousinia qaradaghensis Rech. f.	EN B1ab(iii)+2ab(iii)	x			x		
Cousinia tabrisiana Bunge	NE				x		
Cousinia takhtajanii Tamanian	EN B1ab(i,ii,iii)+ 2ab(i,ii,iii)	x					
Cousinia tenuifolia C. A. Mey. ex DC.	NE	x	x		x		
Cousinia urumiensis Bornm.	NE				x		
Cousinia wilhelminae Rech. f.	VU				x		
Cousinia woronowii Bornm.	VU B1ab(i,ii,iii)+ 2ab(i,ii,iii); C2a(i)						x
Crepis alfredii Bornm.	NE				x		
Crepis alikperi Tamamsch.	DD		x				
Crepis caucasica C. A. Mey., s. str.	LC			x		x	
Crepis ciliata K. Koch	NE	x		x	x		
Crepis conyzifolia (Gouan) A. Kern. subsp. *djimilensis* (K. Koch) Lamond [*Crepis pontica* C. A. Mey.]	NE			x			x
Crepis pontica C. A. Mey. [*Crepis conyzifolia* (Gouan) A. Kern. subsp. *djimilensis* (K. Koch) Lamond]	NE			x			x
Crepis praemorsa (L.) Tausch. subsp. *caucasigena* (Czerep.) Menitsky	LC					x	

TAXON	RED LIST ASSESSMENT	AR	AZ	GE	IR	RU	TU
Crepis sancta (L.) Babc. subsp. *glaucescens* (K. Koch) Menitsky [*Lagoseris glaucescens* (K. Koch) Sosn.]	NE	x	x			x	
Crepis sonchifolia (M. Bieb.) C. A. Mey.	NE	x	x	x		x	
Dolichorrhiza caucasica (M. Bieb.) Galushko [*Senecio caucasicus* (M. Bieb.) DC.]	NE	x		x		x	
Dolichorrhiza correvoniana (Albov) Galushko [*Senecio correvonianus* Albov]	NT			x		x	
Dolichorrhiza persica (Boiss.) B. Nord.	NE				x		
Dolichorrhiza renifolia (C. A. Mey.) Galushko [*Senecio renifolius* (C. A. Mey.) Sch. Bip.]	NT			x		x	
Doronicum balansae Cavill.	NE	x					x
Doronicum dolichotrichum Cavill. [*Doronicum macrophyllum* Fisch. ex Hornem.]	NE	x		x		x	x
Doronicum macrolepis Freyn & Sint.	VU B1ab(iii)+2ab(iii)						x
Doronicum macrophyllum Fisch. ex Hornem. [*Doronicum dolichotrichum* Cavill.]	NE	x		x		x	x
Doronicum tobeyi J. R. Edm.	CR B1ab(ii,iii)						x
Doronicum wendelboi J. R. Edm.	NE				x		
Echinops foliosus Sommier & Levier	VU D2			x			
Echinops galaticus Freyn subsp. *ossicus* (K. Koch) Gagnidze	DD			x			
Echinops polyacanthus Iljin [*Echinops pungens* Trautv. var. *polycanthus* (Iljin) Hedge]	NE	x					x
Echinops pungens Trautv. var. *polycanthus* (Iljin) Hedge [*Echinops polyacanthus* Iljin]	NE	x					x
Echinops pungens Trautv. var. *transcaucasicus* (Iljin) Hedge [*Echinops transcaucasicus* Iljin]	NE	x	x	x			x
Echinops sintenisii Freyn	EN B1ab(iii)+2ab(iii)						x
Echinops sphaerocephalus L. subsp. *cirsifolius* (K. Koch) Gagnidze	DD			x			

TAXON	RED LIST ASSESSMENT	AR	AZ	GE	IR	RU	TU
Echinops transcaucasicus Iljin [*Echinops pungens* Trautv. var. *transcaucasicus* (Iljin) Hedge]	NE	x	x	x			x
Echinops viridifolius Iljin	CR (PE) B1ab(iii)+ 2ab(iii)					x	
Erigeron pseudoelongatus Botsch.	LC					x	
Erigeron schalbusi Vierh.	EN B1ab(iii)+2ab(iii)					x	
Galatella eldarica Kem.-Nath.	DD			x			
Galatella linosyris (L.) Rchb. f. subsp. *fominii* (Kem.-Nath.) Tzvelev	DD					x	
Galatella pontica (Lipsky) Novopokr. & Bogdan	EN B1ab(ii,iii,v)+ 2ab(ii,iii,v)			x		x	
Grossheimia ahverdovii (Gabrieljan) Gabrieljan	NE	x					
Grossheimia caroli-henrici (Gabrieljan & Dittrich) Gabrieljan	CR B1ab(ii,iii)+2ab(ii,iii)	x					
Grossheimia macrocephala (Muss.-Puschk. ex Willd.) Sosn. & Takht. [*Centaurea macrocephala* Muss.-Puschk. ex Willd.]	NE	x	x	x	x		x
Grossheimia polyphylla (Ledeb.) Holub [*Centaurea polyphylla* Ledeb. ex Nordm.]	NT			x		x	
Helichrysum artvinense P. H. Davis & Kupicha	CR B1ab(ii,iii)+2ab(ii,iii)						x
Hieracium abakurae Schelk. & Zahn	DD			x			
Hieracium acuminatifolium (Litv. & Zahn) Üksip	DD			x		x	
Hieracium adenobrachion (Litv. & Zahn) Üksip	EN B2ab(ii,iii)					x	
Hieracium amblyolepis Boiss.	NE						x
Hieracium argillaceoides (Litw. & Zahn) Üksip	NE						x
Hieracium asterodermum (Woronow & Zahn) Üksip	NE						x
Hieracium atrocephalum Schmalh.	NT					x	
Hieracium azerbaijanense Lack	NE				x		

TAXON	RED LIST ASSESSMENT	AR	AZ	GE	IR	RU	TU
Hieracium biebersteinii Litv. & Zahn	DD			x		x	
Hieracium brittatense Üksip	DD			x			
Hieracium bupleurifolium (Wimm. & Grab.) Tausch ex Juksip subsp. *chaetothyrsoides* (Zahn) V. J. Nikolajev	NE					x	
Hieracium callichlorum Litv. & Zahn	NE					x	
Hieracium caloprasinum (Zahn) Üksip	NE						x
Hieracium caucasiense Arv.-Touv.	EN B1ab(ii,v)+2ab(ii,v)					x	
Hieracium chlorochromum Sosn. & Zahn	DD			x			
Hieracium chloroprenanthes Litv. & Zahn	NE					x	x
Hieracium chromolepium (Zahn) Kem.-Nath.	DD			x			
Hieracium cincinnatum Fr.	NE	x	x	x			
Hieracium concinnidens (Zahn) Üksip	DD			x			
Hieracium elisabethae Kem.-Nath.	DD			x			
Hieracium endaurovae Üksip	NE		x	x		x	
Hieracium erythrocarpoides (Litv. & Zahn) Kem.-Nath.	DD			x			
Hieracium gentiliforme (Zahn) P. D. Sell & C. West	NE						x
Hieracium gigantellum Litv. & Zahn	DD			x		x	
Hieracium glomerellum (Zahn) Üksip ex Sennikov	NE					x	
Hieracium gudergomiense Üksip	DD					x	
Hieracium haematoglossum Koslovsky & Zahn	DD			x			
Hieracium hoppeanum Schult.	NE				x		
Hieracium hylogeton (Koslovsky & Zahn) Üksip	DD			x			
Hieracium incaniforme Litv. & Zahn	NE		x	x		x	
Hieracium karagoellense (Zahn) P. D. Sell & C. West	NE						x
Hieracium kharthlicum Kem.-Nath.	DD			x			

TAXON	RED LIST ASSESSMENT	AR	AZ	GE	IR	RU	TU
Hieracium kubanicum (Litv. & Zahn) Üksip ex Sennikov	NT			x		x	
Hieracium laetevirens Sommier & Levier	DD			x			
Hieracium lazicum Boiss. & Balansa	NE						x
Hieracium lazistanum Avr.-Touv.	NE						x
Hieracium leptogrammoides Üksip	NE			x		x	x
Hieracium leptophytomorphum (Litv. & Zahn) Üksip	NT					x	
Hieracium leptoprenanthes Litv. & Zahn	NT			x		x	
Hieracium leucothyrsum (Litv. & Zahn) Üksip ex Sennikov	NE					x	
Hieracium litwinowianum Zahn	NE	x		x		x	
Hieracium longiscapum (Boiss. & Kotschy ex Nägeli & Peter) Zahn	NE					x	x
Hieracium macrolepis Boiss.	NE		x	x		x	
Hieracium macrolepis Boiss. subsp. *simplicicaule* (Sommier & Levier) Zahn [*Hieracium simplicicaule* (Sommier & Levier) Peter]	NE					x	
Hieracium macrolygodes (Zahn) Üksip	DD			x			
Hieracium macrophyllopodum (Zahn) Üksip	NE						x
Hieracium managettae Freyn	NE						x
Hieracium microtum Boiss.	NE					x	x
Hieracium pannoniciforme Litv. & Zahn	LC			x		x	
Hieracium polygonifolium Gottsche & Coşkunç.	NE						x
Hieracium pseudoconstrictum Zahn	DD					x	
Hieracium pseudosvaneticum Peter	LC			x		x	
Hieracium raddeanum Zahn	LC			x		x	
Hieracium retroversilobatum (Schelk. & Zahn) Üksip	DD			x			
Hieracium rigidellum Litv. & Zahn	NT					x	
Hieracium rizense Gottsche & Coşkunç.	NE						x

TAXON	RED LIST ASSESSMENT	AR	AZ	GE	IR	RU	TU
Hieracium ruprechtii Boiss.	NE		x	x		x	
Hieracium sarykamyschense Üksip	NE						x
Hieracium schelkownikowii Zahn	DD		x				
Hieracium simplicicaule (Sommier & Levier) Peter [*Hieracium macrolepis* Boiss. subsp. *simplicicaule* (Sommier & Levier) Zahn]	NE					x	
Hieracium spodocephalum Gottsche & Coşkunç.	NE						x
Hieracium subsimplex Sommier & Levier	DD			x			
Hieracium svaneticiforme (Litv. & Zahn) Kem.-Nath.	NE		x	x		x	
Hieracium syreistschikovii Zahn	NE						x
Hieracium tamderense Hub.-Mor.	NE						x
Hieracium teberdense (Litv. & Zahn) Üksip	NE					x	x
Hieracium terekianum (Litv. & Zahn) Üksip ex V. J. Nikolajev	NE			x	x	x	
Hieracium tersundagense Gottsche & Coşkunç.	NE						x
Hieracium tschkhubianischwilii Kem.-Nath.	DD			x			
Hieracium turfosum Kem.-Nath.	DD			x			
Inula acaulis Schott & Kotschy ex Boiss.	NE	x	x	x			x
Inula grandiflora Willd.	DD		x	x		x	
Inula magnifica Lipsky	LC			x		x	
Inula mariae Bordz.	NE	x	x	x			x
Jurinea akinfievii Nemirova	CR B1ab(i,ii,iii,iv,v)					x	
Jurinea alata (Desf.) Cass.	VU B1ab(ii,iii)+2ab(ii,iii)					x	
Jurinea annae Sosn.	NE					x	
Jurinea armeniaca Sosn.	NE	x	x				
Jurinea bellidioides Boiss.	VU D2					x	
Jurinea blanda (M. Bieb.) C. A. Mey.	NE	x	x	x		x	

TAXON	RED LIST ASSESSMENT	AR	AZ	GE	IR	RU	TU
Jurinea brachypappa Nemirova	VU D2					x	
Jurinea cartaliniana Boiss.	DD			x		x	
Jurinea ciscaucasica (Sosn.) Iljin	NT					x	
Jurinea coronopifolia Sommier & Levier	VU B1ab(iii)+2ab(iii)					x	
Jurinea dolomitica Galushko	NE					x	
Jurinea elegans (Steven) DC.	NE	x	x	x			
Jurinea exuberans (Trautv.) Sosn.	VU D2			x			
Jurinea filicifolia Boiss.	NT			x		x	
Jurinea galushkoi Nemirova	VU C2a(i)					x	
Jurinea iljinii Grossh.	EN B1ab(ii,iii)+2b(ii,iii)					x	
Jurinea leptoloba DC.	NE				x		
Jurinea levieri Albov	DD			x		x	
Jurinea praetermissa Galushko & Nemirova	EN B1ab(i,ii,iii)+ 2ab(i,ii,iii)	x					
Jurinea prokhanovii Nemirova	CR B1ab(ii,iii)+2ab(ii,iii)					x	
Jurinea pseudoiljinii Galushko & Nemirova	EN B1ab(ii)+2ab(ii)					x	
Jurinea pumila Albov	NT			x		x	
Jurinea ruprechtii Boiss.	EN B1ab(ii,iii)					x	
Jurinea sosnowskyi Grossh.	EN B2ab(iii,v)					x	
Jurinea spectabilis Fisch. & C. A. Mey.	NE	x	x				
Jurinea venusta Iljin	NE					x	
Jurinea woronowii Iljin	VU B1ab(iii,v)+2ab(iii,v); C2a(i)					x	
Jurinella squarrosa (Fisch. & C. A. Mey.) Iljin	NE	x	x	x			
Kemulariella abchasica (Kem.-Nath.) Tamamsch.	VU B1ab(ii,iii)+2ab(ii,iii)			x		x	
Kemulariella caucasica (Willd.) Tamamsch.	LC			x		x	x
Kemulariella colchica (Albov) Tamamsch.	VU B2ab(iii)			x		x	x
Kemulariella rosea (Steven ex M. Bieb.) Tamamsch. [*Aster roseus* Steven]	NE		x	x		x	

TAXON	RED LIST ASSESSMENT	AR	AZ	GE	IR	RU	TU
Kemulariella tugana (Albov) Tamamsch.	LC			x			
Lactuca azerbaijanica Rech. f.	CR				x		
Lagoseris glaucescens (K. Koch) Sosn. [*Crepis sancta* (L.) Babc. subsp. *glaucescens* (K. Koch) Menitsky]	NE	x	x			x	
Lamyropsis charadzeae Kimer. [*Cirsium charadzeae* (Kimer.) Czerep.]	DD			x		x	
Lamyropsis sinuata (Trautv.) Dittrich [*Cirsium sinuatum* (Trautv.) Boiss. & Kotschy]	NE		x	x		x	
Lapsana pinnatisecta (Sommier & Levier) Ter-Chatsch.	DD			x			
Ligularia subsagittata Pojark.	NE	x	x	x		x	
Mycelis deltoidea (M. Bieb.) Sennikov [*Cicerbita deltoidea* (M. Bieb.) Beauverd]	NE		x	x		x	
Nardosmia fominii (Bordz.) Kuprian. [*Petasites fominii* Bordz.]	NE			x		x	
Oligochaeta tomentosa Czerep.	NE	x	x		x		
Onopordum cinereum Grossh.	LC	x	x				
Petasites fominii Bordz. [*Nardosmia fominii* (Bordz.) Kuprian.]	LC			x		x	
Petasites georgicus Manden.	DD			x			
Picris canescens (Steven) V. N. Vassil. [*Picris laevis* C. A. Mey.]	LC		x			x	
Picris laevis C. A. Mey. [*Picris canescens* (Steven) V. N. Vassil.]	NE		x			x	
Podospermum alpigenum K. Koch	EN B1ab(iii)+2ab(iii)		x	x			
Podospermum grigoraschvilii Sosn.	EN B2ab(iii)			x			
Podospermum idae Sosn.	VU D2			x			
Podospermum lachnostegium Woronow [*Scorzonera lachnostegia* (Woronow) Lipsch.]	DD	x				x	
Podospermum schischkinii (Lipsch. & Vassilcz.) Kuth.	EN B1ab(iii)+2ab(iii)					x	
Prenanthes abietina (Boiss. & Balansa) Kirp.	LC			x		x	x

TAXON	RED LIST ASSESSMENT	AR	AZ	GE	IR	RU	TU
Prenanthes cacaliifolia (M. Bieb.) Beauverd [*Cicerbita macrophylla* (Willd.) Wallr.]	NE	x	x	x		x	x
Prenanthes petiolata (K. Koch) Sennikov [*Cicerbita petiolata* (K. Koch) Gagnidze]	DD			x		x	
Psephellus abchasicus Albov	VU B1ab(ii,iii)+2ab(ii,iii)			x		x	
Psephellus adjaricus (Albov) Grossh.	VU D2			x			
Psephellus alexeenkoi Alieva	NT					x	
Psephellus barbeyi Albov	VU B1ab(ii)+2ab(ii)			x		x	
Psephellus bellis (Trautv.) Wagenitz [*Centaurea bella* Trautv.]	NE	x		x			
Psephellus boissieri (Sosn.) Sosn.	EN B1ab(iii)					x	
Psephellus buschiorum Sosn.	NE	x		x		x	
Psephellus carthalinicus Sosn.	LC	x		x			
Psephellus circassicus (Albov) Galushko	NT			x		x	
Psephellus ciscaucasicus (Sosn) Galushko	VU B1ab(iii)					x	
Psephellus colchicus Sosn.	LC			x		x	
Psephellus cronquistii (Takht. & Gabrieljan) Gabrieljan	EN B1ab(i,ii,iii)+ 2ab(i,ii,iii)	x					
Psephellus daghestanicus Sosn.	NE	x	x	x		x	
Psephellus dealbatus (Willd.) K. Koch	NE	x	x	x		x	x
Psephellus debedicus (Gabrieljan) Gabrieljan	EN B1ab(i,ii,iii)+ 2ab(i,ii,iii)	x					
Psephellus galushkoi Alieva	CR B1ab(iii,v)+2ab(iii,v); C2a(ii)					x	
Psephellus geghamensis (Gabrieljan) Gabrieljan	EN B2ab(i,ii,iii)	x					
Psephellus holophyllus Soczava & Lipatova	LC			x		x	
Psephellus hymenolepis (Trautv.) Boiss.	VU B1ab(ii,iii)		x			x	
Psephellus kacheticus Rehmann ex Boiss.	DD			x			
Psephellus karabaghensis Sosn.	LC	x	x				
Psephellus kolakovskii (Sosn.) Greuter	DD			x			

TAXON	RED LIST ASSESSMENT	AR	AZ	GE	IR	RU	TU
Psephellus leucophyllus (M. Bieb.) C. A. Mey.	LC					x	
Psephellus maleevii Sosn.	EN B1ab(ii)+2ab(ii)					x	
Psephellus manakyanii (Gabrieljan) Gabrieljan	EN B1ab(i,ii,iii)+ 2ab(i,ii,iii)	x					
Psephellus meskheticus (Sosn.) Gabrieljan	NT	x		x			
Psephellus pambakensis Sosn.	LC	x					
Psephellus paucilobus (Trautv.) Boiss.	EN B1ab(iii)					x	
Psephellus popovae (Gabrieljan) Gabrieljan	NE	x	x		x		
Psephellus prokhanovii Galushko	LC					x	
Psephellus pseudoandinus Galushko & Alieva	DD					x	
Psephellus salviifolius Boiss.	NT					x	
Psephellus somcheticus Sosn.	LC	x		x			
Psephellus sosnowskyi Chukhr.	DD			x			
Psephellus taochius Sosn. [*Centaurea taochia* (Sosn.) Sosn.]	EN B2ab(iii,v)	x					x
Psephellus transcaucasicus Sosn.	DD	x	x				
Psephellus troitzkyi Sosn.	EN B2ab(iii)					x	
Psephellus zangezuri Sosn.	EN B2ab(i,ii,iii)	x					
Psephellus zuvandicus Sosn. subsp. *gegharkunikensis* (Gabrieljan) Gabrieljan	CR B2ab(iii)	x					
Psephellus zuvandicus Sosn. subsp. *jelenewskyi* (Gabrieljan) Gabrieljan	EN B1ab(iii)+2ab(iii)	x					
Psephellus zuvandicus Sosn. subsp. *zuvandicus*	NE	x	x				
Ptarmica ptarmicifolia (Willd.) Galushko [*Achillea ptarmicifolia* (Willd.) Rupr.]	LC			x		x	
Pyrethrum aromaticum Tzvelev [*Pyrethrum daghestanicum* (Rupr. ex Boiss.) Flerow; *Tanacetum daghestanicum* (Rupr. ex Boiss.) K. Bremer & Humphries]	DD		x	x		x	

TAXON	RED LIST ASSESSMENT	AR	AZ	GE	IR	RU	TU
Pyrethrum carneum M. Bieb.	NE		x	x			
Pyrethrum chamaemelifolium (Sommier & Levier) Sosn. [*Tanacetum coccineum* (Willd.) Grierson subsp. *chamaemelifolium* (Sommier & Levier) Grierson]	LC	x	x	x			x
Pyrethrum daghestanicum (Rupr. ex Boiss.) Flerow [*Pyrethrum aromaticum* Tzvelev; *Tanacetum daghestanicum* (Rupr. ex Boiss.) K. Bremer & Humphries]	DD		x	x		x	
Pyrethrum demetrii Manden. [*Pyrethrum divaricatum* Sosn.]	DD		x	x			
Pyrethrum divaricatum Sosn. [*Pyrethrum demetrii* Manden.]	DD		x	x			
Pyrethrum glanduliferum Sommier & Levier	DD			x			
Pyrethrum komarovii Sosn. [*Tanacetum zangezuricum* Chandjian]	NT	x	x				
Pyrethrum kubense Grossh.	DD		x				
Pyrethrum leptophyllum Steven ex M. Bieb. [*Tanacetum leptophyllum* (Steven ex M. Bieb.) Sch. Bip.]	NE		x	x		x	
Pyrethrum marionii Albov	LC			x			
Pyrethrum meyerianum Sosn.	VU D2		x				
Pyrethrum ordubadense Manden.	NT		x				
Pyrethrum peucedanifolium (Sosn.) Manden. [*Tanacetum peucedanifolium* (Sosn.) K. Bremer & Humphries]	DD			x		x	
Pyrethrum punctatum (Desr.) Bordz. ex Grossh. & Schischk.	LC		x	x			
Pyrethrum sericeum (Adams) M. Bieb. [*Tanacetum sericeum* (Adams) Sch. Bip.]	NE			x			x
Pyrethrum sorbifolium Boiss. [*Tanacetum sorbifolium* (Boiss.) Grierson]	NE			x			x
Rhaponticoides hajastana (Tzvelev) M. V. Agab. & Greuter	EN B1ab(i,ii,iii,iv)+2ab(i,ii,iii,iv)	x					

TAXON	RED LIST ASSESSMENT	AR	AZ	GE	IR	RU	TU
Rhaponticoides tamanianiae (M. V. Agab.) M. V. Agab. & Greuter	CR B2ab(ii,iii,v)	x					
Scorzonera aragatzi Kuth.	NT	x	x				
Scorzonera biebersteinii Lipsch.	DD	x	x	x			
Scorzonera charadzeae Papava	DD			x		x	
Scorzonera czerepanovii Kamelin [*Scorzonera lanata* (L.) Hoffm.]	VU B2ab(v)		x	x			
Scorzonera filifolia Boiss.	DD		x			x	
Scorzonera gorovanica Nazarova	VU B1ab(i,ii,iii)+ 2ab(i,ii,iii)	x	x				
Scorzonera grossheimii Lipsch. & Vassilcz.	EN B1ab(i,ii,iii,iv)+ B2ab(i,ii,iii,iv)		x				
Scorzonera ketzkhowelii Sosn. ex Grossh.	CR B1ab(iii)			x			
Scorzonera kirpitschnikovii Lipsch.	NT		x				
Scorzonera kozlowskyi Sosn. ex Grossh.	CR B1ab(iii)			x			
Scorzonera lachnostegia (Woronow) Lipsch. [*Podospermum lachnostegium* Woronow]	NE	x				x	
Scorzonera lanata (L.) Hoffm. [*Scorzonera czerepanovii* Kamelin]	VU B2ab(v)		x	x			
Scorzonera pulchra Lomak.	DD		x				
Scorzonera safievii Grossh.	EN B1ab(i,ii,iii,iv)+ 2ab(i,ii,iii,iv)	x					
Scorzonera seidlitzii Boiss.	NE	x	x	x			x
Scorzonera sosnowskyi Lipsch. [*Scorzonera dzhawakhetica* Sosn. ex Grossh.]	NE			x			x
Scorzonera szovitsii DC.	NE	x			x		
Senecio buschianus Sosn.	VU B1ab(iii)					x	
Senecio caucasicus (M. Bieb.) DC. [*Dolichorrhiza caucasica* (M. Bieb.) Galushko]	NE	x		x		x	
Senecio caucasigenus Schischk. [*Tephroseris caucasigena* (Schischk.) Czerep.]	DD		x			x	x

TAXON	RED LIST ASSESSMENT	AR	AZ	GE	IR	RU	TU
Senecio cladobotrys Ledeb. [*Tephroseris cladobotrys* (Ledeb.) Griseb. & Schenk]	LC			x		x	
Senecio correvonianus Albov [*Dolichorrhiza correvoniana* (Albov) Galushko]	NT			x		x	
Senecio integrifolius (L.) Clairv. subsp. *karsianus* V. A. Matthews	EN B1ab(i,ii,iii)+ 2ab(i,ii,iii)						x
Senecio kolenatianus C. A. Mey.	NE	x		x		x	
Senecio kubensis Grossh. [*Senecio massagetovii* Schischk.]	DD		x	x		x	
Senecio lapsanoides DC.	NE		x	x		x	
Senecio lazicus Boiss. & Balansa	EN B1ab(i,ii,iii)+ 2ab(i,ii,iii)						x
Senecio lipskyi Lomak.	NT		x				
Senecio lorentii Hochst.	NE	x	x	x			x
Senecio macrophyllus M. Bieb.	LC			x		x	
Senecio massagetovii Schischk. [*Senecio kubensis* Grossh.]	DD		x	x		x	
Senecio ovatifolius Boiss.	EN B1ab(i,ii,iii)+ 2ab(i,ii,iii)						x
Senecio pandurifolius K. Koch	NT			x		x	x
Senecio platyphylloides Sommier & Levier	LC			x		x	
Senecio platyphyllus DC. var. *glandulosus* V. A. Matthews	VU B1ab(i,ii,iii)						x
Senecio pojarkoviae Schischk.	NE		x	x		x	
Senecio renifolius (C. A. Mey.) Sch. Bip. [*Dolichorrhiza renifolia* (C. A. Mey.) Galushko]	LC			x		x	
Senecio rhombifolius (Adams ex Willd.) Sch. Bip. [*Adenostyles macrophylla* (M. Bieb.) Czerep.]	NE	x	x	x		x	
Senecio schischkinianus Sofieva	VU B2ab(iii)					x	
Senecio similiflorus Kolak.	VU D2			x			
Senecio sosnovskyi Sofieva	NE		x	x		x	

TAXON	RED LIST ASSESSMENT	AR	AZ	GE	IR	RU	TU
Senecio subfloccosus Schischk. [*Tephroseris subfloccosa* (Schischk.) Czerep.]	DD		x	x		x	
Senecio trapezuntinus Boiss.	CR B2ab(ii,iii)						x
Serratula caucasica Boiss.	NT		x			x	
Solidago caucasica Kem.-Nath.	NE		x	x		x	
Solidago turfosa Woronow ex Grossh.	EN B2ab(iii)			x			
Sonchus araraticus Nazarova & Barsegyan	CR B1ab(i,ii,iii)+ 2ab(i,ii,iii)	x					
Sonchus ketzkhovelii Schchian	DD			x			
Sonchus sosnowskyi Schchian	DD	x					
Sosnowskya amblyolepis (Ledeb.) Takht.	NE		x	x			
Stemmacantha pulchra (Fisch. & C. A. Mey.) Dittrich	NE		x	x	x	x	
Tanacetum akinfiewii (T. Alex.) Tzvelev	CR B1ab(ii)+2ab(ii); C2a(ii)					x	
Tanacetum aucherianum Sch. Bip.	EN B1ab(v)+2ab(v)						x
Tanacetum coccineum (Willd.) Grierson subsp. *chamaemelifolium* (Sommier & Levier) Grierson [*Pyrethrum chamaemelifolium* (Sommier & Levier) Sosn.]	LC	x	x	x			x
Tanacetum daghestanicum (Rupr. ex Boiss.) K. Bremer & Humphries [*Pyrethrum aromaticum* Tzvelev; *Pyrethrum daghestanicum* (Rupr. ex Boiss.) Flerow]	DD		x	x		x	
Tanacetum leptophyllum (Steven ex M. Bieb.) Sch. Bip. [*Pyrethrum leptophyllum* Steven ex M. Bieb.]	NE		x	x		x	
Tanacetum oltense (Sosn.) Grierson	DD						x
Tanacetum oxystegium (Sosn.) Grierson	CR C2a(i)						x
Tanacetum peucedanifolium (Sosn.) K. Bremer & Humphries [*Pyrethrum peucedanifolium* (Sosn.) Manden.]	DD			x		x	
Tanacetum salsugineum Podlech	NE				x		

TAXON	RED LIST ASSESSMENT	AR	AZ	GE	IR	RU	TU
Tanacetum sericeum (Adams) Sch. Bip. [*Pyrethrum sericeum* (Adams) M. Bieb.]	NE			x			x
Tanacetum silaifolium (DC.) Sch. Bip. [*Chamaemelum silaifolium* (Steven) Trautv.]	DD		x			x	
Tanacetum sorbifolium (Boiss.) Grierson [*Pyrethrum sorbifolium* Boiss.]	NE			x			x
Tanacetum tenuissimum (Trautv.) Grossh.	NT		x				
Tanacetum tricholobum (Sosn. ex Manden.) Chandjian	NT					x	
Tanacetum zangezuricum Chandjian [*Pyrethrum komarovii* Sosn.]	NT	x	x				
Taraxacum anatolicum Soest	CR B1ab(i,ii,iii)+ 2ab(i,ii,iii)						x
Taraxacum bessarabicum (Hornem.) Hand.-Mazz. subsp. *gumusanicum* Soest	DD						x
Taraxacum ciscaucasicum Schischk.	NT					x	
Taraxacum confusum Schischk.	NE	x	x	x		x	
Taraxacum desertorum Schischk.	LC		x			x	
Taraxacum divulsiforme R. Doll	DD					x	
Taraxacum grossheimii Schischk.	NE	x	x	x		x	
Taraxacum praticola Schischk.	DD		x	x			
Taraxacum tenuisectum Sommier & Levier	LC		x			x	
Taraxacum voronovii Schischk.	EN			x			
Tephroseris caucasigena (Schischk.) Czerep. [*Senecio caucasigenus* Schischk.]	DD		x			x	x
Tephroseris cladobotrys (Ledeb.) Griseb. & Schenk [*Senecio cladobotrys* Ledeb.]	LC			x		x	
Tephroseris karjaginii (Sofieva) Holub	LC			x		x	
Tephroseris subfloccosa (Schischk.) Czerep. [*Senecio subfloccosus* Schischk.]	DD		x	x		x	
Tomanthea daralaghezica (Fomin) Takht.	VU B1ab(i,ii,iii)+ 2ab(i,ii,iii)	x	x				

TAXON	RED LIST ASSESSMENT	AR	AZ	GE	IR	RU	TU
Tragopogon armeniacus Kuth.	EN B1ab(i,ii,iii)+ 2ab(i,ii,iii)	x					
Tragopogon brevirostris DC.	NT		x			x	
Tragopogon colchicus Albov	VU D2			x			
Tragopogon daghestanicus (Artemczuk) Kuth.	LC					x	
Tragopogon fibrosus Freyn & Sint. ex Freyn	DD						x
Tragopogon filifolius Rehmann ex Boiss.	LC			x		x	
Tragopogon karjaginii Kuth.	EN B1ab(v)		x				
Tragopogon kemulariae Kuth.	NE	x	x	x	x		
Tragopogon ketzkhovelii Kuth.	VU D2			x			
Tragopogon leonidae Kuth.	DD					x	
Tragopogon macropogon C. A. Mey.	DD		x				
Tragopogon makaschwilii Kuth.	VU D2			x			
Tragopogon marginatus Boiss. & Buhse	NE	x	x	x	x		
Tragopogon meskheticus Kuth.	VU D2			x			
Tragopogon nachitschevanicus Kuth.	EN B1ab(ii,iii)+2ab(ii,iii)		x				
Tragopogon otschiaurii Kuth.	VU D2			x			
Tragopogon rezaiyensis Rech. f.	NE				x		
Tragopogon segetus Kuth.	NT	x					
Tragopogon serotinus Sosn.	NE	x	x	x			
Tragopogon sosnowskyi Kuth.	LC	x	x				
Tragopogon tuberosus K. Koch	NE	x	x	x		x	
Tripleurospermum colchicum (Manden.) Pobed.	DD			x			
Tripleurospermum fissurale (Sosn.) E. Hossain	EN B1ab(i,ii,iii)+ 2ab(i,ii,iii); C1						x
Tripleurospermum heterolepis (Freyn & Sint.) Bornm.	DD						x
Tripleurospermum nathaliae Zernov	NT					x	
Tripleurospermum szowitzii (DC.) Pobed.	VU D2			x			

TAXON	RED LIST ASSESSMENT	AR	AZ	GE	IR	RU	TU
BERBERIDACEAE							
Berberis iberica Steven & Fisch. ex DC.	NE	x	x	x		x	
Epimedium circinatocucullatum Sosn.	EN B2ab(ii,iii); C2a(i)			x		x	
Epimedium colchicum (Boiss.) Trautv. [*Epimedium pinnatum* Fisch. ex DC. subsp. *colchicum* (Boiss.) N. Busch.]	NT			x		x	x
Epimedium pinnatum Fisch. ex DC. subsp. *colchicum* (Boiss.) N. Busch. [*Epimedium colchicum* (Boiss.) Trautv.]	NT			x		x	x
Gymnospermium smirnovii (Trautv.) Takht.	EN B1ab(ii,iii,v)+ 2ab(ii,iii,v)			x			
BETULACEAE							
Alnus barbata C. A. Mey. [*Alnus glutinosa* (L.) Gaertn. subsp. *barbata* (C. A. Mey.) Yalt.]	NE		x	x		x	x
Alnus glutinosa (L.) Gaertn. subsp. *barbata* (C. A. Mey.) Yalt. [*Alnus barbata* C. A. Mey.]	NE		x	x		x	x
Betula browicziana Güner	EN B1ab(i,ii,iii,iv)+ 2ab(i,ii,iii,iv)						x
Betula medwediewii Regel	VU B1ab(iii,v)			x			x
Betula megrelica Sosn.	VU B1ab(iii,v)+2b(iii,v)			x			
Betula raddeana Trautv.	NT		x	x		x	
Carpinus geoktschaica Radde-Fomina [*Carpinus schuschaensis* H. J. P. Winkl.]	NE	x	x				
Carpinus schuschaensis H. J. P. Winkl. [*Carpinus geoktschaica* Radde-Fomina]	NE	x	x				
Corylus abchasica (Kem.-Nath.) Kem.-Nath.	DD			x			
Corylus cervorum Petrov	DD		x				
Corylus colchica Albov	VU B2ab(iii)			x			
Corylus egrissiensis Kem.-Nath.	DD			x			
Corylus iberica Wittm. ex Kem.-Nath.	DD		x	x			
Corylus imeretica Kem.-Nath.	DD			x			
Corylus kachetica Kem.-Nath.	DD			x			

TAXON	RED LIST ASSESSMENT	AR	AZ	GE	IR	RU	TU
BORAGINACEAE							
Aipyanthus pulcher (Willd. ex Roem. & Schult.) Kolak. [*Arnebia echioides* (L.) DC.; *Arnebia pulchra* (Roem. & Schult.) J. R. Edm.; *Macrotomia echioides* (L.) Boiss.]	NE	x	x	x		x	x
Alkanna cordifolia K. Koch	VU B1ab(i,ii,iii,iv)						x
Arnebia echioides (L.) DC. [*Aipyanthus pulcher* (Willd. ex Roem. & Schult.) Kolak.; *Arnebia pulchra* (Roem. & Schult.) J. R. Edm.; *Macrotomia echioides* (L.) Boiss.]	NE	x	x	x		x	x
Arnebia pulchra (Roem. & Schult.) J. R. Edm. [*Aipyanthus pulcher* (Willd. ex Roem. & Schult.) Kolak.; *Arnebia echioides* (L.) DC.; *Macrotomia echioides* (L.) Boiss.]	NE	x	x	x		x	x
Cynoglossum holosericeum Steven	NE		x			x	x
Eritrichium caucasicum (Albov) Grossh.	NT			x		x	
Eritrichium gracillimum Rech. f.	NE				x		
Heliotropium gypsaceum Rech. f. & Riedl	CR				x		
Heliotropium styligerum Trautv.	LC			x		x	
Heliotropium szovitsii (Steven) Bunge	NE	x	x		x		
Heliotropium tzvelevii T. N. Popova	VU B1ab(i,ii,iii)+ 2ab(i,ii,iii)	x	x				
Lindelofia kandavanensis Bornm. & Gauba	NE				x		
Macrotomia echioides (L.) Boiss. [*Aipyanthus pulcher* (Willd. ex Roem. & Schult.) Kolak.; *Arnebia echioides* (L.) DC.; *Arnebia pulchra* (Roem. & Schult.) J. R. Edm.]	NE	x	x	x		x	x
Myosotis amoena (Rupr.) Boiss.	NE	x	x	x		x	x
Myosotis anomala Riedl	NE				x		
Myosotis daralaghezica T. N. Popova	EN B2ab(i,ii,iii)	x					
Myosotis lazica Popov	NT			x			x
Myosotis schistosa A. P. Khokhr.	DD					x	
Nonea alpestris (Steven) G. Don	NE		x	x		x	

TAXON	RED LIST ASSESSMENT	AR	AZ	GE	IR	RU	TU
Nonea anchusoides Boiss. & Buhse	NE				x		
Nonea armeniaca (Kusn.) Grossh.	LC	x		x			x
Nonea daghestanica Kusn.	DD		x			x	
Nonea decurrens (C. A. Mey.) G. Don	NE		x	x		x	
Nonea intermedia Ledeb.	NE			x		x	x
Nonea karsensis Popov	CR B2ab(i,ii,iii)						x
Nonea polychroma Selvi & Bigazzi	NE	x					
Nonea pulmonarioides Boiss. & Balansa	EN B1ab(i,ii,iii)+ 2ab(i,ii,iii)						x
Nonea setosa (Lehm.) Roem. & Schult.	NE		x	x		x	
Nonea versicolor (Steven) Sweet	NE	x		x		x	x
Omphalodes cappadocica (Willd.) DC.	LC			x		x	x
Omphalodes caucasica Brand.	LC			x		x	
Omphalodes kusnetzovii Kolak.	DD			x			
Omphalodes lojkae Sommier & Levier	LC			x		x	
Omphalodes rupestris Rupr. ex Boiss.	NE		x	x		x	
Onosma arcuata Riedl	CR B2ab(i,ii,iii); C2a(i)						x
Onosma bilabiatum Boiss. & Buhse	NE				x		
Onosma caucasica E. G. Levin ex Popov	NE			x		x	x
Onosma caucasica E. G. Levin ex Popov subsp. *oligotricha* T. N. Popova	DD					x	
Onosma circinnata Riedl	VU B1ab(i,ii,iii,iv)						x
Onosma gracilis Trautv.	NT	x	x				x
Onosma levinii T. N. Popova	DD		x			x	
Onosma nigricaulis Riedl	CR B1ab(i,ii,iii)						x
Onosma obtusifolia Hausskn. & Sint. ex Riedl	CR B2ab(i,ii,iii)						x
Onosma sharifii Riedl	NE				x		
Paracaryum artvinense R. R. Mill	EN B1ab(i,ii,iii)+ 2ab(i,ii,iii)						x
Paracaryum laxiflorum Trautv.	NE	x	x	x			x
Paracynoglossum imeretinum (Kusn.) Popov	VU B2ab(iii)			x			

TAXON	RED LIST ASSESSMENT	AR	AZ	GE	IR	RU	TU
Rindera media (Turrill) Riedl	NE				x		
Rindera regia (S. G. Gmel.) Kusn.	NE				x		
Symphytum abchasicum Trautv.	DD			x		x	
Symphytum armeniacum Bucknall	DD						x
Symphytum ciscaucasicum Gvin.	NE					x	
Symphytum grandiflorum DC.	LC			x		x	
Symphytum hajastanum Gvin.	EN B2ab(i,ii,iii,iv)	x					
Symphytum ibericum Steven ex M. Bieb.	DD			x			x
Symphytum longipetiolatum Wickens	VU B1ab(i,ii,iii,iv)						x
Symphytum podcumicum Yu. M. Frolov	EN B1ab(iii,v)+2ab(iii,v); C2a(i)					x	
Symphytum savvalense Kurtto	CR B1ab(i,ii,iii)+ 2ab(i,ii,iii)						x
Symphytum sylvaticum Boiss. subsp. *sylvaticum*	CR B1ab(i,ii,iii)						x
Symphytum sylvaticum Boiss. var. *hordokopii* (Kurtto) R. R. Mill	DD						x
Symphytum sylvaticum Boiss. var. *sepulcrale*	VU B1ab(i,ii,iii)						x
Trigonocaryum involucratum (Steven) Medw.	NE		x	x		x	
BRASSICACEAE							
Aethionema diastrophis Bunge	LC	x	x				x
Aethionema edentulum N. Busch	NE	x	x		x		
Aethionema levandowskyi N. Busch	NT		x				
Aethionema pulchellum Boiss. & A. Huet	DD	x	x				
Aethionema sintenisii Hausskn. & Bornm.	CR B2ab(i,ii,iii)						x
Alliaria brachycarpa M. Bieb.	NE		x	x		x	
Alyssum andinum Rupr.	EN B2ab(ii,v)		x			x	
Alyssum artwinense N. Busch	EN B1ab(i,ii,iii,iv)+ 2ab(i,ii,iii,iv)						x

TAXON	RED LIST ASSESSMENT	AR	AZ	GE	IR	RU	TU
Alyssum dagestanicum Rupr.	LC					x	
Alyssum gehamense Fed.	NE	x	x			x	x
Alyssum hajastanum Avet.	EN B2ab(i,ii,iii)	x					
Apterigia pumila (Steven) Galushko	NE		x	x		x	
Arabis armena N. Busch	NT	x	x				
Arabis brachycarpa Rupr.	NE	x	x	x		x	x
Arabis christianii N. Busch [*Arabis mollis* Steven]	NE	x	x			x	
Arabis colchica Kolak.	NT			x			
Arabis farinacea Rupr.	DD		x			x	
Arabis kazbegi Mtskhvet.	NT			x			
Arabis mollis Steven [*Arabis christianii* N. Busch]	NE	x	x			x	
Arabis nordmanniana Rupr.	LC			x		x	
Arabis sachokiana N. Busch	VU B2ab(ii); C2a(i)			x		x	
Barbarea grandiflora N. Busch	EN B1ab(ii,iii,v)					x	
Barbarea ketzkhovelii Mardal.	NT			x			
Barbarea lutea Coode & Cullen	EN B1ab(i,ii,iii)+ 2ab(i,ii,iii)						x
Callothlaspi abchasicum F. K. Mey.	DD			x			
Cardamine microphylla (Willd.) O. E. Schulz [*Dentaria microphylla* Willd.]	LC			x		x	x
Cardamine seidlitziana Albov	DD			x		x	
Cardaria propinqua (Fisch. & C. A. Mey.) N. Busch [*Lepidium propinquum* Fisch. & C. A. Mey.]	NE	x	x	x	x	x	
Carpoceras tatianae (Bordz.) Grossh.	NT	x	x				
Chorispora iberica (M. Bieb.) DC.	NE	x	x	x	x	x	x
Clypeola raddeana Albov	EN A3c; C2a(i)			x			x
Crambe armena N. Busch	EN B1ab(iii)+2ab(iii); D	x	x				
Crambe cordifolia Steven	EN B1ab(iii,v)c(iv); C2b					x	
Crambe gibberosa Rupr.	EN B2ab(iii,v)		x			x	
Dentaria bipinnata C. A. Mey.	LC			x		x	

TAXON	RED LIST ASSESSMENT	AR	AZ	GE	IR	RU	TU
Dentaria microphylla Willd. [*Cardamine microphylla* (Willd.) O. E. Schulz]	NE			x		x	x
Draba bruniifolia Steven subsp. *armeniaca* Coode & Cullen	VU B1ab(iii)						x
Draba bryoides DC. [*Draba rigida* Willd. var. *bryoides* (DC.) Boiss.]	NE	x	x	x		x	x
Draba elisabethae N. Busch	LC					x	
Draba hispida Willd.	NE	x		x		x	x
Draba imbricata C. A. Mey.	DD					x	
Draba imeretica (Rupr.) Rupr. [*Draba terekemensis* Yıld.]	EN B1ab(iii)+2ab(iii)			x			x
Draba incompta Steven	DD		x			x	
Draba longisiliqua Schmalh. ex Akinf.	NT		x			x	
Draba meskhetica Chinth.	VU D2			x			
Draba mingrelica Schischk. ex Grossh.	NT			x			
Draba mollissima Steven	NT		x			x	
Draba narmanensis Yıld.	CR B2ab(ii,iii)						x
Draba ossetica (Rupr.) Sommier & Levier	VU B2ab(iii)			x		x	
Draba rigida Willd. var. *bryoides* (DC.) Boiss. [*Draba bryoides* DC.]	NE	x	x	x		x	x
Draba scabra C. A. Mey.	NE	x	x	x		x	
Draba siliquosa M. Bieb.	NE	x	x	x	x	x	x
Draba subsecunda Sommier & Levier	VU B2ab(v)			x		x	
Draba supranivalis Rupr.	LC			x		x	
Draba terekemensis Yıld. [*Draba imeretica* (Rupr.) Rupr.]	EN B1ab(iii)+2ab(iii)			x			x
Erucastrum takhtajanii V. I. Dorof.	NE	x					
Erysimum argyrocarpum N. Busch	NE		x		x		
Erysimum armeniacum (Sims) J. Gay	NE	x				x	x
Erysimum babadagense Prima	VU B1ab(ii,iii)		x			x	
Erysimum brachycarpum Boiss.	DD	x	x				
Erysimum brevistylum Sommier & Levier	DD					x	
Erysimum callicarpum Lipsky	VU B2ab(ii,iii)c(ii)					x	

TAXON	RED LIST ASSESSMENT	AR	AZ	GE	IR	RU	TU
Erysimum caspicum N. Busch	VU D2		x				
Erysimum caucasicum Trautv.	LC			x			
Erysimum collinum (M. Bieb.) Andrz. ex DC.	NE	x	x	x		x	
Erysimum contractum Sommier & Levier	VU D2			x			
Erysimum deflexum Cullen	CR B1ab(i,ii,iii)+ 2ab(i,ii,iii)						x
Erysimum feodorovii Kassumov	NE	x	x				
Erysimum ibericum (Adams) DC.	NE	x	x	x	x	x	
Erysimum krynitzkii Bordz.	NE	x	x	x		x	
Erysimum leptocarpum J. Gay	EN B2ab(i,ii,iii)						x
Erysimum lilacinum Steinb.	EN B1ab(iii)+2ab(iii); D	x	x				
Erysimum meyerianum (Rupr.) N. Busch	NT					x	
Erysimum strictisiliquum N. Busch	NT		x				
Erysimum subnivale Prima	NT			x			
Erysimum substrigosum (Rupr.) N. Busch	NE	x	x	x		x	
Erysimum szowitsianum Boiss.	NE	x	x	x	x		
Erysimum wagifii Kassumov	EN B1ab(iii)+2ab(iii)	x	x				
Eunomia rotundifolia C. A. Mey.	NE	x	x	x		x	
Hesperis adzharica Tzvelev [*Hesperis matronalis* L. subsp. *adzharica* (Tzvelev) Cullen]	NE	x		x		x	x
Hesperis ciscaucasica F. Dvořák	DD					x	
Hesperis matronalis L. subsp. *adzharica* (Tzvelev) Cullen [*Hesperis adzharica* Tzvelev]	NE	x		x		x	x
Hesperis pseudocinerea V. I. Dorof.	CR B1ab(iii)+2ab(iii)					x	
Hesperis transcaucasica Tzvelev	LC	x					x
Hornungia angustilimbata V. I. Dorof.	CR B1ab(iii,v)c(iv); C2a(ii)b					x	
Hymenolobus puberulus (Rupr.) N. Busch	NT					x	
Isatis arnoldiana N. Busch	DD	x					
Isatis brachycarpa C. A. Mey.	NE	x	x			x	

TAXON	RED LIST ASSESSMENT	AR	AZ	GE	IR	RU	TU
Isatis bungeana Seidl.	LC	x		x			
Isatis buschiorum Hovh.	DD	x					
Isatis caucasica (Rupr.) N. Busch	LC			x		x	
Isatis iberica Steven	NE	x	x	x			
Isatis karjaginii Schischk.	EN B1ab(iii)+2ab(iii); D	x	x				
Isatis kozlowskyi Grossh.	DD			x			x
Isatis latisiliqua Steven	NT		x			x	
Isatis nummularia Trautv. [*Sameraria nummularia* Bornm.]	NE		x		x		
Isatis ornithorhynchus N. Busch	DD	x	x				x
Isatis reticulata C. A. Mey.	DD			x		x	
Isatis sevangensis N. Busch	EN B1ab(i,ii)+2ab(i,ii)	x					
Lepidium propinquum Fisch. & C. A. Mey. [*Cardaria propinqua* (Fisch. & C. A. Mey.) N. Busch]	NE	x	x	x	x	x	
Malcolmia meyeri Boiss. [*Zuvanda meyeri* (Boiss.) Askerova]	EN B1ab(iii,v)+2ab(iii,v)	x	x				
Matthiola boissieri Grossh.	LC		x				
Matthiola caspica (N. Busch) Grossh.	NT		x			x	
Matthiola daghestanica (Conti) N. Busch	VU B1ab(ii,iii)		x			x	
Murbeckiella huetii (Boiss.) Rothm.	NE	x	x	x	x	x	x
Neotorularia ledebouri (Boiss.) V. E. Avet. [*Torularia contortuplicata* (Stephan) O. E. Schulz]	NE		x	x		x	
Pachyphragma macrophyllum (Hoffm.) N. Busch	NE	x	x	x		x	x
Peltariopsis drabicarpa (Boiss.) N. Busch	NE				x		
Peltariopsis grossheimii N. Busch	CR B1ab(iii)+2ab(iii)	x	x				
Pseudocamelina szowitsii (Boiss.) N. Busch	CR				x		
Pseudovesicaria digitata (C. A. Mey.) Rupr.	NT	x	x	x		x	
Raphanus rostratus DC.	NT		x				
Rorippa ampullicarpa V. I. Dorof.	NT		x			x	

TAXON	RED LIST ASSESSMENT	AR	AZ	GE	IR	RU	TU
Rorippa spasskajae V. I. Dorof.	CR B2ab(iii,v)	x					
Sameraria glastifolia (Fisch. & C. A. Mey.) Boiss.	VU B1ab(iii)+2ab(iii)	x	x				x
Sameraria nummularia Bornm. [*Isatis nummularia* Trautv.]	NE		x		x		
Sameraria odontophora Bordz.	EN B2ab(i,ii,iii)	x					
Sisymbrium elatum K. Koch	DD			x			x
Sisymbrium erucastrifolium (Rupr.) Trautv.	NE		x	x		x	
Sisymbrium lipskyi N. Busch	LC			x		x	
Sobolewskia caucasica (Rupr.) N. Busch	NE		x	x		x	
Sobolewskia truncata N. Busch	DD		x			x	
Sterigmostemum acanthocarpum (Fisch. & C. A. Mey.) Grossh.	VU B1ab(iii)+2ab(iii)	x	x				
Sterigmostemum contortuplicatum (Boiss.) Kuntze	NE				x		
Thlaspi freynii N. Busch.	NE	x					
Thlaspi huetii Boiss.	NE	x		x		x	x
Thlaspi maassoumii Mozaff.	VU				x		
Thlaspi macranthum (Lipsky) N. Busch	NE	x		x		x	
Thlaspi orbiculatum Steven	DD			x		x	
Thlaspi pumilum (Steven) Ledeb.	NE		x	x		x	
Thlaspi rostratum N. Busch	CR B1ab(i,ii,iii,v)+ 2ab(i,ii,iii,v)		x				
Thlaspi sintenisii Hausskn. ex Bornm.	VU B1ab(i,ii,iii)						x
Thlaspi szowitsianum Boiss.	NE	x	x			x	
Thlaspi tenue (Boiss. & Buhse) Hedge	NE				x		
Thlaspi zangezuricum Tzvelev	CR B2ab(v); D	x					
Torularia contortuplicata (Stephan) O. E. Schulz [*Neotorularia ledebouri* (Boiss.) V. E. Avet.]	NE		x	x		x	
Torularia eldarica Grossh.	NE		x	x			
Zuvanda meyeri (Boiss.) Askerova [*Malcolmia meyeri* Boiss.]	EN B1ab(iii,v)+2ab(iii,v)	x	x				

TAXON	RED LIST ASSESSMENT	AR	AZ	GE	IR	RU	TU
CAESALPINIACEAE							
Gleditsia caspia Desf.	NE		x		x		
CAMPANULACEAE							
Asyneuma campanuloides (M. Bieb. ex Sims) Bornm.	NE	x	x	x		x	
Campanula aghrica Kit Tan & Sorger	CR B1ab(i,ii,iii)+2ab(i,ii,iii)						x
Campanula albovii Kolak.	DD			x			
Campanula alliariifolia Willd.	NE	x	x	x		x	x
Campanula andina Rupr.	VU B1ab(iii)					x	
Campanula ardonensis Rupr.	CR B1ab(ii,v)					x	
Campanula argunensis Rupr. [*Campanula saxifraga* M. Bieb. subsp. *argunensis* (Rupr.) Ogan.]	NE		x	x		x	
Campanula armazica Charadze	EN B2ab(iii)			x			
Campanula autraniana Albov	EN B1ab(ii,iii)+2ab(ii,iii)					x	
Campanula bayerniana Rupr. [*Campanula elegantissima* Grossh.; *Campanula takhtadzhianii* Fed.]	NE	x	x		x		
Campanula bellidifolia Adams	NT			x		x	
Campanula besenginica Fomin	EN B2ab(v)					x	
Campanula biebersteiniana Roem. & Schult. [*Campanula tridentata* Schreb. subsp. *biebersteiniana* (Roem. & Schult.) Ogan.]	NE		x	x		x	
Campanula bzybica Jabr.-Kolak.	VU D2			x			
Campanula calcarata Sommier & Levier	DD					x	
Campanula calcarea (Albov) Charadze	LC			x			
Campanula caucasica M. Bieb.	NE	x	x			x	
Campanula charadzeae Grossh. [*Campanula sibirica* L. subsp. *charadzeae* (Grossh.) Ogan.]	NE		x	x		x	
Campanula choruhensis Kit Tan & Sorger	EN B1ab(i,ii,iii)+2ab(i,ii,iii)						x
Campanula ciliata Steven	NE		x	x		x	

TAXON	RED LIST ASSESSMENT	AR	AZ	GE	IR	RU	TU
Campanula circassica Fomin	LC			x		x	
Campanula collina Sims subsp. *collina*	NE	x	x	x		x	x
Campanula collina Sims subsp. *sphaerocarpa* (Kolak.) Ogan. [*Campanula sphaerocarpa* Kolak.]	LC			x		x	
Campanula crispa Lam.	NE	x	x	x			x
Campanula czerepanovii Fed.	CR B2ab(ii,iii,v); C2a(ii)					x	
Campanula daghestanica Fomin	EN B1ab(ii,iii)					x	
Campanula daralaghezica (Grossh.) Kolak. & Serdyuk.	DD		x				
Campanula dolomitica E. A. Busch	VU					x	
Campanula doluchanovii Charadze	DD		x	x			
Campanula dzaaku Albov	VU D2			x			
Campanula dzyschrica Kolak.	VU D2			x		x	
Campanula elegantissima Grossh. [*Campanula bayerniana* Rupr.; *Campanula takhtadzhianii* Fed.]	NE	x	x		x		
Campanula engurensis Charadze	VU D2			x			
Campanula fedorovii Charadze	NE		x	x			
Campanula fominii Grossh.	VU D2		x				
Campanula fondervisii Albov	VU D2			x			
Campanula gilliatii Milne-Redh. & Turrill	NE				x		
Campanula glomerata L. subsp. *caucasica* (Trautv.) Ogan.	NE	x	x			x	
Campanula glomerata L. subsp. *maleevii* (Fed.) Ogan.	NT					x	
Campanula glomerata L. subsp. *oblongifolia* (K. Koch) Fed. [*Campanula oblongifolia* (K. Koch) Charadze]	NE	x	x	x			
Campanula glomerata L. subsp. *oblongifolioides* (Galushko) Ogan.	LC					x	
Campanula grossheimii Charadze	DD			x			
Campanula hieracioides Kolak.	DD			x			

TAXON	RED LIST ASSESSMENT	AR	AZ	GE	IR	RU	TU
Campanula hohenackeri Fisch. & C. A. Mey. [*Campanula sibirica* L. subsp. *hohenackeri* (Fisch. & C. A. Mey.) Damboldt]	NE	x	x	x		x	x
Campanula hypopolia Trautv.	NE		x	x		x	
Campanula imeretina Rupr.	NT			x			
Campanula irinae A. I. Kuth.	VU D2			x			
Campanula jadvigae Kolak.	DD			x			
Campanula kachethica Kantsch.	CR B2ab(ii,iii)			x			
Campanula kadargavanica Amirkh. & Komzha	EN B1ab(i)					x	
Campanula kantschavelii Zagareli	DD			x			
Campanula kemulariae Fomin	VU D2			x			
Campanula kluchorica Kolak.	DD			x			
Campanula kolakovskyi Charadze	VU D2			x			
Campanula kolenatiana C. A. Mey. ex Rupr.	NE		x	x		x	
Campanula komarovii Maleev	EN B1ab(iii,v)					x	
Campanula kryophila Rupr.	EN B1ab(ii); C2a(i)					x	
Campanula lactiflora M. Bieb. [*Gadellia lactiflora* (M. Bieb.) Shulkina]	NE			x	x	x	x
Campanula latiloba A. DC. subsp. *rizeensis* Güner	EN B1ab(i,ii,iii,iv)+ 2ab(i,ii,iii,iv)						x
Campanula letschchumensis Kem.-Nath.	VU D2			x			
Campanula lezgina (F. N. Alex.) Kolak. & Serdyuk.	EN B2ab(ii,iii)		x			x	
Campanula longistyla Fomin	NE			x		x	
Campanula macrochlamys Boiss. & A. Huet	NE		x				x
Campanula makaschvilii E. A. Busch	VU D2			x			
Campanula massalskyi Fomin	EN B2ab(iii)	x					x
Campanula megrelica Manden. & Kuth.	VU D2			x			
Campanula meyeriana Rupr. [*Campanula saxifraga* M. Bieb. subsp. *meyeriana* (Rupr.) Ogan.]	EN B2ab(ii)		x			x	

TAXON	RED LIST ASSESSMENT	AR	AZ	GE	IR	RU	TU
Campanula minsteriana Grossh.	NE	x	x				
Campanula mirabilis Albov	DD			x			
Campanula oblongifolia (K. Koch) Charadze [*Campanula glomerata* L. subsp. *oblongifolia* (K. Koch) Fed.]	NE	x	x	x			
Campanula ochroleuca Kem.-Nath.	DD			x			
Campanula ossetica M. Bieb.	VU B1ab(ii,iii)+2ab(ii,iii)			x		x	
Campanula panjutinii Kolak.	DD			x			
Campanula paradoxa Kolak.	VU D2			x			
Campanula pendula M. Bieb. [*Symphyandra pendula* (M. Bieb.) A. DC.]	NT			x		x	
Campanula petrophila Rupr.	NE		x	x		x	
Campanula pontica Albov	VU D2			x			x
Campanula radchensis Charadze	VU D2			x			
Campanula raddeana Trautv.	VU D2			x			
Campanula sarmatica Ker Gawl. subsp. *ramosissima* (Sommier & Levier) Ogan.	DD					x	
Campanula sarmatica Ker Gawl. subsp. *sarmatica*	NE		x	x		x	
Campanula sarmatica Ker Gawl. subsp. *woronowii* (Charadze) Ogan.	EN B1ab(i,iii)+2ab(i,iii)					x	
Campanula savalanica Fed.	CR				x		
Campanula saxifraga M. Bieb., s. str.	NT			x		x	
Campanula saxifraga M. Bieb. subsp. *argunensis* (Rupr.) Ogan. [*Campanula argunensis* Rupr.]	NE		x	x		x	
Campanula saxifraga M. Bieb. subsp. *meyeriana* (Rupr.) Ogan. [*Campanula meyeriana* Rupr.]	NE		x			x	
Campanula schistosa Kolak.	VU D2			x			
Campanula sclerophylla (Kolak.) Czerep.	EN B1ab(iii,v)+2ab(iii,v)					x	
Campanula seraglio Kit Tan & Sorger	CR B1ab(i,ii)+2ab(i,ii)						x
Campanula sibirica L. subsp. *charadzeae* (Grossh.) Ogan. [*Campanula charadzeae* Grossh.]	NE		x	x		x	

TAXON	RED LIST ASSESSMENT	AR	AZ	GE	IR	RU	TU
Campanula sibirica L. subsp. *ciscaucasica* (Charadze) Ogan.	LC					x	
Campanula sibirica L. subsp. *hohenackeri* (Fisch. & C. A. Mey.) Damboldt [*Campanula hohenackeri* Fisch. & C. A. Mey.]	NE	x	x	x		x	x
Campanula songutica Amirkh.	EN B1ab(v)+2ab(v)					x	
Campanula sosnowskyi Charadze	DD			x			
Campanula sphaerocarpa Kolak. [*Campanula collina* Sims subsp. *sphaerocarpa* (Kolak.) Ogan.]	LC			x		x	
Campanula suanetica Rupr.	VU D2			x			
Campanula symphytifolia (Albov) Kolak.	DD			x			
Campanula takhtadzhianii Fed. [*Campanula bayerniana* Rupr.; *Campanula elegantissima* Grossh.]	NE	x	x		x		
Campanula trautvetteri Grossh. ex Fed.	DD		x	x			
Campanula tridentata Schreb. subsp. *biebersteiniana* (Roem. & Schult.) Ogan. [*Campanula biebersteiniana* Roem. & Schult.]	NE		x	x		x	
Campanula troegerae Damboldt	CR A3c; C2a(i)						x
Campanula zangezura (Lipsky) Kolak. & Serdyuk. [*Symphyandra zangezura* Lipsky]	NT	x	x				
Campanula zeyensis Amirkh. & Tavasiev	EN B1ab(ii,iii,v)+ 2ab(ii,iii,v)					x	
Gadellia lactiflora (M. Bieb.) Shulkina [*Campanula lactiflora* M. Bieb.]	NE			x	x	x	x
Muehlbergella oweriniana (Rupr.) Feer	CR B1ab(ii,iii,v)					x	
Symphyandra antiqua Kolak.	VU D2			x			
Symphyandra lazica Boiss. & Balansa	CR B2ab(ii)			x			x
Symphyandra pendula (M. Bieb.) A. DC. [*Campanula pendula* M. Bieb.]	NE			x		x	
Symphyandra zangezura Lipsky [*Campanula zangezura* (Lipsky) Kolak. & Serdyuk.]	NT	x	x				

TAXON	RED LIST ASSESSMENT	AR	AZ	GE	IR	RU	TU
CAPPARACEAE							
Cleome circassica Tzvelev	EN B2ab(ii,iii)					x	
Cleome daghestanica (Rupr.) Tzvelev	LC			x		x	
CAPRIFOLIACEAE							
Lonicera buschiorum Pojark.	LC			x		x	
Lonicera steveniana Fisch. ex Pojark.	LC			x		x	
Sambucus tigranii Troitzky	EN B1ab(i,ii,iii)+ 2ab(i,ii,iii)	x		x			
CARYOPHYLLACEAE							
Allochrusa bungei Boiss.	NE	x	x		x		x
Allochrusa persica (Boiss.) Boiss.	NE				x		
Allochrusa takhtajanii Gabrieljan & Dittrich	CR B1ab(iii,v)+2ab(iii,v); C2a(ii); D	x					
Allochrusa versicolor (Fisch. & C. A. Mey.) Boiss.	NE	x	x		x		x
Arenaria blepharophylla Boiss. var. *parviflora* (Fenzl) McNeill [*Arenaria graminea* C. A. Mey.]	NE		x				x
Arenaria graminea C. A. Mey. [*Arenaria blepharophylla* Boiss. var. *parviflora* (Fenzl) McNeill]	NE		x				x
Arenaria holostea M. Bieb. subsp. *macrantha* (Schischk.) McNeill	NT	x					x
Arenaria lychnidea M. Bieb. [*Eremogone lychnidea* (M. Bieb.) Rupr.]	LC			x		x	x
Arenaria scariosa Boiss.	EN						x
Arenaria szowitsii Boiss.	NE				x		
Bufonia takhtajanii Nersesian	CR B2ab(ii,iii,v)	x					
Cerastium argenteum M. Bieb.	LC			x			
Cerastium dagestanicum Schischk.	VU B1ab(ii,iii)		x			x	
Cerastium hemschinicum Schischk.	DD			x		x	
Cerastium holosteum Fisch. ex Hornem.	NE	x	x	x		x	
Cerastium holosteum Fisch. ex Hornem. var. *meyerianum* (Rupr.) I. V. Sokolova ex Czerep.	NT					x	

TAXON	RED LIST ASSESSMENT	AR	AZ	GE	IR	RU	TU
Cerastium kasbek Parrot	NT			x		x	x
Cerastium lazicum Boiss.	EN B1ab(i,ii,iii)+ 2ab(i,ii,iii)						x
Cerastium multiflorum C. A. Mey.	NT			x		x	
Cerastium oreades Schischk.	NT			x		x	
Cerastium polymorphum Rupr.	LC			x		x	
Cerastium ponticum Albov	VU C2ab(i)			x		x	
Cerastium salatavicum Rupr.	DD			x			
Cerastium svanicum Charadze	VU D2			x			
Cerastium szowitsii Boiss.	LC	x	x		x		
Cerastium undulatifolium Sommier & Levier	VU B2ab(ii,iii)			x		x	
Charesia akinfievii (Schmalh.) E. A. Busch [*Elisanthe akinfievii* (Schmalh.) Grossh.]	VU B1ab(ii)+2ab(ii)			x		x	
Dianthus abchasicus Gvinian.	VU D2			x			
Dianthus acantholimonoides Schischk.	VU					x	
Dianthus agrostolepis Rech. f.	NE				x		
Dianthus andronakii Woronov ex Schischk.	CR B2a(iii,iv); D						x
Dianthus awaricus Charadze	NT					x	
Dianthus azkurensis Sosn.	VU D2			x			
Dianthus bicolor Adams [*Dianthus preobrashenskii* Klokov]	LC	x	x				
Dianthus caucaseus Smith ex Sims	LC			x		x	
Dianthus charadzeae Gagnidze & Gvinian.	VU D2			x			
Dianthus daghestanicus Charadze	NT			x		x	
Dianthus elbrusense Charadze	VU C2a(i)					x	
Dianthus fragrans Adams	LC		x			x	
Dianthus grossheimii Schischk.	EN B1ab(ii)+2ab(ii)	x					
Dianthus hyrcanicus Rech. f.	NE				x		
Dianthus imereticus (Rupr.) Schischk.	NT			x		x	

TAXON	RED LIST ASSESSMENT	AR	AZ	GE	IR	RU	TU
Dianthus inamoenus Schischk.	NE	x	x	x			x
Dianthus jaroslavii Galushko	DD					x	
Dianthus kubanensis Schischk.	VU B1ab(iii)c(i,ii,iv); C2a(i)					x	
Dianthus kusnezovii Marcow.	NT			x		x	
Dianthus oschtenicus Galushko	NT					x	
Dianthus preobrashenskii Klokov [*Dianthus bicolor* Adams]	LC	x	x				
Dianthus raddeanus Vierh.	LC	x	x				x
Dianthus rudbaricus Assadi	NE				x		
Dianthus ruprechtii Schischk. ex Grossh.	NE		x	x		x	
Dianthus sahandicus Assadi	NE				x		
Dianthus schemachensis Schischk.	LC		x				x
Dianthus seidlitzii Boiss.	CR				x		
Dianthus subulosus Freyn & Conrath	NE	x	x	x			
Dianthus talyschensis Boiss. & Buhse	CR				x		
Dianthus transcaucasicus Schischk.	DD			x			
Dianthus vladimiri Galushko	EN B2ab(ii,iii)		x			x	
Elisanthe akinfievii (Schmalh.) Grossh. [*Charesia akinfievii* (Schmalh.) E. A. Busch]	VU B1ab(ii)+2ab(ii)			x		x	
Eremogone lychnidea (M. Bieb.) Rupr. [*Arenaria lychnidea* M. Bieb.]	LC			x		x	x
Gypsophila acutifolia Fisch. ex Spreng.	NE		x	x		x	
Gypsophila bazorganica Rech. f.	CR				x		
Gypsophila capitata M. Bieb.	LC		x			x	
Gypsophila glandulosa (Boiss.) Walp.	LC			x		x	x
Gypsophila imbricata Rupr.	VU B2ab(i,iii); C2a(i)					x	
Gypsophila iranica Barkoudah	CR				x		
Gypsophila lipskyi Schischk.	DD	x	x				
Gypsophila meyeri Rupr.	VU B2ab(ii,iii); C2a(i)					x	
Gypsophila paniculata L. var. *araratica* Hub.-Mor.	CR B1ab(i,ii,iii)+ 2ab(i,ii,iii)						x

TAXON	RED LIST ASSESSMENT	AR	AZ	GE	IR	RU	TU
Gypsophila perfoliata L. var. *araratica* Kit Tan	NE	x			x		x
Gypsophila robusta Grossh.	VU B2ab(iii,v)		x	x			
Gypsophila silenoides Rupr.	NE			x		x	x
Gypsophila simulatrix Bornm. & Woronow	EN B2ab(i,ii,iii)						x
Gypsophila steupii Schischk.	EN B2ab(ii,iii,v); C2a(i)			x		x	
Gypsophila stevenii Fisch. ex Schrank	NE	x	x	x			
Gypsophila szovitsii Fisch. & C. A. Mey. ex Fenzl	VU B2ab(ii,iii,iv)	x	x		x		
Gypsophila takhtadzhanii Schischk. ex Ikonn.	DD	x					
Gypsophila tenuifolia M. Bieb.	NE	x	x	x	x		x
Gypsophila wilhelminae Rech. f.	NE				x		
Melandrium balansae Boiss.	DD			x		x	
Minuartia abchasica Schischk.	VU B2ab(ii,iii)			x		x	
Minuartia acuminata Turrill	NE				x		
Minuartia armena Schischk. ex Ikonn.	DD	x					
Minuartia biebersteinii (Rupr.) Schischk.	LC			x		x	
Minuartia brotherana (Trautv.) Woronow	NT			x		x	
Minuartia buschiana Schischk.	LC			x		x	
Minuartia circassica (Albov) Woronow ex Grossh.	NE			x		x	x
Minuartia colchica Charadze	LC			x		x	
Minuartia daralagexica Schischk. ex Ikonn.	DD	x					
Minuartia inamoena (C. A. Mey.) Woronow ex Grossh.	NE			x		x	
Minuartia rhodocalyx (Albov) Woronow ex Grossh.	VU B1ab(iii)+2ab(iii)			x		x	
Minuartia sabalanica Assadi & Mostafavi	CR				x		
Minuartia subuniflora (Albov) Woronow ex Grossh.	NT			x			

TAXON	RED LIST ASSESSMENT	AR	AZ	GE	IR	RU	TU
Minuartia trautvetterana Sosn. & Charadze	DD			x		x	
Minuartia woronowii Schischk.	VU B2ab(ii,iii)	x		x			x
Petrocoma hoefftiana (Fisch. ex C. A. Mey.) Rupr.	VU B2ab(ii,iii); C2a(i)					x	
Saponaria cerastoides Fisch. & C. A. Mey.	NE		x	x			x
Saponaria picta Boiss.	EN B2ab(i,ii,iii)						x
Saponaria prostrata Willd. subsp. *calvertii*	NE			x			x
Silene alexeji Kolak.	DD			x			
Silene alpicola Schischk.	VU D1			x		x	
Silene boissieri Panjut.	DD			x			
Silene caucasica (Bunge) Boiss. [*Silene tatjanae* Schischk.]	NE	x	x	x		x	x
Silene chloropetala Rupr.	VU B1ab(iii)		x			x	
Silene chustupica Nersesian	CR B2ab(iii,v)	x					
Silene daghestanica Rupr.	NE			x		x	
Silene depressa M. Bieb.	VU B1ab(ii,iii)+2ab(ii,iii)	x	x				
Silene eremitica Boiss.	NE	x	x		x		x
Silene gynodioica Ghaz. subsp. *gynodioica*	NE				x		
Silene humilis C. A. Mey.	NE		x	x		x	
Silene ikonnikovii Lazkov	DD					x	
Silene ispirensis A. P. Khokhr.	CR B1ab(i,ii,iii)+ 2ab(i,ii,iii)						x
Silene kubanensis Sommier & Levier	VU B1ab(v)+2ab(v); C2a(i)					x	
Silene lacera Sims	LC			x		x	
Silene lazica Boiss.	NE			x			x
Silene linearifolia Otth ex DC.	NE		x	x		x	
Silene lineata Boiss. & Buhse	NE				x		
Silene lipskyi Lazkov	DD					x	
Silene lychnidea C. A. Mey.	LC			x		x	
Silene marcowiczii Schischk.	NE		x	x	x	x	

TAXON	RED LIST ASSESSMENT	AR	AZ	GE	IR	RU	TU
Silene praestans Schischk.	EN B2ab(v)		x				
Silene prilipkoana Schischk.	NE		x		x		
Silene pygmaea Adams	NE		x	x		x	
Silene scythicina Coode & Cullen	EN B1ab(i,ii,iii)+ 2ab(i,ii,iii); C2a(i)						x
Silene solenantha Trautv.	NE		x	x		x	
Silene talyschensis Schischk.	VU D2		x				
Silene tatjanae Schischk. [*Silene caucasica* (Bunge) Boiss.]	NE	x	x	x		x	x
Stellaria anagalloides Rupr.	DD			x			
CELASTRACEAE							
Euonymus leiophloea Steven	NE			x		x	x
CHENOPODIACEAE							
Anabasis eugeniae Iljin	VU D2		x				
Atriplex fominii Iljin	NT		x			x	
Axyris caucasica (Sommier & Levier) Lipsky	LC					x	
Gamanthus pilosus (Pall.) Bunge	NE				x		
Hablitzia tamnoides M. Bieb.	NE	x	x	x	x	x	
Halimocnemis azarbaijanensis Assadi	NE				x		
Salsola daghestanica (Turcz. ex Bunge) Lipsky	EN B1ab(iii)					x	
Salsola flavovirens Iljin	DD		x				
Salsola futilis Iljin	DD		x		x		
Salsola nodulosa (Moq.) Iljin	NE	x	x	x			x
Salsola persica Bunge ex Boiss.	NE				x		
Salsola tamamschjaniae Iljin	VU B2ab(i,ii,iv)	x	x		x		
CISTACEAE							
Helianthemum dagestanicum Rupr.	EN B1ab(iii,v)+2ab(iii,v)					x	
Helianthemum georgicum Juz. & Pozd.	VU B1ab(iii)+2ab(iii)			x			
CONVOLVULACEAE							
Convolvulus ruprechtii Boiss.	EN B1ab(iii)					x	

TAXON	RED LIST ASSESSMENT	AR	AZ	GE	IR	RU	TU
CORNACEAE							
Swida armasica (Sanadze) Gvin.	CR B1ab(iii)			x			
Swida iberica (Woronow) Pojark. ex Grossh.	NE	x	x	x			
Swida koenigii (C. K. Schneid.) Pojark. ex Grossh.	VU B2ab(iii)			x		x	
CRASSULACEAE							
Chiastophyllum oppositifolium (Ledeb. ex Nordm.) A. Berger [*Umbilicus oppositifolius* (Ledeb. ex Nordm.) Ledeb.]	NE	x		x		x	
Hylotelephium caucasicum (Grossh.) H. Ohba [*Sedum caucasicum* (Grossh.) Boriss.]	DD	x	x	x		x	
Prometheum pilosum (M. Bieb.) H. Ohba [*Sedum pilosum* M. Bieb.]	NE	x	x		x	x	x
Rosularia sempervivum (M. Bieb.) A. Berger	NE	x			x	x	x
Sedum abchasicum Kolak. ex V. V. Byalt	NT			x		x	
Sedum caucasicum (Grossh.) Boriss. [*Hylotelephium caucasicum* (Grossh.) H. Ohba]	NT	x	x	x		x	
Sedum euxinum 't Hart & Alpinar	CR B2ab(ii,iii,iv)						x
Sedum gracile C. A. Mey.	NE	x	x	x	x	x	x
Sedum involucratum M. Bieb.	NE		x	x		x	
Sedum lenkoranicum Grossh.	NT		x				
Sedum oppositifolium Sims [*Sedum spurium* M. Bieb.]	NE	x	x	x		x	x
Sedum pilosum M. Bieb. [*Prometheum pilosum* (M. Bieb.) H. Ohba]	NE	x	x		x	x	x
Sedum spurium M. Bieb. [*Sedum oppositifolium* Sims]	NE	x	x	x		x	x
Sedum stevenianum Rouy & E. G. Camus	NE		x	x		x	
Sedum subulatum (C. A. Mey.) Boiss.	LC		x				
Sempervivum annae Gurgen.	DD			x			

TAXON	RED LIST ASSESSMENT	AR	AZ	GE	IR	RU	TU
Sempervivum artvinense Muirhead	NE						x
Sempervivum brevipetalum Kit Tan & Sorger	DD						x
Sempervivum caucasicum Rupr. ex Boiss.	NE		x	x		x	
Sempervivum caucasicum Rupr. ex Boiss. var. *borissovae* (Wale) Gurgen.	NE		x	x		x	
Sempervivum charadzeae Gurgen.	DD			x			
Sempervivum davisii Muirhead subsp. *davisii*	NE					x	
Sempervivum ermanicum Gurgen.	DD			x			
Sempervivum furseorum Muirhead	EN B2ab(i,ii,iii)						x
Sempervivum glabrifolium Boriss.	EN B1ab(i,ii,iii,iv)+ 2ab(i,ii,iii,iv)						x
Sempervivum globiferum L. subsp. *aghricum* Kit Tan & Sorger	CR B2ab(i,ii,iii)						x
Sempervivum minus Turril ex Wale var. *glabrum* Wale	EN B1ab(i,ii,iii)+ 2ab(i,ii,iii)						x
Sempervivum minus Turril ex Wale var. *minus*	LC						x
Sempervivum pumilum M. Bieb.	NE		x	x		x	
Sempervivum sosnowskyi Ter-Chatsch.	VU 2ab(i,ii,iv)			x			x
Sempervivum staintonii Muirhead	EN B1ab(i,ii,iii)+ 2ab(i,ii,iii)						x
Sempervivum transcaucasicum Muirhead	LC	x	x	x			x
Umbilicus oppositifolius (Ledeb. ex Nordm.) Ledeb. [*Chiastophyllum oppositifolium* (Ledeb. ex Nordm.) A. Berger]	NE	x		x		x	
DIPSACACEAE							
Cephalaria anatolica Schchian	CR B1ab(i,ii,iii)+ 2ab(i,ii,iii)						x
Cephalaria armeniaca Bordz.	NT	x	x	x			
Cephalaria balkharica E. A. Busch	NT			x		x	

TAXON	RED LIST ASSESSMENT	AR	AZ	GE	IR	RU	TU
Cephalaria brevipalea (Sommier & Levier) Litv.	NT					x	
Cephalaria calcarea Albov	NT			x		x	
Cephalaria dagestanica Bobrov	NT					x	
Cephalaria gigantea (Ledeb.) Bobrov	NE	x	x	x		x	
Cephalaria grossheimii Bobrov [*Cephalaria kotschyi* Boiss. & Hohen.]	NE	x	x				
Cephalaria kotschyi Boiss. & Hohen. [*Cephalaria grossheimii* Bobrov]	NE	x	x				
Cephalaria media Litv.	NE	x	x	x		x	x
Cephalaria microdonta Bobrov	NE		x	x		x	
Cephalaria nachiczevanica Bobrov	CR B1ab(iii)+2ab(iii)	x	x				
Cephalaria sosnowskyi Kolak.	DD			x			
Cephalaria velutina Bobrov	NT		x			x	
Scabiosa adzharica Schchian	NT			x			
Scabiosa amoena J. Jacq. f.	NE		x	x	x		
Scabiosa caucasica M. Bieb.	NE	x		x	x	x	x
Scabiosa colchica Steven	NT			x			
Scabiosa correvoniana Sommier & Levier	LC			x		x	
Scabiosa georgica Sulak.	DD		x	x	x		
Scabiosa gumbetica Boiss.	LC					x	
Scabiosa imeretica (Sommier & Levier) Sulak.	NT			x			
Scabiosa olgae Albov	VU B1ab(ii,iii)+2ab(ii,iii)			x		x	
Scabiosa owerinii Boiss.	NE		x	x		x	
Scabiosa rupestris M. Bieb.	EN B2ab(ii,iii)					x	
Scabiosa schimperiana Boiss. & Buhse	LC				x		
Scabiosa sosnowskyi Sulak.	LC			x		x	
EMPETRACEAE				x		x	
Empetrum androgynum V. N. Vassil. var. *caucasicum* V. N. Vassil. [*Empetrum caucasicum* Juz.]	NE	x	x	x		x	

TAXON	RED LIST ASSESSMENT	AR	AZ	GE	IR	RU	TU
Empetrum caucasicum Juz. [*Empetrum androgynum* V. N. Vassil. var. *caucasicum* V. N. Vassil.]	NE	x	x	x		x	
ERICACEAE							
Arctostaphylos caucasica Lipsch.	LC			x		x	
Epigaea gaultherioides (Boiss. & Balansa) Takht.	VU B2ab(iii)			x			x
Rhododendron caucasicum Pall.	NE	x	x	x		x	x
Rhododendron smirnowii Trautv. ex Regel	VU B1ab(iii)			x			x
Rhododendron ungernii Trautv. ex Regel	VU B1ab(iii)			x			x
Rhodothamnus sessilifolius P. H. Davis	CR B1ab(i,ii,iii)+2ab(i,ii,iii)						x
EUPHORBIACEAE							
Euphorbia abchasica Woronow	DD			x			
Euphorbia ardonensis Galushko	EN B2ab(ii,iii)					x	
Euphorbia aristata Schmalh.	EN B1ab(i,ii,iii)+2ab(i,ii,iii)					x	
Euphorbia armena Prokh.	DD			x			x
Euphorbia buschiana Grossh.	EN B1ab(ii,iii)+2ab(ii,iii)					x	
Euphorbia czerepanovii Geltman	DD			x		x	
Euphorbia daghestanica Geltman	NT					x	
Euphorbia dubovikiae Oudejans	DD					x	
Euphorbia eugeniae Prokh.	EN B2ab(i,ii,iii)			x		x	
Euphorbia glaberrima K. Koch	LC	x	x	x		x	
Euphorbia grossheimii Prokh.	EN B1ab(iii)+2ab(iii)	x	x				
Euphorbia hyrcana Grossh.	NT		x				
Euphorbia iberica Boiss. [*Euphorbia vedica* Ter-Chatsch.]	NE	x		x			
Euphorbia macroceras Fisch. & C. A. Mey.	LC	x	x	x		x	x
Euphorbia normannii Schmalh. ex Lipsky	VU D2					x	
Euphorbia panjutinii Grossh.	DD			x		x	

TAXON	RED LIST ASSESSMENT	AR	AZ	GE	IR	RU	TU
Euphorbia procera M. Bieb.	LC	x	x	x		x	
Euphorbia sahendi Bornm.	CR				x		
Euphorbia scripta Sommier & Levier	LC			x		x	
Euphorbia wittmannii Boiss.	NT			x			x
Leptopus colchicus (Fisch. & C. A. Mey. ex Boiss.) Pojark.	VU B2ac(i,ii,iii)			x		x	
FABACEAE							
Amoria bobrovii (Khalilov) Roskov	CR B2ab(i,ii,iii,iv,v)		x				
Amoria egrissica (Mikheev & Magulaev) Roskov	NE		x	x		x	
Amoria elizabethae (Grossh.) Roskov [*Trifolium elizabethae* Grossh.]	DD		x			x	
Amoria ruprechtii (Tamamsch. & Fed.) Roskov [*Trifolium ruprechtii* Tamamsch. & Fed.]	NE	x	x	x		x	
Amoria talyschensis (Khalilov) Roskov	VU D2		x				
Anthyllis caucasica (Grossh.) Juz. [*Anthyllis variegata* Boiss. ex Grossh.]	NE	x	x	x		x	
Anthyllis grossheimii Chinth.	NE		x	x			
Anthyllis irenae Juz.	DD			x			
Anthyllis variegata Boiss. ex Grossh. [*Anthyllis caucasica* (Grossh.) Juz.]	NE	x	x	x		x	
Astracantha denudata (Steven) Podlech [*Astragalus denudatus* Steven]	NE		x	x		x	
Astragalus achundovii Grossh. ex Fed.	NE	x	x		x		
Astragalus acmophylloides Grossh.	CR B2ab(i,ii,iii)						x
Astragalus acutifolius Bunge	NE				x		
Astragalus aduncus Willd.	NE	x	x				
Astragalus adzharicus Popov	VU D2			x			x
Astragalus agassii Manden.	EN D	x					
Astragalus aharicus Maassoumi & Podlech	EN				x		
Astragalus albanicus Grossh.	EN B1ab(i,ii,iii,iv,v)+2ab(i,ii,iii,iv,v)		x				

TAXON	RED LIST ASSESSMENT	AR	AZ	GE	IR	RU	TU
Astragalus alexandrii Charadze	DD		x				
Astragalus ammodendroides Bornm. ex Podlech & Zarre	NE				x		
Astragalus andreji-sytinii Podlech [*Astragalus eugenii* Grossh.]	EN B1ab(ii)		x			x	
Astragalus ansinii A. Uzun, Terzioğlu & S. Palabas-Uzun	CR B2ab(ii,iv)						x
Astragalus arasbaranensis Maassoumi & Ranjbar	EN				x		
Astragalus argillosus Manden.	VU D2			x			
Astragalus arnacatha M. Bieb.	EN B1ab(ii,iii)					x	
Astragalus aspindzicus Manden. & Chinth.	VU D2			x			
Astragalus atenicus Ivan.	VU D2			x			
Astragalus azizii Maassoumi	NE				x		
Astragalus bachmarensis Grossh.	NE			x			x
Astragalus bakuensis Bunge	CR B1ab(i,ii,iii,iv,v)		x				
Astragalus balkaricus Sytin	NT					x	
Astragalus basilicus Maassoumi & Podlech	NE				x		
Astragalus beckerianus Trautv.	NE	x	x			x	
Astragalus biebersteinii Bunge	NE		x			x	
Astragalus birangiae Maassoumi	CR				x		
Astragalus brachypetalus Trautv.	DD		x				x
Astragalus brachytropis (Steven) C. A. Mey.	NE	x	x	x		x	
Astragalus bungeanus Boiss.	NE		x	x		x	
Astragalus buschiorum Galushko	NT		x			x	
Astragalus bylowae Elenevsky	CR B2ab(ii,iii)	x					
Astragalus candolleanus Boiss.	LC	x	x				
Astragalus canosus Maassoumi	NE				x		
Astragalus captiosus Boriss.	NE		x	x		x	
Astragalus caspicus M. Bieb.	NE		x			x	

TAXON	RED LIST ASSESSMENT	AR	AZ	GE	IR	RU	TU
Astragalus caucasicus Pall.	NE		x	x		x	x
Astragalus chalilovii Grossh. ex Fed.	VU D2		x				
Astragalus charadzeae Grossh.	EN B1ab(i,ii,iii)+ 2ab(i,ii,iii)					x	
Astragalus chehreganii Podlech & Zarre	NE				x		
Astragalus choicus Bunge	NE				x		
Astragalus chordorrhizus Bunge	NE			x		x	
Astragalus ciceroides Sosn.	LC					x	
Astragalus circassicus Grossh.	EN B1ab(iv,v); C2a(i)			x		x	
Astragalus clavatus DC.	DD						x
Astragalus coarctatus Trautv.	DD						x
Astragalus coelestis Boiss.	DD	x					
Astragalus conspicuus Boriss.	NT	x	x				
Astragalus cuscutae Bunge	EN B1ab(v)+2ab(v)		x				
Astragalus cyri Fomin ex Grossh.	EN B1ab(iii,v)+2ab(iii,v)			x			
Astragalus daghestanicus Grossh.	CR B1ab(iii,v); C2a(ii)					x	
Astragalus dasypoda Rupr. ex Boiss.	DD			x			
Astragalus demetrii Charadze	NE					x	
Astragalus denudatus Steven [*Astracantha denudata* (Steven) Podlech]	NE		x	x		x	
Astragalus dianat-nejadii Ghahrem.	NE				x		
Astragalus divaricatus Bunge	NE		x		x		
Astragalus doghrunensis Maassoumi & Podlech	NE				x		
Astragalus doluchanovii Manden.	VU D2			x			
Astragalus dzhebrailicus Grossh. [*Astragalus schuschaensis* Grossh.]	EN B1ab(iii)+2ab(iii)	x	x		x		
Astragalus elegans Bunge	NE				x		
Astragalus eliasianus Kit Tan & Sorger	CR B1ab(i,ii,iii)+ 2ab(i,ii,iii); C2a(ii); D						x
Astragalus erinaceus Fisch. & C. A. Mey.	NE				x		x

TAXON	RED LIST ASSESSMENT	AR	AZ	GE	IR	RU	TU
Astragalus erivanensis Bornm. & Woronow	VU B1ab(iii)+2ab(iii)	x	x				
Astragalus eugenii Grossh. [*Astragalus andreji-sytinii* Podlech]	NE		x			x	
Astragalus euoplus Trautv.	EN B1ab(iii)+2ab(iii)	x	x				
Astragalus fabaceus M. Bieb.	NE	x			x	x	
Astragalus fissuralis T. Alex.	VU D2					x	
Astragalus flavirubens Al. Fed., Fed. & Rzazade	DD		x				
Astragalus flexicaulis Sosn.	DD					x	
Astragalus freynii Albov	NT			x		x	
Astragalus frickii Bunge	LC			x		x	x
Astragalus gezeldarensis Grossh.	DD		x				
Astragalus goktschaicus Grossh. [*Astragalus kosmaljanicus* Rzazade]	NE	x	x	x	x		x
Astragalus gudrathi Al. Fed., Fed. & Rzazade	DD		x				
Astragalus haesitabundus Lipsky	NT		x			x	
Astragalus hajastanus Grossh.	LC	x	x				
Astragalus hendelanicus Maassoumi	NE				x		
Astragalus hirtulus Ledeb.	NT			x			
Astragalus holophyllus Boriss.	VU B1ab(i,ii,iii)+ 2ab(i,ii,iii)	x					
Astragalus holosemius Bunge	NE				x		
Astragalus horridissimus Širj. & Bornm.	EN				x		
Astragalus husseinovii Rzazade	DD		x		x		
Astragalus hymenocystis Fisch. & C. A. Mey.	NE				x		x
Astragalus hyrcanus Pall.	VU B1ab(iii)		x			x	
Astragalus hystrix Fisch. & C. A. Mey.	NE				x		
Astragalus igniarius Popov	EN B1ab(i,ii,iii,iv,v)+ 2ab(i,ii,iii,iv,v)		x				
Astragalus ilachchiensis Ranjbar & Ghahrem.	CR				x		

TAXON	RED LIST ASSESSMENT	AR	AZ	GE	IR	RU	TU
Astragalus imbricatus Boriss.	DD						x
Astragalus inoblitus Maassoumi	NE				x		
Astragalus insidiosus Boriss.	NT		x				
Astragalus kabristanicus Grossh.	LC		x				
Astragalus kadschorensis Bunge	NE	x	x	x			
Astragalus karabaghensis Bunge	EN B1ab(iii)+2ab(iii)	x	x		x		
Astragalus karakuschensis Gontsch.	EN B1ab(i,ii,iii)+ 2ab(i,ii,iii)	x	x				
Astragalus karjaginii Boriss.	EN B1ab(i,ii,iii)+ 2ab(i,ii,iii)	x	x				
Astragalus karl-heinzii Maassoumi	NE				x		
Astragalus kazbekii Charadze	NE		x	x		x	
Astragalus kemulariae Grossh.	LC			x			
Astragalus ketzkhovelii Charadze	DD			x			
Astragalus khajiboulaghensis Maassoumi	NE				x		
Astragalus kiamaky-daghensis Maassoumi & Podlech	NE				x		
Astragalus kochianus Sosn.	NE	x	x				
Astragalus kosmaljanicus Rzazade [*Astragalus goktschaicus* Grossh.]	NE	x	x	x	x		x
Astragalus kozlovskyi Grossh.	NT			x			
Astragalus kubensis Grossh.	VU D2		x				
Astragalus lasioglottis Steven ex M. Bieb.	NT					x	x
Astragalus latifolius Lam.	VU B1ab(i,ii,iii)+ 2ab(i,ii,iii)	x	x		x		
Astragalus leonidae Manden.	VU D2			x			
Astragalus leushanensis Maassoumi	NE				x		
Astragalus levieri Freyn ex Sommier & Levier	NE		x	x		x	
Astragalus longivexillatus Podlech & Ekiçi	CR B2ab(ii,iii)						x
Astragalus lunatus Pall.	NE		x		x	x	
Astragalus lussiae Rzazade	CR B1ab(v)+2ab(v)		x				

TAXON	RED LIST ASSESSMENT	AR	AZ	GE	IR	RU	TU
Astragalus magnificus Kolak.	VU D2			x			
Astragalus marandicus Podlech	NE				x		
Astragalus maraziensis Rzazade	EN B1ab(i,ii,iii,iv,v)+ 2ab(i,ii,iii,iv,v)		x				
Astragalus massalskyi Grossh. ex Fed.	NE	x	x	x			
Astragalus maximus Willd. var. *dasysemius* D. F. Chamb. & V. A. Matthews	EN B1ab(i,ii,iii)+ 2ab(i,ii,iii); C2a(i)						x
Astragalus meskheticus Manden.	VU D2			x			
Astragalus meyeri Boiss.	NE	x	x		x		x
Astragalus mirus Sosn.	DD						x
Astragalus montis-aquilis Grossh.	EN B1ab(i,ii,iii)+ 2ab(i,ii,iii)	x	x				
Astragalus montismishoudaghi Sheikh Akbari Mehr, Ghorbani & Maassoumi	CR				x		
Astragalus mozaffarianii Maassoumi	NE				x		
Astragalus nachitschevanicus Rzazade	CR B1ab(i,ii,iii)c(i)+ 2ab(i,ii,iii)c(i)		x				
Astragalus neochaldoranicus Podlech & Maassoumi	NE				x		
Astragalus neomobayenii Maassoumi	NE				x		
Astragalus nigrocalycinus Podlech	CR B1ab(i,ii,iii)+ 2ab(i,ii,iii)						x
Astragalus oltensis Grossh.	DD						x
Astragalus olurensis Podlech	CR B1ab(i,ii,iii)+ 2ab(i,ii,iii)						x
Astragalus ordubadensis Grossh.	EN B1ab(i,ii,iii)+ 2ab(i,ii,iii)	x	x		x		
Astragalus oreades C. A. Mey.	NE		x	x		x	x
Astragalus owirensis Ali ex Podlech	CR B2ab(iii,v)						x
Astragalus paradoxus Bunge	NE	x	x		x		x
Astragalus parvulus Bornm.	NE				x		
Astragalus pauperiflorus Bornm.	NE				x		
Astragalus pereshkhoranicus Maassoumi & Gharem.	NE				x		

TAXON	RED LIST ASSESSMENT	AR	AZ	GE	IR	RU	TU
Astragalus pinetorum Boiss. subsp. *alamutensis* Maassoumi & Podlech	NE				x		
Astragalus pinetorum Boiss. subsp. *declinatus* Podlech	NE				x		
Astragalus pinetorum Boiss. subsp. *pinetorum*	NE				x		
Astragalus polyphyllus Bunge	NT		x			x	
Astragalus prilipkoanus Grossh.	NE	x	x		x		
Astragalus pseudopersicus Maassoumi & Podlech	NE				x		
Astragalus pseudoutriger Grossh.	NE	x	x	x	x		x
Astragalus psoraloides Lam.	NE	x	x		x	x	x
Astragalus punctatus Bunge	NE				x		
Astragalus qoturensis Podlech	NE				x		
Astragalus raddeanus Regel	LC			x			
Astragalus recognitus Fisch.	NE				x		
Astragalus refractus C. A. Mey.	NT	x	x		x		
Astragalus regelii Trautv.	NT		x				
Astragalus rollovii Grossh.	NE				x		
Astragalus rostratus C. A. Mey.	NE		x		x		
Astragalus rubrocalycinus Maassoumi & Podlech	NE				x		
Astragalus ruprechtii Bunge	NE		x				
Astragalus rzaevii Grossh.	NE	x	x		x		
Astragalus sachokianus Grossh.	CR B1ab(v)+2ab(v)		x				
Astragalus sahendi Buhse ex Fisch.	NE				x		
Astragalus salatavicus Bunge	CR B1ab(ii,iii)+2ab(ii,iii); C2a(ii)					x	
Astragalus sangesuricus Boriss.	EN B2ab(i,ii,iii)	x	x				
Astragalus sanguinolentus M. Bieb.	NE	x	x			x	
Astragalus savellanicus Podlech	NE				x		
Astragalus schachbuzensis Rzazade	CR B2ab(v)		x				
Astragalus schemachensis Karjagin	NE		x		x		

TAXON	RED LIST ASSESSMENT	AR	AZ	GE	IR	RU	TU
Astragalus schischkinii Grossh.	VU D2			x			
Astragalus schuschaensis Grossh. [*Astragalus dzhebrailicus* Grossh.]	EN B1ab(iii)+2ab(iii)	x	x		x		
Astragalus sevangensis Grossh.	NE	x	x				
Astragalus shagalensis Grossh.	VU B1ab(i,ii,iii)+ 2ab(i,ii,iii)	x					
Astragalus shelkovnikovi Grossh.	EN B2ab(iii)	x	x				
Astragalus sonamerensis Schischk.	DD						x
Astragalus sosnowskyi Grossh. [*Astragalus tanae* Sosn.]	NE	x		x	x		x
Astragalus sphaerocephalus Manden.	NE		x	x			
Astragalus supinus Bunge	NE		x	x		x	
Astragalus szovitsii Fisch. & C. A. Mey.	LC	x	x		x		x
Astragalus takhtadzhjanii Grossh.	NE	x	x	x	x		
Astragalus tanae Sosn. [*Astragalus sosnowskyi* Grossh.]	NE	x		x	x		x
Astragalus tatlii Peşmen	CR B1ab(i,ii,iii)+ 2ab(i,ii,iii)						x
Astragalus terekensis Al. Fed., Fed. & Rzazade	NE		x				
Astragalus theodorii Grossh.	NE		x				
Astragalus trabzonicus Ekici, Aytaç & Akan	CR B1ab(i,ii,iii)+ 2ab(i,ii,iii)						x
Astragalus tshegemensis Galushko	VU B1ab(ii,iii)					x	
Astragalus uraniolimneus Boiss.	LC	x	x		x		x
Astragalus vardziae Charadze & Chinth.	VU D2			x			
Astragalus vavilovii Tamamsch. & Fed.	NT	x		x			
Astragalus vedicus Takht.	NT	x	x				
Astragalus viridis Bunge	NE	x	x		x		x
Astragalus voronovianus (Boriss.) Boriss.	DD						x
Astragalus wagneri Bunge	NE	x	x		x		
Astragalus xiphidium Bunge	NE	x	x	x	x	x	x
Astragalus zangelanus Grossh.	NE		x		x		

TAXON	RED LIST ASSESSMENT	AR	AZ	GE	IR	RU	TU
Astragalus zuvanticus Grossh.	NT		x				
Chamaecytisus colchicus (Albov) Portenier	NE					x	
Chamaecytisus hirsutissimus (K. Koch) Czerep. [*Cytisus hirsutissimus* K. Koch]	LC			x		x	
Chesneya elegans Fomin	EN B1ab(i,ii,iii)+ 2ab(i,ii,iii)						x
Cicer balcaricum Galushko	DD					x	
Cicer caucasicum Bornm.	VU D2			x			
Cicer minutum Boiss. & Hohen.	NE					x	
Colutea acutifolia Shap.	DD					x	
Colutea armena Boiss. & A. Huet	NE		x			x	x
Colutea komarovii Takht.	NT	x	x				
Colutea uniflora Beck	NE				x		
Coronilla balansae (Boiss.) Grossh. [*Coronilla orientalis* Mill. var. *balansae* (Boiss.) Hrabětova]	NE			x		x	x
Coronilla orientalis Mill. var. *balansae* (Boiss.) Hrabětova [*Coronilla balansae* (Boiss.) Grossh.]	NE			x		x	x
Cytisus caucasicus Grossh.	DD			x			
Cytisus hirsutissimus K. Koch [*Chamaecytisus hirsutissimus* (K. Koch) Czerep.]	LC			x		x	
Galega orientalis Lam.	NE	x	x	x		x	
Genista abchasica Sachokia	EN B1ab(iii)+2ab(iii)			x		x	
Genista adzharica Popov	CR 2ab(iii,v)			x			
Genista angustifolia Schischk.	VU B2ab(ii)					x	
Genista compacta Schischk.	NT					x	
Genista humifusa L.	DD			x		x	
Genista kolakowskyi Sachokia	DD			x		x	
Genista mingrelica Albov	VU D2			x			
Genista patula M. Bieb.	NE		x	x		x	
Genista sachokiana A. I. Kuth.	NT			x			

TAXON	RED LIST ASSESSMENT	AR	AZ	GE	IR	RU	TU
Genista suanica Schischk. ex Grossh.	VU B2ab(iv)			x		x	
Genista transcaucasica Schischk.	NE	x	x	x			
Hedysarum caucasicum M. Bieb.	DD	x	x	x		x	
Hedysarum daghestanicum Rupr. ex Boiss.	EN B1ab(v)					x	
Hedysarum elegans Boiss. & A. Huet	VU B2ab(iii,v)	x	x	x			x
Hedysarum huetii Boiss.	DD						x
Hedysarum ibericum M. Bieb.	NE	x	x	x	x		
Hedysarum marandense Mozaff., Akrami & Maassoumi	EN				x		
Hedysarum orumiehense Ranjbar	VU				x		
Hedysarum papillosum Boiss.	NE				x		
Hedysarum sericeum M. Bieb.	NE	x	x	x	x	x	x
Lathyrus alamutensis Mozaff., Ahvazi & Charkhch.	NE				x		
Lathyrus colchicus Lipsky	NE					x	
Lathyrus cyaneus (Steven) K. Koch	NE	x	x	x	x	x	x
Lathyrus karsianus P. H. Davis	EN B1ab(i,ii,iii)+ 2ab(i,ii,iii)						x
Lathyrus woronowii Bornm.	CR C2a(i)						x
Lotus armeniacus Kit Tan & Sorger	CR C2a(i)						x
Lotus caucasicus Kuprian. ex Juz.	NE	x	x	x		x	
Lupinaster polyphyllus (C. A. Mey.) Latsch. [*Trifolium polyphyllum* C. A. Mey.]	LC			x		x	x
Medicago caucasica Vassilcz.	NE		x	x		x	
Medicago daghestanica Rupr. ex Boiss.	NT		x			x	
Medicago difalcata Sinskaya subsp. *orientalicaucasica* Sinskaya	DD					x	
Medicago dzawakhetica Bordz.	NE	x		x			
Medicago glutinosa M. Bieb. [*Medicago virescens* Grossh.]	NE		x	x		x	
Medicago gunibica Vassilcz.	NT					x	
Medicago hemicoerulea Sinskaya	LC					x	

TAXON	RED LIST ASSESSMENT	AR	AZ	GE	IR	RU	TU
Medicago hemicycla Grossh.	NE	x	x	x		x	
Medicago hemicycla Grossh. f. *pseudodzawakhetica* Troitzky [*Medicago hemicycla* Grossh. subsp. *medidaghestanica* Sinskaya]	DD			x		x	
Medicago hemicycla Grossh. subsp. *medidaghestanica* Sinskaya [*Medicago hemicycla* Grossh. f. *pseudodzawakhetica* Troitzky]	DD			x		x	
Medicago virescens Grossh. [*Medicago glutinosa* M. Bieb.]	NE		x	x		x	
Melilotus hirsutus Lipsky	LC			x		x	
Onobrychis angustifolia Chinth.	NT			x			
Onobrychis atropatana Boiss. var. *grandiflora* Aktoklu	CR B1ab(i,ii)+2ab(i,ii)						x
Onobrychis aucheri Boiss. subsp. *aucheri*	NE				x		
Onobrychis biebersteinii Sirj.	NE	x	x	x		x	
Onobrychis bobrovii Grossh.	LC		x			x	
Onobrychis cyri Grossh.	NE	x	x	x		x	
Onobrychis daghestanica Grossh.	DD					x	
Onobrychis depauperata Boiss.	CR				x		
Onobrychis grossheimii Kolak. ex Fed.	NT			x			
Onobrychis hajastana Grossh.	EN	x	x				x
Onobrychis hamata Vassilcz.	VU B1ab(v)+2ab(v)					x	
Onobrychis heterophylla C. A. Mey.	NT		x				
Onobrychis iberica Grossh.	NE	x	x	x		x	
Onobrychis kachetica Boiss. & Buhse	LC			x			
Onobrychis kemulariae Chinth.	VU D2			x			
Onobrychis kluchorica Chinth.	DD					x	
Onobrychis komarovii Grossh.	NE		x	x			
Onobrychis lasistanica Širj.	EN B1ab(i,ii,iii)+2ab(i,ii,iii)						x
Onobrychis majorovii Grossh.	NT					x	
Onobrychis meschetica Grossh.	LC			x			

TAXON	RED LIST ASSESSMENT	AR	AZ	GE	IR	RU	TU
Onobrychis oxytropoides Bunge	NE	x		x		x	
Onobrychis petraea (M. Bieb. ex Willd.) Fisch.	NE	x	x	x		x	
Onobrychis schuschajensis O. D. Agajeva	DD		x				
Onobrychis sosnowskyi Grossh. [*Onobrychis stenostachya* Freyn subsp. *sosnowskyi* (Grossh.) Hedge]	EN B2ab(i,ii,iii)			x			x
Onobrychis stenostachya Freyn subsp. *sosnowskyi* (Grossh.) Hedge [*Onobrychis sosnowskyi* Grossh.]	EN B2ab(i,ii,iii)			x			x
Onobrychis subacaulis Boiss.	NE	x	x		x		x
Onobrychis szovitsii Boiss.	CR				x		
Onobrychis takhtajanii Sytin	EN B1ab(i,ii,iii)+ 2ab(i,ii,iii)	x					
Onobrychis transcaucasica Grossh.	NE	x	x	x			
Onobrychis vaginalis C. A. Mey.	DD		x				
Orobus ciliatidentatus (Czefr.) Avazneli	NE		x	x			
Oxytropis armeniaca Sosn. ex Mulk.	NT	x		x			
Oxytropis cyanea M. Bieb.	DD	x	x	x		x	
Oxytropis dasypoda Rupr. ex Boiss.	NE		x	x		x	
Oxytropis kubanensis Leskov	NT					x	
Oxytropis lazica Boiss.	NT	x		x			x
Oxytropis meyeri Bunge ex Boiss.	EN B2ab(iii)					x	
Oxytropis owerinii Bunge	NE		x	x		x	
Oxytropis takhti-soleimanii Vassilcz.	CR				x		
Psoralea acaulis Steven ex M. Bieb.	LC			x		x	x
Trifolium elizabethae Grossh. [*Amoria elizabethae* (Grossh.) Roskov]	NE		x			x	
Trifolium fontanum Bobrov	NE		x	x		x	
Trifolium polyphyllum C. A. Mey. [*Lupinaster polyphyllus* (C. A. Mey.) Latsch.]	LC			x		x	x
Trifolium raddeanum Trautv.	VU D2					x	

TAXON	RED LIST ASSESSMENT	AR	AZ	GE	IR	RU	TU
Trifolium ruprechtii Tamamsch. & Fed. [*Amoria ruprechtii* (Tamamsch. & Fed.) Roskov]	NE	x	x	x		x	
Trifolium sintenisii Freyn [*Trifolium stipitatum* Boiss. & Balansa]	NE			x			x
Trifolium stipitatum Boiss. & Balansa [*Trifolium sintenisii* Freyn]	NE			x			x
Trigonella biflora Griseb.	DD		x				
Vicia abbreviata Fisch. ex Spreng.	NT			x		x	
Vicia akhmaganica Kazar.	DD			x			
Vicia alpestris Steven	NE	x	x	x		x	x
Vicia antiqua Grossh.	NE	x	x	x			
Vicia caucasica Ekutim.	NE		x	x		x	
Vicia ciliatula Lipsky	NE	x	x	x			
Vicia dadianorum Sommier & Levier	NE						x
Vicia erzurumica Demirkuş & Erik	CR B2ab(i,ii,iii)						x
Vicia grossheimii Ekutim.	NE	x	x	x		x	
Vicia hololasia Woronow	VU B1ab(ii,iii)		x			x	
Vicia iberica Grossh.	NE		x	x			
Vicia persica Boiss.	NE				x		
Vicia quadrijuga P. H. Davis	CR B1ab(i,ii,iii)+2ab(i,ii,iii)						x
Vicia rafigae Tamamsch.	VU D2		x				
Vicia rechingeri Chrtková	NE				x		
Vicia semiglabra Rupr. ex Boiss.	NE		x			x	x
Vicia sosnowskyi Ekutim.	DD			x		x	
FAGACEAE							
Quercus imeretina Steven ex Woronow	VU B2ab(iii)			x			
Quercus pontica K. Koch	VU B2ab(iii)			x			x
FUMARIACEAE							
Corydalis alexeenkoana N. Busch	DD			x		x	
Corydalis alpestris C. A. Mey.	NE	x	x	x		x	x
Corydalis conorhiza Ledeb.	NE			x		x	x

TAXON	RED LIST ASSESSMENT	AR	AZ	GE	IR	RU	TU
Corydalis emanuelii C. A. Mey. [*Corydalis pallidiflora* (Rupr.) N. Busch.]	LC			x		x	
Corydalis heteropetala Ochiauri	DD			x		x	
Corydalis pallidiflora (Rupr.) N. Busch. [*Corydalis emanuelii* C. A. Mey.]	LC			x		x	
Corydalis tarkiensis Prokh.	VU D2					x	
Corydalis teberdensis A. P. Khokhr.	LC					x	
Corydalis vittae Kolak.	VU D2			x			
Fumaria daghestanica Mikhailova	DD					x	
GENTIANACEAE							
Comastoma dechyanum (Sommier & Levier) Holub	NT			x		x	
Gentiana angulosa M. Bieb.	NE		x	x		x	
Gentiana grossheimii Doluch.	EN B2ab(iii)					x	
Gentiana kolakovskyi Doluch.	NT			x			
Gentiana lagodechiana (Kusn.) Grossh.	NE		x	x		x	
Gentiana oschtenica (Kusn.) Woronow	LC			x		x	
Gentiana owerinii (Kusn.) Grossh.	DD					x	
Gentiana paradoxa Albov	VU B1ab(i)+2ab(i)			x		x	
Gentiana rhodocalyx Kolak.	DD			x			
Gentianella biebersteinii (Bunge) Holub	LC			x		x	
Gentianella lipskyi (Kusn.) Holub	CR B2ab(ii)					x	
Gentianella poretzkyi Tzvelev	LC					x	
Gentianella promethea (Juz.) Holub	NE	x	x			x	
Swertia iberica Fisch. & C. A. Mey.	NE	x	x	x		x	x
GERANIACEAE							
Erodium anthemidifolium M. Bieb.	DD			x			
Erodium fumarioides Steven	NT		x			x	
Erodium hendrikii Alpinar	CR B1ab(i,ii,iii)+ 2ab(i,ii,iii)						x
Erodium schemachense Grossh.	NT		x				
Erodium sosnowskianum Fed.	CR B2ab(i,ii,iii)	x					

TAXON	RED LIST ASSESSMENT	AR	AZ	GE	IR	RU	TU
Erodium stevenii M. Bieb.	EN B1ab(ii,iv)c(iv)+ 2ab(ii,iv)c(iv)					x	
Geranium asphodeloides Burm. f. subsp. *pallens* (M. Bieb.) Woronow [*Geranium pallens* M. Bieb.]	DD			x		x	
Geranium cinereum Cav. var. *lazicum* (Woronow) Güner	DD			x			x
Geranium cinereum Cav. var. *ponticum* P. H. Davis & J. Roberts	VU B1ab(i,ii,iii)						x
Geranium davisianum Peşmen & Güner	CR B2ab(ii)						x
Geranium gracile Ledeb. ex Nordm.	NE			x	x	x	x
Geranium gymnocaulon DC.	NE			x		x	x
Geranium ibericum Cav.	NE	x		x		x	x
Geranium kemulariae Charadze	DD			x			
Geranium montanum Hablitz ex Pall.	NE		x		x		
Geranium pallens M. Bieb. [*Geranium asphodeloides* Burm. f. subsp. *pallens* (M. Bieb.) Woronow]	DD			x		x	
Geranium platypetalum Fisch. & C. A. Mey.	NE	x	x	x	x	x	x
Geranium platypetalum Fisch. & C. A. Mey. var. *albipetalum* Erik & Demirkuş	VU D2						x
Geranium psilostemon Ledeb.	NE			x			x
Geranium renardii Trautv.	LC			x		x	
Geranium ruprechtii (Woronow) Grossh.	NE	x	x	x		x	
GROSSULARIACEAE							
Ribes achurjanii Mulk.	EN B1ab(i,ii,iii)+ 2ab(i,ii,iii)	x					
Ribes armenum Pojark.	EN B1ab(i,ii,iii)+ 2ab(i,ii,iii)	x					
Ribes biebersteinii Berland. ex DC. [*Ribes caucasicum* M. Bieb.]	NE	x	x	x	x	x	x
Ribes caucasicum M. Bieb. [*Ribes biebersteinii* Berland. ex DC.]	NE	x	x	x	x	x	x
HYDRANGEACEAE							
Philadelphus caucasicus Koehne	NE	x	x	x		x	x

TAXON	RED LIST ASSESSMENT	AR	AZ	GE	IR	RU	TU
HYPERICACEAE							
Hypericum atropatanum Rzazade	NE	x	x				
Hypericum eleonorae Elenevsky	CR B2ab(ii,iii)	x					
Hypericum fissurale Woronow	CR B2ab(ii,iii); C1						x
Hypericum formosissimum Takht.	NE	x	x				
Hypericum grossheimii Kem.-Nath. ex Fed. [*Hypericum pruinatum* Boiss. & Balansa]	NE			x			x
Hypericum karjagini Rzazade	DD		x				
Hypericum maleevii Zernov & Elenevsky	CR B1ab(i,ii)					x	
Hypericum marginatum Woronow	EN B1ab(i,ii,iii)+ 2ab(i,ii,iii)						x
Hypericum nordmannii Boiss.	DD			x			
Hypericum nummularioides Trautv.	NE			x		x	x
Hypericum pruinatum Boiss. & Balansa [*Hypericum grossheimii* Kem.-Nath. ex Fed.]	NE			x			x
Hypericum theodorii Woronow	EN B1ab(v)+2ab(v)		x				
Hypericum xylosteifolium (Spach) N. Robson	NE			x		x	x
LAMIACEAE							
Ballota grisea Pojark. [*Ballota nigra* L. var. *grisea* (Pojark.) Menitsky]	NT		x			x	
Ballota nigra L. var. *grisea* (Pojark.) Menitsky [*Ballota grisea* Pojark.]	NT		x			x	
Betonica abchasica (Grossh.) Chinth.	VU B1ab(iii,v)+2ab(iii,v)			x		x	
Betonica nivea Steven, s. str .	NE		x		x	x	
Betonica orientalis L.	NE	x	x		x	x	x
Betonica ossetica (Bornm.) Chinth.	NT			x		x	
Dracocephalum botryoides Steven	NE	x	x			x	
Eremostachys azerbaijanica Rech. f.	NE				x		
Galeopsis nana Otsch.	NT			x			

TAXON	RED LIST ASSESSMENT	AR	AZ	GE	IR	RU	TU
Hyssopus angustifolius M. Bieb. [*Hyssopus officinalis* L. var. *angustifolius* (M. Bieb.) Benth.]	NE	x		x	x	x	x
Hyssopus officinalis L. var. *angustifolius* (M. Bieb.) Benth. [*Hyssopus angustifolius* M. Bieb.]	NE	x		x	x	x	x
Lamium album L. subsp. *sempervirens* (A. P. Khokhr.) Menitsky [*Lamium sempervirens* A. P. Khokhr.]	NE					x	x
Lamium gundelsheimeri K. Koch	EN B1ab(i,ii)+2ab(i,ii)					x	x
Lamium sempervirens A. P. Khokhr. [*Lamium album* L. subsp. *sempervirens* (A. P. Khokhr.) Menitsky]	NE					x	x
Lamium sulfureum Hausskn. & Sint. ex R. R. Mill	EN B1ab(i,ii,iii)+ 2ab(i,ii,iii)						x
Lamium tschorochense A. P. Khokhr.	CR B2ab(ii,iii)						x
Lamium vreemanii A. P. Khokhr.	EN B2ab(ii,iii)						x
Marrubium goktschaicum N. Pop.	DD		x	x			
Marrubium nanum Knorr.	NT		x				
Marrubium persicum C. A. Mey.	NE				x		
Marrubium plumosum C. A. Mey.	VU B1ab(iii)					x	
Micromeria elliptica K. Koch	EN B1ab(i,ii,iii,iv)+ 2ab(i,ii,iii,iv)						x
Micromeria fruticosa (L.) Druce subsp. *giresunica* P. H. Davis	CR B1ab(i,ii)+2ab(i,ii)						x
Nepeta alaghezi Pojark.	EN B2ab(ii,iii)	x					
Nepeta amoena Stapf	NE		x	x	x		
Nepeta betonicifolia C. A. Mey. subsp. *betonicifolia*	CR B2ab(ii,v)		x				
Nepeta betonicifolia C. A. Mey. subsp. *strictifolia* (Pojark.) Menitsky [*Nepeta grossheimii* Pojark.; *Nepeta strictifolia* (Pojark.) Menitsky]	DD	x	x	x			
Nepeta betonicifolia subsp. *somkhetica* (Kapeller) Menitsky [*Nepeta somkhetica* Kapeller]	NE	x	x			x	
Nepeta biebersteiniana (Trautv.) Pojark.	NT			x		x	

TAXON	RED LIST ASSESSMENT	AR	AZ	GE	IR	RU	TU
Nepeta cyanea Steven	LC					x	
Nepeta daghestanica Pojark. [*Nepeta teucriifolia* Willd. subsp. *daghestanica* (Pojark.) A. L. Budantzev]	NT		x			x	
Nepeta grossheimii Pojark. [*Nepeta betonicifolia* C. A. Mey. subsp. *strictifolia* (Pojark.) Menitsky; *Nepeta strictifolia* (Pojark.) Menitsky]	DD	x	x	x			
Nepeta iberica Pojark.	VU D2			x			
Nepeta komarovii E. A. Busch	NT			x		x	
Nepeta longituba Pojark.	CR B2ab(ii,v)		x				
Nepeta makuensis Jamzad & Mozaff.	CR				x		
Nepeta noraschenica Grossh.	NE	x	x				
Nepeta pogonosperma Jamzad & Assadi	NE				x		
Nepeta sahandica Noroozi & Ajani	CR				x		
Nepeta schischkinii Pojark.	NT		x				
Nepeta somkhetica Kapeller [*Nepeta betonicifolia* subsp. *somkhetica* (Kapeller) Menitsky]	NE	x	x			x	
Nepeta sosnovskyi Askerova	CR B2ab(ii,v)		x				
Nepeta strictifolia (Pojark.) Menitsky [*Nepeta betonicifolia* C. A. Mey. subsp. *strictifolia* (Pojark.) Menitsky; *Nepeta grossheimii* Pojark.]	NT	x	x	x			
Nepeta teucriifolia Willd. subsp. *daghestanica* (Pojark.) A. L. Budantzev [*Nepeta daghestanica* Pojark.]	NE		x			x	
Nepeta trautvetteri Boiss. & Buhse [*Nepeta velutina* Pojark.]	NE	x	x				
Nepeta velutina Pojark. [*Nepeta trautvetteri* Boiss. & Buhse]	NE	x	x				
Nepeta zangezura Grossh.	NE	x	x				
Origanum rotundifolium Boiss.	NE			x			x
Phlomis ghilanensis K. Koch	EN				x		
Phlomis lenkoranica Knorr.	NT		x				
Salvia andreji Pobed.	NT		x				

TAXON	RED LIST ASSESSMENT	AR	AZ	GE	IR	RU	TU
Salvia beckeri Trautv.	VU B1ab(v)		x			x	
Salvia canescens C. A. Mey.	LC					x	
Salvia canescens C. A. Mey. var. *daghestanica* (Sosn.) Menitsky	NE					x	
Salvia compar Trautv. ex Grossh.	LC			x			
Salvia fominii Grossh. [*Salvia limbata* C. A. Mey.]	NE	x	x	x			
Salvia garedji Troitzky	VU B1ab(v)+2ab(v)		x	x			
Salvia grossheimii Sosn.	NE	x	x				
Salvia huberi Hedge	EN B2ab(i,ii)						x
Salvia hypochionaea Boiss.	EN				x		
Salvia kuznetzovii Sosn.	NT		x			x	
Salvia limbata C. A. Mey. [*Salvia fominii* Grossh.]	NE	x	x	x			
Salvia oligophylla Aucher ex Benth.	NE				x		
Salvia pachystachya Trautv.	NE	x	x		x		x
Salvia sahendica Boiss. & Buhse	NE				x		
Satureja atropatana Bunge	NE				x		
Satureja bzybica Woronow	DD			x			
Satureja intermedia C. A. Mey.	NE		x		x		
Satureja isophylla Rech. f.	NE				x		
Satureja montana L. subsp. *subdentata* (Boiss.) Menitsky	LC					x	
Scutellaria albida L. subsp. *colchica* (Rech. f.) J. R. Edm.	NE			x			x
Scutellaria araxensis Grossh. [*Scutellaria orientalis* L. subsp. *araxensis* Fed.]	NE	x			x		
Scutellaria darriensis Grossh.	NT		x				
Scutellaria grossheimiana Juz.	NT		x				
Scutellaria helenae Albov	LC			x			
Scutellaria karatschaica Charadze [*Scutellaria orientalis* L. subsp. *karatschhaica* (Charadze) Menitsky]	EN B1ab(iii)+2ab(iii)			x		x	
Scutellaria karjaginii Grossh.	DD		x				

TAXON	RED LIST ASSESSMENT	AR	AZ	GE	IR	RU	TU
Scutellaria leptostegia Juz.	DD			x			
Scutellaria oreophila Grossh. [*Scutellaria orientalis* L. subsp. *oreophila* (Grossh.) Fed.]	NE		x	x		x	
Scutellaria orientalis L. subsp. *araxensis* Fed. [*Scutellaria araxensis* Grossh.]	NE	x			x		
Scutellaria orientalis L. subsp. *dagestanica* (Grossh.) Fed.	NT					x	
Scutellaria orientalis L. subsp. *granulosa* (Juz.) Fed.	VU B1ab(iii)					x	
Scutellaria orientalis L. subsp. *karatschhaica* (Charadze) Menitsky [*Scutellaria karatschaica* Charadze]	EN B1ab(iii)+2ab(iii)			x		x	
Scutellaria orientalis L. subsp. *karjaginii* (Grossh.) Fed. [*Scutellaria sevanensis* Sosn. ex Grossh.]	NE	x	x				
Scutellaria orientalis L. subsp. *novorossica* (Juz.) Fed.	EN B1ab(iii,iv)c(ii,iii)+ 2ab(iii,iv)c(ii,iii)					x	
Scutellaria orientalis L. subsp. *oreophila* (Grossh.) Fed. [*Scutellaria oreophila* Grossh.]	NE		x	x		x	
Scutellaria orientalis L. subsp. *polyodon* (Juz.) Fed.	NE					x	
Scutellaria orientalis L. subsp. *raddeana* (Juz.) Fed. [*Scutellaria raddeana* Juz.]	NT			x		x	
Scutellaria orientalis L. subsp. *sedelmeyerae* (Juz.) Fed. [*Scutellaria sedelmeyerae* Juz.]	NE	x	x				
Scutellaria orientalis L. subsp. *sosnowskyi* (Takht.) Fed. [*Scutellaria sosnowskyi* Takht.]	NE	x		x			x
Scutellaria pontica K. Koch	NE			x			x
Scutellaria prilipkoana Grossh.	VU D2		x				
Scutellaria raddeana Juz. [*Scutellaria orientalis* L. subsp. *raddeana* (Juz.) Fed.]	NT			x		x	
Scutellaria rhomboidalis Grossh.	CR B1ab(i,ii,v)+ 2ab(i,ii,v)		x				

TAXON	RED LIST ASSESSMENT	AR	AZ	GE	IR	RU	TU
Scutellaria sedelmeyerae Juz. [*Scutellaria orientalis* L. subsp. *sedelmeyerae* (Juz.) Fed.]	NE	x	x				
Scutellaria sevanensis Sosn. ex Grossh. [*Scutellaria orientalis* L. subsp. *karjaginii* (Grossh.) Fed.]	NE	x	x				
Scutellaria sosnowskyi Takht. [*Scutellaria orientalis* L. subsp. *sosnowskyi* (Takht.) Fed.]	NE	x		x			x
Scutellaria tortumensis (Kit Tan & Sorger) A. P. Khokhr.	NE						x
Scutellaria uzundereensis A. P. Khokhr.	CR B1ab(i,ii,iii)+ 2ab(i,ii,iii)						x
Stachys araxina Kapeller	NE	x	x				x
Stachys bayburtensis R. Bhattacharjee & Hub.-Mor.	CR B1ab(i,ii)+2ab(i,ii)						x
Stachys choruhensis Kit Tan & Sorger	CR B1ab(i,ii)+2ab(i,ii)						x
Stachys cretica L. subsp. *trapezuntica* Rech. f.	EN B1ab(i,ii,iii)+ 2ab(i,ii,iii)						x
Stachys fominii Sosn. ex Grossh.	LC		x				
Stachys fruticulosa M. Bieb. subsp. *fruticulosa*	NE		x	x			
Stachys fruticulosa M. Bieb. subsp. *grossheimii* (Kapeller) Menitsky	DD		x				
Stachys paulii Grossh.	VU D2		x				
Stachys rizeensis R. Bhattacharjee	EN B1ab(i,ii,iii)+ 2ab(i,ii,iii)						x
Stachys stchegleewii Sosn.	NT		x				
Stachys sosnowskyi Kapeller	CR B2ab(ii,iii); C2a(i)						x
Stachys talyschensis Kapeller	LC		x				
Teucrium canum Fisch. & C. A. Mey.	NE	x	x			x	
Teucrium chamaedrys L. subsp. *nuchense* (K. Koch) Rech. f. [*Teucrium nuchense* K. Koch]	LC			x		x	
Teucrium nuchense K. Koch [*Teucrium chamaedrys* L. subsp. *nuchense* (K. Koch) Rech. f.]	NE			x		x	

TAXON	RED LIST ASSESSMENT	AR	AZ	GE	IR	RU	TU
Thymus caucasicus Willd. ex Ronniger, s. str. [*Thymus praecox* Opiz subsp. *medwedewii* (Ronninger) Jalas]	NE		x	x		x	x
Thymus collinus M. Bieb., s. str.	NE	x	x	x		x	
Thymus coriifolius Ronninger	NE		x	x			
Thymus daghestanicus Klokov & Des.-Shost.	NE		x			x	
Thymus desjatovae Ronniger	VU D2		x				
Thymus elisabethae Klokov & Des.-Shost.	NT					x	
Thymus grossheimii Ronniger [*Thymus praecox* Opiz subsp. *grossheimii* (Ronniger) Jalas]	NE		x	x		x	x
Thymus hadzhievii Grossh.	VU D2		x				
Thymus helendzhicus Klokov & Des.-Shost.	EN B1ab(iii)			x		x	
Thymus karamarjanicus Klokov & Des.-Shost.	NT		x				
Thymus karjaginii Grossh.	NE		x	x			
Thymus kjapazi Grossh.	LC		x				
Thymus klokovii (Ronninger) Des.-Shost. [*Thymus tiflisiensis* Klokov & Des.-Shost.]	NE		x	x			
Thymus ladjanuricus Kem.-Nath.	VU D2			x			
Thymus majkopensis Klokov & Des.-Shost.	NT					x	
Thymus marandensis Jamzad	VU				x		
Thymus markhotensis Maleev	VU B1ab(ii,iii,v)+2ab(ii,iii,v); C2a(i)					x	
Thymus nummularius M. Bieb., s. str.	NE		x	x		x	
Thymus pastoralis Iljin	VU B2ac(i,ii)		x			x	
Thymus persicus (Ronniger ex Rech. f.) Jalas	NE				x		
Thymus praecox Opiz subsp. *grossheimii* (Ronniger) Jalas [*Thymus grossheimii* Ronniger]	NE		x	x		x	x

TAXON	RED LIST ASSESSMENT	AR	AZ	GE	IR	RU	TU
Thymus praecox Opiz subsp. *medwedewii* (Ronninger) Jalas [*Thymus caucasicus* Willd. ex Ronniger, s. str.]	NE		x	x		x	x
Thymus pseudopulegioides Klokov & Des.-Shost.	NE			x		x	x
Thymus pulchellus C. A. Mey.	EN B2ab(ii,iii,iv,v)					x	
Thymus sosnowskyi Grossh.	LC			x			
Thymus tiflisiensis Klokov & Des.-Shost. [*Thymus klokovii* (Ronninger) Des.-Shost.]	NE		x	x			
Thymus trautvetteri Klokov & Des.-Shost.	NE		x		x		
Ziziphora borzhomica Juz. ex Grossh.	LC			x			
Ziziphora brantii K. Koch	DD	x					x
Ziziphora clinopodioides Lam. subsp. *szowitsii* (Rech. f.) Rech. f.	NE	x			x		
Ziziphora denticulata Juz.	DD		x				
Ziziphora puschkinii Adams	NE		x	x		x	
Ziziphora serpyllacea M. Bieb.	NE	x	x	x		x	
Ziziphora subnivalis Ter-Chatsch.	NE		x	x		x	
Ziziphora woronowii Maleev	EN B1ab(iii,iv)+2ab(iii,iv)			x		x	
LINACEAE							
Linum barsegjanii Gabrieljan & Dittrich	CR B1ab(i,ii,iii)+2ab(i,ii,iii)	x					
Linum hypericifolium Salisb.	NE	x	x	x		x	x
Linum subbiflorum Juz.	NE	x	x				
LYTHRACEAE							
Ammania pubiflora (Koehne) Sosn.	LC		x				
MALVACEAE							
Alcea abchasica Iljin	DD			x			
Alcea assadii Pakravan	CR				x		
Alcea flavovirens (Boiss. & Buhse) Iljin	NE	x			x		x
Alcea grossheimii Iljin	EN B2ab(i,ii,iii)	x					

TAXON	RED LIST ASSESSMENT	AR	AZ	GE	IR	RU	TU
Alcea karsiana (Bordz.) Litv.	DD	x					x
Alcea kusariensis (Iljin ex Grossh.) Iljin	NE		x			x	
Alcea lenkoranica Iljin	NE		x		x		
Alcea rugosa Alef.	NE	x	x	x		x	
Alcea sachsachanica Iljin	DD		x				
Alcea sophiae Iljin	NE	x					
Alcea sosnovskyi Iljin	NE	x					
Alcea transcaucasica (Iljin ex Grossh.) Iljin	NT			x			
Hibiscus ponticus Rupr.	CR C2a(i)			x			
Malva grossheimii Iljin	DD		x				
MORACEAE							
Ficus hyrcana Grossh.	NT		x				
MORINACEAE							
Morina persica L. var. *decussatifolia* Erik & Demirkuş	CR C2a(i)						x
NYMPHAEACEAE							
Nymphaea colchica (Woronow ex Grossh.) Kem.-Nath.	DD			x		x	
OLEACEAE							
Fraxinus coriariifolia Scheele [*Fraxinus excelsior* L. subsp. *coriariifolia* (Scheele) A. E. Murray]	NE		x	x	x	x	x
Fraxinus excelsior L. subsp. *coriariifolia* (Scheele) A. E. Murray [*Fraxinus coriariifolia* Scheele]	NE		x	x	x	x	x
Osmanthus decorus (Boiss. & Balansa) Kasapligil	VU B2ab(iii)			x			x
ONAGRACEAE							
Chamerion colchicum (Albov) Holub [*Epilobium colchicum* Albov]	NE	x	x	x		x	x
Epilobium colchicum Albov [*Chamerion colchicum* (Albov) Holub]	NE	x	x	x		x	x

TAXON	RED LIST ASSESSMENT	AR	AZ	GE	IR	RU	TU
OROBANCHACEAE							
Orobanche armena Tzvelev	EN B2ab(iii,iv)c(ii,iii)						x
Orobanche colorata K. Koch	NE	x	x	x		x	
Orobanche connata K. Koch	DD	x					
Orobanche gamosepala Reut.	NE	x		x		x	x
Orobanche grossheimii Novopokr.	NE	x	x	x		x	
Orobanche ingens (Beck) Tzvelev	DD					x	
Orobanche inulae Novopokr. & Abramov	NE		x	x		x	
Orobanche pulchella (C. A. Mey.) Novopokr.	NE	x		x		x	
Orobanche quadrifida K. Koch	DD			x			
Orobanche raddeana Beck	LC			x		x	
Orobanche schelkovnikovii Tzvelev	DD	x					
Orobanche septemloba (Beck) Tzvelev	DD		x				
PAEONIACEAE							
Paeonia carthalinica Ketsk.	EN B2ab(v)			x			
Paeonia caucasica (Schipcz.) Schipcz.	LC			x		x	
Paeonia lagodechiana Kem.-Nath.	NT			x			
Paeonia macrophylla (Albov) Lomakin	VU B1ab(iii,v)			x		x	
Paeonia majko Ketsk.	DD			x			
Paeonia mlokosewitschii Lomakin	VU B1ab(iii,v)+2ab(iii,v)		x	x		x	
Paeonia ruprechtiana Kem.-Nath.	EN B1ab(iii,v)+2ab(iii,v)			x			
Paeonia steveniana Kem.-Nath.	EN B1ab(iii,v)+2ab(iii,v)			x			
PAPAVERACEAE							
Papaver alberti Mikheev	NT					x	
Papaver bipinnatum C. A. Mey.	NE	x	x		x	x	
Papaver bracteatum Lindl.	NE	x			x	x	
Papaver gabrielianae M. V. Agab.	VU D2	x					
Papaver lasiothrix Fedde	NE				x		x
Papaver lateritium K. Koch subsp. *lateritium*	VU B1ab(i,ii,iii,iv)						x

TAXON	RED LIST ASSESSMENT	AR	AZ	GE	IR	RU	TU
Papaver lateritium K. Koch subsp. *monanthum* (Trautv.) Kadereit [*Papaver monanthum* Trautv.]	NE	x	x	x			x
Papaver lisae N. Busch	EN B1ab(iii)					x	
Papaver maschukense Mikheev	DD					x	
Papaver monanthum Trautv. [*Papaver lateritum* K. Koch subsp. *monanthum* (Trautv.) Kadereit]	NE	x	x	x			x
Papaver orientale L.	NE	x		x		x	x
Papaver orientale L. var. *parviflora* N. Busch [*Papaver paucifoliatum* (Trautv.) Fedde]	NE	x		x		x	x
Papaver paczoskii Mikheev	DD					x	
Papaver paucifoliatum (Trautv.) Fedde [*Papaver orientale* L. var. *parviflora* N. Busch]	NE	x		x		x	x
Papaver roseolum M. V. Agab. & Fragman	EN B2ab(ii,iii)	x					
Papaver schelkovnikovii N. Busch	EN B2ab(iii)		x				
Papaver sjunicicum M. V. Agab.	DD	x					
Papaver talyshense Grossh.	VU D2		x				
PLANTAGINACEAE							
Plantago atrata Hoppe subsp. *circassica* Tzvelev	DD					x	
PLUMBAGINACEAE							
Acantholimon araxanum Bunge	DD	x	x		x		x
Acantholimon atropatanum Bunge	CR				x		
Acantholimon bracteatum (Girard) Boiss.	NE	x	x		x		
Acantholimon densiflorum Assadi	NE				x		
Acantholimon esfandiarii Rech. f. & Schiman-Czeika	CR				x		
Acantholimon fedorovii Tamamsch. & Mirzoeva	CR B1ab(iii)	x	x				
Acantholimon fominii Kusn.	NT		x	x			
Acantholimon gabrieljaniae Mirzoeva	DD	x					

TAXON	RED LIST ASSESSMENT	AR	AZ	GE	IR	RU	TU
Acantholimon glumaceum (Jaub. & Spach) Boiss.	NE	x	x	x			x
Acantholimon karelinii (Stschegl.) Bunge	NE	x	x		x		
Acantholimon lepturoides (Jaub. & Spach) Boiss.	NE	x	x	x			
Acantholimon mishaudaghense Mobayen	NE				x		
Acantholimon sahendicum Boiss. & Buhse	NE	x	x		x		
Acantholimon schemachense Grossh.	VU B1ab(iii)+2ab(iii)		x			x	
Acantholimon tenuiflorum Boiss.	EN B1ab(iii,v)+ 2ab(iii,v)		x				
Acantholimon vedicum Mirzoeva	DD	x					
Acantholimon wilhelminae Rech. f. & Schiman-Czeika	CR				x		
Limonium fischeri (Trautv.) Lincz.	EN B2ab(iii)	x	x				
POLYGALACEAE							
Polygala alata (Tamamsch.) Galushko	NT					x	
Polygala albowii Kem.-Nath.	LC			x			
Polygala alpicola Rupr.	NE	x	x	x		x	
Polygala amoenissima Tamamsch.	DD			x		x	
Polygala caucasica Rupr.	NE	x	x	x		x	
Polygala grossheimii Kem.-Nath.	CR B2ab(v)		x				
Polygala hohenackeriana Fisch. & C. A. Mey.	NE	x	x	x			
Polygala makaschwilii Kem.-Nath.	DD			x		x	
Polygala mariamae Tamamsch.	NT	x		x			
Polygala pseudohospita (Tamamsch.) Tamamsch.	NE	x	x				
Polygala sosnowskyi Kem.-Nath.	NE		x	x		x	
Polygala transcaucasica Tamamsch.	LC	x		x			x
Polygala urartu Tamamsch.	EN B1ab(iii,v)+2ab(iii,v)	x					
POLYGONACEAE							

TAXON	RED LIST ASSESSMENT	AR	AZ	GE	IR	RU	TU
Aconogonon panjutinii (Charkev.) Soják [*Polygonum panjutinii* Charkev.]	LC			x		x	
Atraphaxis angustifolia Jaub. & Spach	EN B1ab(iii)		x				
Atraphaxis aucheri Jaub. & Spach	NE				x		
Atraphaxis suaedifolia Jaub. & Spach	CR				x		
Calligonum bakuense Litv.	EN B1ab(i,ii,iii,iv,v)+ 2ab(i,ii,iii,iv,v)		x				
Persicaria imeretina (Charkev.) Soják	NE					x	
Polygonum caspicum Kom.	CR B1ab(i,ii,iii,v)+ 2ab(i,ii,iii,v)		x				
Polygonum dshawachischwilii Charkev.	DD			x			
Polygonum panjutinii Charkev. [*Aconogonon panjutinii* (Charkev.) Soják]	LC			x		x	
Polygonum tiflisiense Kom.	DD		x	x			
Rumex caucasicus Rech. f.	DD						x
PRIMULACEAE							
Androsace intermedia Ledeb.	NE			x			x
Androsace raddeana Sommier & Levier	NE	x	x	x		x	
Cyclamen abchasicum (Medw. ex Kusn.) Kolak. [*Cyclamen vernum* Sweet var. *abchasicum* (Medw. ex Kusn.) Menitsky]	NT			x		x	
Cyclamen colchicum (Albov) Albov	VU D2			x			
Cyclamen parviflorum Pobed.	LC						x
Cyclamen parviflorum Pobed. var. *subalpinum* Grey-Wilson	DD						x
Cyclamen vernum Sweet var. *abchasicum* (Medw. ex Kusn.) Menitsky [*Cyclamen abchasicum* (Medw. ex Kusn.) Kolak.]	NT			x		x	
Cyclamen vernum Sweet var. *vernum*	NE	x	x	x		x	
Primula abchasica Sosn.	DD			x			
Primula amoena M. Bieb.	NE	x	x	x		x	
Primula bayernii Rupr.	NT			x		x	
Primula cordifolia Rupr.	NE	x	x	x		x	

TAXON	RED LIST ASSESSMENT	AR	AZ	GE	IR	RU	TU
Primula darialica Rupr.	VU B2ab(ii,iii,v)			x		x	
Primula elatior (L.) Hill subsp. *pseudoelatior* (Kusn.) Sm. & Forrest [*Primula pseudoelatior* Kusn.]	NE			x		x	x
Primula farinifolia Rupr.	LC			x		x	
Primula juliae Kusn.	EN B1ab(ii,iii)		x	x		x	
Primula kusnetzovii Fed.	DD			x		x	
Primula longipes Freyn & Sint.	VU B1ab(i,ii,iv)						x
Primula luteola Rupr.	NT			x		x	
Primula megaseifolia Boiss. & Balansa ex Boiss.	NE			x			x
Primula meyeri Rupr.	NE			x		x	x
Primula ossetica Kusn.	DD					x	
Primula pseudoelatior Kusn. [*Primula elatior* (L.) Hill subsp. *pseudoelatior* (Kusn.) Sm. & Forrest]	NE			x		x	x
Primula renifolia Volgunov	EN B1ab(ii,iii)+2ab(ii,iii)					x	
Primula ruprechtii Kusn. ex Lipsky	NE	x	x	x		x	
Primula saguramica Gavr.	DD			x			
Primula vulgaris subsp. *woronowii* (Losinsk.) Fed. [*Primula woronowii* Losinsk.]	NE	x	x	x	x		
Primula woronowii Losinsk. [*Primula vulgaris* subsp. *woronowii* (Losinsk.) Fed.]	NE	x	x	x	x		
Sredinskya grandis (Trautv.) Fed.	NT			x		x	
RAFFLESIACEAE							
Cytinus ruber (Fourr.) Fritsch	NE			x			
RANUNCULACEAE							
Aconitum cymbulatum (Schmalh.) Lipsky subsp. *fastigiatum* (N. Busch) Vorosch.	DD					x	
Aconitum cymbulatum (Schmalh.) Lipsky	LC			x		x	
Aconitum nasutum Fisch. ex Rchb. subsp. *anfalovii* Vorosch.	NE					x	x

TAXON	RED LIST ASSESSMENT	AR	AZ	GE	IR	RU	TU
Aconitum nasutum Fisch. ex Rchb. subsp. *nasutum*	NE	x	x	x		x	x
Aconitum nasutum Fisch. ex Rchb. var. *pubiceps* Rupr. [*Aconitum pubiceps* (Rupr.) Trautv.]	DD			x		x	
Aconitum nasutum Fisch. ex Rchb. subsp. *tuscheticum* (N. Busch) Luferov [*Aconitum tuscheticum* (N. Busch) N. Busch]	DD			x		x	
Aconitum orientale Mill. var. *eriocarpum* Akinf. ex N. Busch	DD					x	
Aconitum orientale Mill. var. *ochroleucum* (Willd.) Rapaics	DD					x	
Aconitum orientale Mill. var. *orientale*	NE	x		x		x	x
Aconitum orientale Mill. var. *ponticum* Rapaics	DD					x	
Aconitum pubiceps (Rupr.) Trautv. [*Aconitum nasutum* Fisch. ex Rchb. var. *pubiceps* Rupr.]	DD			x		x	
Aconitum tuscheticum (N. Busch) N. Busch [*Aconitum nasutum* Fisch. ex Rchb. subsp. *tuscheticum* (N. Busch) Luferov]	NE			x		x	
Adonis cyllenea Boiss., Heldr. & Orph. var. *paryadrica* Boiss.	DD						x
Anemone fasciculata L.	NE	x		x		x	x
Anemone kusnetzowii Woronow ex Grossh.	CR B1ab(i,ii,v)+ 2ab(i,ii,v)		x				
Anemone speciosa Adams ex Pritz.	NE		x	x		x	
Aquilegia caucasica (Ledeb.) Rupr. [*Aquilegia olympica* Boiss.]	NE			x	x	x	
Aquilegia colchica Kem.-Nath.	EN B1ab(iii,v)+2ab(iii,v)			x			
Aquilegia gegica Jabr.-Kolak.	VU D2			x			
Aquilegia kubanica I. M. Vassiljeva	DD					x	
Aquilegia olympica Boiss. [*Aquilegia caucasica* (Ledeb.) Rupr.]	NE			x	x	x	
Delphinium arcuatum N. Busch	DD		x			x	

TAXON	RED LIST ASSESSMENT	AR	AZ	GE	IR	RU	TU
Delphinium bracteosum Sommier & Levier	NE		x	x		x	
Delphinium buschianum Grossh.	NE	x	x				x
Delphinium caucasicum C. A. Mey.	NE		x	x		x	
Delphinium charadzeae Kem.-Nath. & Gagnidze	DD			x		x	
Delphinium crispulum Rupr.	NT					x	
Delphinium darginicum Dimitrova	DD					x	
Delphinium dasycarpum Steven ex DC.	DD			x		x	
Delphinium dzavakhischwilii Kem.-Nath.	DD			x			
Delphinium elisabethae N. Busch	VU D2			x			
Delphinium fedorovii Dimitrova	NE			x		x	
Delphinium flexuosum M. Bieb.	NE	x	x	x		x	x
Delphinium foetidum Lomakin	NE	x	x				
Delphinium freynii Conrath [*Delphinium schmalhausenii* Albov]	NE	x		x		x	x
Delphinium iris Ilarslan & Kit Tan	CR B1ab(i,ii)+2ab(i,ii)						x
Delphinium ironorum N. Busch	EN B1ab(iii)+2ab(iii)			x			
Delphinium linearilobum (Trautv.) N. Busch	NE	x	x	x			
Delphinium macropogon Prokh.	DD				x	x	
Delphinium megalanthum Nevski	NT					x	
Delphinium munzianum P. H. Davis & Kit Tan	CR B2ab(ii,iii)						x
Delphinium ochroleucum Steven ex DC.	DD	x	x	x		x	
Delphinium osseticum N. Busch	CR B2ab(iii)			x			
Delphinium prokhanovii Dimitrova	DD		x			x	
Delphinium pyramidatum Albov	NE		x	x		x	
Delphinium schmalhausenii Albov [*Delphinium freynii* Conrath]	NE	x		x		x	x
Delphinium speciosum M. Bieb.	NE		x	x		x	
Delphinium szowitsianum Boiss.	NE	x	x		x		x

TAXON	RED LIST ASSESSMENT	AR	AZ	GE	IR	RU	TU
Delphinium thamarae Kem.-Nath.	EN B1ab(iii)+2ab(iii)			x			
Delphinium tomentellum N. Busch	VU B2ab(ii,iii)		x			x	
Ficaria ledebourii Grossh. & Schischk.	NE		x	x		x	
Ficaria varia Otsch.	VU D2			x			
Helleborus abchasicus A. Braun	DD			x		x	
Helleborus caucasicus A. Braun	DD	x		x		x	
Pulsatilla albana (Steven) Bercht. & J. Presl var. *andina* Rupr.	DD		x			x	
Pulsatilla armena (Boiss.) Rupr.	NE	x	x				
Pulsatilla aurea (Sommier & Levier) Juz.	LC			x		x	
Pulsatilla georgica Rupr.	EN B1ab(iii,v)+2ab(iii,v)			x			
Pulsatilla violacea Rupr. subsp. *violacea*	NE	x		x		x	
Ranunculus abchasicus Freyn	NT			x		x	
Ranunculus arachnoideus C. A. Mey.	NE		x	x		x	
Ranunculus baidarae Rupr.	NE	x	x	x		x	
Ranunculus balkaricus N. Busch	LC			x		x	
Ranunculus buhsei Boiss.	NE			x	x	x	x
Ranunculus caucasicus M. Bieb. subsp. *caucasicus*	DD			x	x	x	x
Ranunculus caucasicus M. Bieb. subsp. *subleiocarpus* (Sommier & Levier) P. H. Davis [*Ranunculus raddeanus* Regel]	NE	x		x		x	x
Ranunculus crassifolius (Rupr.) Grossh.	NE		x	x		x	
Ranunculus dissectus M. Bieb. subsp. *aragazii* (Grossh.) Bulany & Derv.-Sok.	CR B1ab(iii,v)	x					
Ranunculus dissectus M. Bieb. subsp. *rigidulus* (Boiss.) P. H. Davis	DD						x
Ranunculus dolosus Fisch. & C. A. Mey.	NE				x		
Ranunculus elegans K. Koch [*Ranunculus grandiflorus* L.]	NE	x	x	x			
Ranunculus georgicus Kem.-Nath.	NE					x	
Ranunculus grandiflorus L. [*Ranunculus elegans* K. Koch]	NE	x	x	x			

TAXON	RED LIST ASSESSMENT	AR	AZ	GE	IR	RU	TU
Ranunculus grossheimii Kolak. [*Ranunculus suukensis* N. Busch]	NT			x		x	
Ranunculus helenae Albov	VU B1ab(ii,iii,v)+ 2ab(ii,iii,v); C2a(i)			x		x	
Ranunculus lojkae Sommier & Levier	LC			x		x	
Ranunculus migaricus Kem.-Nath. & Chinth.	NT			x			
Ranunculus obesus Trautv.	NE	x		x			x
Ranunculus raddeanus Regel [*Ranunculus caucasicus* M. Bieb. subsp. *subleiocarpus* (Sommier & Levier) P. H. Davis]	NE	x		x		x	x
Ranunculus renzii Iranshahr & Rech. f.	CR				x		
Ranunculus sahendicus Boiss. & Buhse	NE				x		
Ranunculus sommieri Albov	DD			x			
Ranunculus subtilis Trautv.	LC			x		x	
Ranunculus suukensis N. Busch [*Ranunculus grossheimii* Kolak.]	NT			x		x	
Ranunculus svaneticus Rupr.	DD			x			
Ranunculus tempskyanus Freyn & Sint.	CR B2ab(ii,iii)						x
Ranunculus trisectilis Ovcz.	NE			x	x	x	x
Ranunculus vermirrhizus A. P. Khokhr.	CR B1ab(i,ii,iii)+ 2ab(i,ii,iii)						x
Ranunculus villosus DC. subsp. *villosus*	NE	x	x			x	
Thalictrum triternatum Rupr.	LC			x		x	
RESEDACEAE							
Reseda globulosa Fisch. & C. A. Mey.	NE	x	x	x		x	
RHAMNACEAE							
Rhamnus cordata Medw.	NT			x			
Rhamnus depressa Grubov	NE	x	x	x		x	x
Rhamnus imeretina Booth, Petz. & Kirchn.	NE			x		x	x
Rhamnus microcarpa Boiss.	NE			x		x	x
Rhamnus pallasii Fisch. & C. A. Mey.	NE	x	x	x	x	x	x

TAXON	RED LIST ASSESSMENT	AR	AZ	GE	IR	RU	TU
Rhamnus spathulifolia Fisch. & C. A. Mey.	NE	x	x	x	x	x	x
ROSACEAE							
Alchemilla abchasica Buser	NE			x		x	x
Alchemilla adelodictya Juz.	DD			x			
Alchemilla akdoganica Kalheber	NE						x
Alchemilla alexandri Juz.	DD			x			
Alchemilla amardica Rothm.	NE				x		
Alchemilla ancerensis Kalheber	NE						x
Alchemilla aurata Juz.	DD			x			
Alchemilla ayazii Kalheber	NE						x
Alchemilla ayderensis Kalheber	NE						x
Alchemilla bakurianica Sosn.	DD			x			
Alchemilla barbatiflora Juz.	NE			x			x
Alchemilla basakii Kalheber	NE						x
Alchemilla beyazoglii Kalheber	NE						x
Alchemilla capillacea Juz.	DD			x			
Alchemilla cartalinica Juz.	DD			x			
Alchemilla caucasica Buser	NE	x	x	x	x		
Alchemilla chlorosericea (Buser) Juz.	LC			x		x	
Alchemilla cimilensis Kalheber	NE						x
Alchemilla circassica Juz.	DD					x	
Alchemilla condensa A. Froehner	NE				x		
Alchemilla daghestanica Juz.	NE					x	x
Alchemilla debilis Juz.	DD			x		x	
Alchemilla depexa Juz.	DD			x		x	
Alchemilla divaricans Buser	DD			x		x	
Alchemilla diversipes Juz.	NE	x	x	x		x	
Alchemilla dombaica Juz.	DD					x	
Alchemilla dura Buser	NE	x		x		x	x
Alchemilla dzhavakhetica Juz.	DD		x				
Alchemilla elevitensis Kalheber	NE						x

TAXON	RED LIST ASSESSMENT	AR	AZ	GE	IR	RU	TU
Alchemilla elisabethae Juz.	LC			x		x	
Alchemilla epidasys Rothm. [*Alchemilla pseudomollis* Juz.]	DD			x		x	
Alchemilla epipsila Juz.	NE	x	x	x			x
Alchemilla erectilis Juz.	DD			x			
Alchemilla erythropoda Juz.	NE		x		x		
Alchemilla farinosa A. Froehner	NE				x		
Alchemilla frondosa Juz.	DD					x	
Alchemilla georgica Juz.	NE	x		x			
Alchemilla gigantodus A. Froehner	NE				x		x
Alchemilla grandidens Juz.	DD			x			
Alchemilla hayirlioglii Kalheber	NE						x
Alchemilla hemsinica Kalheber	NE						x
Alchemilla hessii Rothm.	NE				x		
Alchemilla hirtipedicellata Juz.	NE			x			x
Alchemilla holotricha Juz.	NE		x	x		x	x
Alchemilla hypochlora Juz.	DD			x			
Alchemilla hypotricha Juz.	DD			x			
Alchemilla hyrcana (Buser) Juz.	NE	x	x		x		
Alchemilla ikizdereensis Kalheber	NE						x
Alchemilla impolita Juz.	DD			x			
Alchemilla indurata Juz.	DD			x			
Alchemilla insignis Juz.	DD			x			
Alchemilla jaroschenkoi Grossh.	DD		x				
Alchemilla kackarensis Kalheber	NE						x
Alchemilla kozlowskii Juz.	NE		x	x			
Alchemilla laeta Juz.	LC			x		x	
Alchemilla languida Buser	NE	x		x		x	
Alchemilla microdictya Juz.	DD			x			
Alchemilla microdonta Juz.	NE		x	x		x	
Alchemilla minusculiflora Buser	NE			x			x

TAXON	RED LIST ASSESSMENT	AR	AZ	GE	IR	RU	TU
Alchemilla obtegens Juz.	DD			x			
Alchemilla orduensis Pawł.	NE						x
Alchemilla orthotricha Rothm.	NE					x	x
Alchemilla oxysepala Juz.	NE			x		x	x
Alchemilla pascualis Juz.	DD			x			
Alchemilla pectiniloba A. Froehner	NE				x		
Alchemilla plicatissima A. Froehner	NE				x		x
Alchemilla procerrima A. Froehner	NE						x
Alchemilla pseudomollis Juz. [*Alchemilla epidasys* Rothm.]	DD			x		x	
Alchemilla psilocaula Juz.	NT					x	
Alchemilla pycnotricha Juz.	NE		x	x			
Alchemilla raddeana (Buser) Juz.	NE	x	x				
Alchemilla rechingeri Rothm.	NE				x		
Alchemilla rigida Buser	NE	x		x	x	x	
Alchemilla rizensis Pawł.	NE						x
Alchemilla sedelmeyeriana Juz.	NE	x	x	x			
Alchemilla sericata Rchb.	NE	x	x	x	x	x	
Alchemilla sintenisii Rothm.	NE						x
Alchemilla smirnovii Juz.	NE	x	x				x
Alchemilla speciosa Buser	NE			x		x	x
Alchemilla stellulata Juz.	DD			x		x	
Alchemilla subcrenatiformis Juz.	DD			x			
Alchemilla suberectipila Juz.	DD			x			
Alchemilla subsplendens Buse	DD			x			
Alchemilla surculosa A. Froehner	NE				x		
Alchemilla tamarae Juz.	DD			x		x	
Alchemilla tephroserica (Buser) Juz.	DD			x		x	
Alchemilla tiryalensis Pawł.	NE						x
Alchemilla trabzonica Hayırlıoğlu-Ayaz & Beyazoğlu	NE						x
Alchemilla transcaucasica Rothm.	NE						x

TAXON	RED LIST ASSESSMENT	AR	AZ	GE	IR	RU	TU
Alchemilla tredecimloba Buser	DD			x		x	
Alchemilla undecimloba Juz.	NE			x		x	
Alchemilla urceolata Juz.	NT			x		x	
Alchemilla valdehirsuta Buser	NE	x	x	x	x		
Alchemilla venosa Juz.	NE	x	x	x			x
Alchemilla woronowii Juz.	DD			x			
Alchemilla ziganadagensis Pawł.	NE						x
Amygdalus georgica Desf.	VU			x			
Amygdalus nairica Fed. & Takht.	EN B1ab(i,ii,iii,v)+ 2ab(i,ii,iii,v)	x					
Cotoneaster armenus Pojark.	VU B1ab(iii,v)+2ab(iii,v)	x					
Cotoneaster melanocarpus Lodd., G. Lodd. & W. Lodd var. *daghestanicus* (Zinserl.) Gladkova	DD					x	
Cotoneaster morulus Pojark.	NE	x	x	x		x	x
Cotoneaster saxatilis Pojark.	DD	x	x	x		x	
Cotoneaster soczavianus Pojark.	NE			x		x	
Cotoneaster transcaucasicus Pojark.	VU B2ab(ii,iii)	x	x				x
Crataegus armena Pojark.	EN B1ab(iii,v)+2ab(iii,v)	x					
Crataegus caucasica K. Koch	NE	x	x	x		x	
Crataegus daghestanica Gladkova	DD					x	
Crataegus eriantha Pojark.	NE	x	x			x	
Crataegus turcica Dönmez	CR B2ab(ii,iii); D						x
Crataegus zangezura Pojark.	EN B1ab(iii,v)+2ab(iii,v)	x					
Dryas caucasica Juz.	NE		x	x		x	
Geum latilobum Sommier & Levier	DD			x		x	
Potentilla adscharica Sommier & Levier	NE	x	x	x	x	x	x
Potentilla alexeenkoi Lipsky	NE		x		x	x	
Potentilla brachypetala Fisch. & C. A. Mey. ex Lehm.	NE	x		x		x	
Potentilla buchneri Kit Tan & Sorger [*Potentilla lomakinii* Grossh.]	NE	x	x				x
Potentilla camillae Kolak.	DD			x		x	

TAXON	RED LIST ASSESSMENT	AR	AZ	GE	IR	RU	TU
Potentilla caucasica Juz.	DD		x	x		x	
Potentilla divina Albov	LC			x		x	
Potentilla doddsii P. H. Davis	EN B1ab(i,ii,iii)+ 2ab(i,ii,iii)						x
Potentilla elatior Willd. ex Schlecht.	NE	x		x		x	x
Potentilla foliosa Sommier & Levier ex R. Keller	DD	x		x		x	
Potentilla gilanica (Th. Wolf) Th. Wolf	NE				x		
Potentilla imerethica Gagnidze & Sokhadze	DD			x			
Potentilla kemulariae Kapeller & A. I. Kuth.	VU D2			x			
Potentilla lazica Boiss. & Balansa	LC	x	x	x			x
Potentilla lomakinii Grossh. [*Potentilla buchneri* Kit Tan & Sorger]	NE	x	x				x
Potentilla nordmanniana Ledeb.	NT			x		x	
Potentilla orbiculata Th. Wolf	NE					x	x
Potentilla oweriniana Boiss.	NE					x	x
Potentilla petraea Willd. ex Schlecht.	NE				x		
Potentilla porphyrantha Juz.	NE	x	x		x		
Potentilla raddeana (Th. Wolf) Juz.	NE	x	x				
Potentilla ruprechtii Boiss.	NE		x	x		x	x
Potentilla savvalensis Pawł.	VU D2						x
Potentilla seidlitziana Bien.	CR B2ab(iii)	x					
Potentilla sommieri Siegfr. & R. Keller	DD			x			
Potentilla sosnowskyi Kapeller	VU D2			x			
Potentilla sphenophylla Th. Wolf	EN B1ab(iv)c(ii)+ 2ab(iv)c(ii)					x	
Potentilla subpalmata Ledeb.	CR B1ab(i,ii,iii)+ 2ab(i,ii,iii)						x
Potentilla svanetica Siegfr. & R. Keller	NT			x		x	
Potentilla szovitsii Th. Wolf	NE		x		x		
Pyrus browiczii Mulk.	CR B2ab(iii)	x					

TAXON	RED LIST ASSESSMENT	AR	AZ	GE	IR	RU	TU
Pyrus chosrovica Gladkova	DD	x					
Pyrus complexa Rubtzov	VU B1ab(iii,v)+2ab(iii,v)	x	x				
Pyrus daralaghezi Mulk.	EN B2ab(iii)	x					
Pyrus demetrii Kuth.	EN B2ab(iii,v)	x		x			
Pyrus elata Rubtzov	EN B2ab(v)	x					
Pyrus eldarica Grossh.	CR A2abc; C1		x	x			
Pyrus fedorovii Kuth.	NE	x		x			
Pyrus georgica Kuth.	NE	x	x	x		x	
Pyrus gergerana Gladkova	CR B1ab(iii)+2ab(iii); D	x					
Pyrus grossheimii Fed.	NE	x	x		x	x	
Pyrus hajastana Mulk.	EN B1ab(i,ii,v)+ 2ab(i,ii,v)	x					
Pyrus hyrcana Fed.	NE		x		x		
Pyrus ketzkhovelii Kuth.	EN B2ab(iii,v)	x		x			
Pyrus medvedevii Rubtzov	NE	x	x				
Pyrus nutans Rubtzov	EN B1ab(iii,v)+2ab(iii,v); D	x	x				
Pyrus oxyprion Woronow	NE	x	x	x			x
Pyrus raddeana Woronow	NT	x	x				
Pyrus sachokiana Kuth.	NT			x		x	
Pyrus salicifolia Pall.	NE	x	x	x		x	x
Pyrus sosnovskyi Fed.	EN B1ab(i,ii,iii)+ 2ab(i,ii,iii)	x					
Pyrus takhtadzhianii Fed.	NE	x	x	x			
Pyrus tamamschianiae Fed.	EN B1ab(i,ii,iii)+ 2ab(i,ii,iii)	x					
Pyrus theodorovii Mulk.	EN B1ab(v)+2ab(v)	x					
Pyrus voronovii Rubtzov	CR B1ab(i,ii,v)	x					
Pyrus vsevolodii Heideman	VU D1	x	x				
Pyrus zangezura Maleev	NT	x	x				
Rosa abutalybovii Gadzh.	NT		x				
Rosa adenophylla Galushko	NT					x	
Rosa alexeenkoi Crép. ex Juz.	EN B1ab(v)+2ab(v)		x				

TAXON	RED LIST ASSESSMENT	AR	AZ	GE	IR	RU	TU
Rosa altidaghestanica Husseinov	DD					x	
Rosa arensii Juz. & Galushko	DD					x	
Rosa atropatana Sosn.	NE	x	x				
Rosa awarica Husseinov	DD					x	
Rosa azerbajdzhanica Novopokr. & Rzazade	EN B2ab(ii,iii,iv,v)		x				
Rosa balcarica Galushko	NT					x	
Rosa baxanensis Galushko	DD					x	
Rosa borissovae Chrshan.	DD					x	
Rosa brotherorum Chrshan.	NT		x			x	
Rosa buschiana Chrshan.	NE		x	x		x	
Rosa cziragensis Husseinov	DD					x	
Rosa darginica Husseinov	DD					x	
Rosa didoensis Boiss.	DD			x		x	
Rosa dolichocarpa Galushko	CR D					x	
Rosa doluchanovii Manden.	DD			x			
Rosa elasmacantha Trautv.	LC					x	
Rosa elongata Galushko	NT					x	
Rosa ermanica Manden.	DD			x			
Rosa gadzhievii Chrshan. & Iskend.	NT		x				
Rosa galushkoi Demurova	VU B1ab(iii)			x		x	
Rosa hirtissima Lonacz.	NE			x		x	x
Rosa hracziana Tamamsch.	DD	x	x				
Rosa irinae Demurova	NT					x	
Rosa irysthonica Manden.	DD			x			
Rosa isaevii Gadzh. & Iskend.	DD		x				
Rosa jaroshenkoi Gadzh. & Iskend.	DD		x				
Rosa kamelinii Husseinov	DD					x	
Rosa karjaginii Sosn.	NT		x				
Rosa kazarjanii Sosn.	DD	x					
Rosa khasautensis Galushko	DD					x	

TAXON	RED LIST ASSESSMENT	AR	AZ	GE	IR	RU	TU
Rosa komarovii Sosn.	LC		x				
Rosa kossii Galushko	NT					x	
Rosa kozlowskii Chrashan.	NE		x	x			
Rosa marschalliana Sosn.	NE		x	x		x	
Rosa nisami Sosn.	NT		x				
Rosa obtegens Galushko	DD					x	
Rosa oplisthes Boiss.	LC			x		x	
Rosa ossethica Manden.	DD			x			
Rosa oxyodon Boiss.	LC			x		x	
Rosa oxyodontoides Galushko	DD					x	
Rosa praetermissa Galushko	DD					x	
Rosa prilipkoana Sosn.	NE		x	x			
Rosa prokhanovii Galushko	NT					x	
Rosa pubicaulis Galushko	NT					x	
Rosa pulverulenta M. Bieb.	NE		x	x			
Rosa ruprechtii Boiss.	DD		x				
Rosa sachokiana Jarosch.	NE	x	x				
Rosa sjunikii Jarosch.	DD	x	x				
Rosa sosnovskyana Tamamsch.	EN B2ab(ii)	x	x				
Rosa sosnovskyi Chrshan.	EN B2ab(ii,iii,iv,v)		x				
Rosa subbuschiana Husseinov	DD					x	
Rosa tchegemensis Galushko	EN B2ab(i,ii,iii)					x	
Rosa teberdensis Chrshan.	LC			x		x	
Rosa terscolensis Galushko	DD					x	
Rosa tlaratensis Husseinov	DD					x	
Rosa transcaucasica Manden.	DD			x			
Rosa tscherekensis Galushko	DD					x	
Rosa tuschetica Boiss.	NE		x	x		x	
Rosa uniflora Galushko	NT					x	
Rosa usischensis Husseinov	DD					x	
Rosa valentinae Galushko	DD					x	

TAXON	RED LIST ASSESSMENT	AR	AZ	GE	IR	RU	TU
Rosa zakatalensis Gadzh.	NT		x				
Rosa zangezura Jarosch.	VU B1ab(ii)+2ab(ii)	x	x				
Rosa zaramagensis Demurova	EN VU B2ab(ii,iii)					x	
Rosa zuvandica Gadzh.	DD		x				
Rubus abchaziensis Sudre	DD			x			
Rubus adscharicus Sanadze	DD			x			
Rubus armeniacus Focke	NE	x	x				
Rubus buschii Grossh.	DD		x				
Rubus cartalinicus Juz.	NE	x		x			
Rubus caucasicus Focke	NE	x	x	x		x	x
Rubus caucasigenus (Sudre) Juz.	DD			x			
Rubus charadzeae Sanadze	DD			x			
Rubus cyri Juz.	DD			x			
Rubus discernendus Sudre	DD			x			
Rubus dolichocarpus Juz.	LC		x	x			
Rubus georgicus Focke	LC		x	x			
Rubus hyrcanus Juz.	DD		x				
Rubus ibericus Juz.	DD		x	x		x	
Rubus juzepczukii Sanadze	DD			x			
Rubus kachethicus Sanadze	DD			x			
Rubus ketzkhovelii Sanadze	DD			x			
Rubus kudagorensis Sanadze	DD			x			
Rubus lepidulus (Sudre) Juz.	DD			x			
Rubus leptostemon Juz.	DD			x			
Rubus longipetiolatus Sanadze	DD			x			
Rubus miszczenkoi Juz.	DD			x			
Rubus moschus Juz.	DD			x			
Rubus nakeralicus Sanadze	DD			x			
Rubus ochthodes Juz.	DD			x			
Rubus ossicus Juz.	DD			x			
Rubus piceetorum Juz.	NE	x		x			

TAXON	RED LIST ASSESSMENT	AR	AZ	GE	IR	RU	TU
Rubus platyphylloides Sanadze	DD			x			
Rubus platyphyllos K. Koch	NE			x		x	x
Rubus ponticus (Focke) Juz.	DD			x			
Rubus semicaucasicus Sudre	DD					x	
Rubus takhtadjanii Mulk.	CR B1ab(i,ii)	x					
Rubus woronowii (Sudre) Sudre	DD			x			
Rubus zangezurus Mulk.	EN B1ab(ii)+2ab(ii)	x					
Sorbus albovii Zinserl.	DD					x	
Sorbus armeniaca Hedl.	NE	x	x				
Sorbus buschiana Zinserl.	NT			x		x	
Sorbus caucasica Zinserl.	NE	x	x	x		x	x
Sorbus caucasica Zinserl. var. *yaltirikii* Gökşin	EN B1ab(i,ii,v)+ 2ab(i,ii,v)						x
Sorbus caucasigena Kom. ex Gatsch.	DD		x	x			
Sorbus colchica Zinserl.	LC			x		x	
Sorbus fedorovii Zaik.	DD			x			
Sorbus hajastana Gabrieljan	NE	x				x	
Sorbus kuznetzovii Zinserl.	NE	x	x			x	
Sorbus migarica Zinserl.	LC			x		x	
Sorbus subfusca (Ledeb.) Boiss.	LC	x	x	x		x	x
Sorbus velutina (Albov) C. K. Schneid.	NT			x		x	
Spiraea kotschyana (Boiss.) Gladkova	DD					x	
Woronowia speciosa (Albov) Juz.	NE			x		x	
RUBIACEAE							
Asperula abchasica V. I. Krecz.	LC			x		x	
Asperula accrescens Klokov	DD					x	
Asperula affinis Boiss. & A. Huet	NE	x					
Asperula alpina M. Bieb.	LC		x			x	
Asperula biebersteinii V. I. Krecz.	LC					x	
Asperula caucasica Pobed. [*Asperula taurina* L. subsp. *caucasica* (Pobed.) Ehrend.]	NE	x		x		x	x

TAXON	RED LIST ASSESSMENT	AR	AZ	GE	IR	RU	TU
Asperula cristata (Sommier & Levier) V. I. Krecz.	LC			x		x	
Asperula diminuta Klokov	EN B1ab(ii,iii)+2ab(ii,iii)		x			x	
Asperula glomerata (M. Bieb.) Griseb. subsp. *hirsutiuscula* (Pobed.) Ehrend. & Schönb.-Tem. [*Asperula hirsutiuscula* Pobed.]	NE B2ab(i,ii,iii,iv,v)		x		x		
Asperula gracilis C. A. Mey. [*Asperula ovatifolia* Boiss. & A. Huet]	NE			x		x	x
Asperula hirsutiuscula Pobed. [*Asperula glomerata* (M. Bieb.) Griseb. subsp. *hirsutiuscula* (Pobed.) Ehrend. & Schönb.-Tem.]	NE B2ab(i,ii,iii,iv,v)		x		x		
Asperula intersita Klokov	NE					x	
Asperula kemulariae Manden.	NE		x	x		x	
Asperula lipskyana V. I. Krecz.	EN B1ab(ii,iii)+2ab(ii,iii); C2a(i)					x	
Asperula microphylla Boiss.	CR				x		
Asperula ovatifolia Boiss. & A. Huet [*Asperula gracilis* C. A. Mey.]	NE			x		x	x
Asperula pedicellata Klokov	NT					x	
Asperula prostrata (Adams) K. Koch subsp. *albovii* (Manden.) Mikheev	DD			x		x	
Asperula prostrata (Adams) K. Koch subsp. *prostrata*	NE	x	x	x		x	
Asperula sherardioides Jaub. & Spach	NE				x		
Asperula taurina L. subsp. *caucasica* (Pobed.) Ehrend. [*Asperula caucasica* Pobed.]	NE	x		x		x	x
Asperula virgata Hub.-Mor. ex Ehrend. & Schönb.-Tem.	CR B2ab(ii,iii)						x
Crucianella gilanica Trin. subsp. *gilanica*	NE		x		x		
Crucianella gilanica Trin. subsp. *suleimanica* Ehrend. & Schönb.-Tem.	NE				x		
Crucianella gilanica Trin. subsp. *transcaucasica* (Ehrend.) T. N. Popova & Takht.	NE	x			x		

TAXON	RED LIST ASSESSMENT	AR	AZ	GE	IR	RU	TU
Crucianella suaveolens C. A. Mey.	NE				x		
Cruciata sosnowskyi (Manden.) Pobed. [*Galium sosnowskyi* Manden.]	EN B2ab(iii)	x	x				
Cruciata taurica (Pall. ex Willd.) Ehrend. subsp. *armeniaca* (Mikheev) Ehrend.	NE	x			x		
Cruciata valentinae (Galushko) Galushko	NT		x			x	
Gaillonia szowitsii DC. [*Neogaillonia szowitsii* (DC.) Lincz.]	NE	x	x		x		
Galium anfractum Sommier & Levier	LC			x		x	
Galium atropatanum Grossh.	NE	x	x				
Galium azerbayjanicum Ehrend. & Schonb.-Tem.	NE		x		x		
Galium brachyphyllum Roem. & Schult.	LC		x			x	
Galium bullatum Lipsky	VU B1ab(iii)+2ab(iii)	x	x				
Galium buschiorum Mikheev	DD		x			x	
Galium caspicum Steven	NE		x		x		
Galium chloroleucum Fisch. & C. A. Mey.	NE		x		x		
Galium consanguineum Boiss. subsp. *majmechense* (Bordz.) Mikheev [*Galium majmechense* Bordz.]	NE	x	x	x		x	
Galium czerepanovii Pobed.	NE	x	x		x		x
Galium fistulosum Sommier & Levier	LC		x			x	
Galium flaviflorum (Trautv.) Mikheev	LC		x			x	
Galium hyrcanicum C. A. Mey.	NT		x				
Galium kiapazi Manden.	VU D2	x	x				
Galium kuetzingii Boiss. & Buhse	NE		x		x		
Galium majmechense Bordz. [*Galium consanguineum* Boiss. subsp. *majmechense* (Bordz.) Mikheev]	NE	x	x	x		x	
Galium oshtenicum Ehrend. & Schanzer ex Mikheev	LC					x	
Galium praemontanum Mardal.	EN D			x			
Galium sosnowskyi Manden. [*Cruciata sosnowskyi* (Manden.) Pobed.]	EN B2ab(iii)	x	x				

TAXON	RED LIST ASSESSMENT	AR	AZ	GE	IR	RU	TU
Galium subuliferum Sommier & Levier	LC		x	x			
Galium tortumense Ehrend. & Schönb.-Tem.	CR B1ab(i,ii,iii)+2ab(i,ii,iii)						x
Galium valantioides M. Bieb.	NE	x	x	x		x	
Galium vartanii Grossh.	DD		x			x	
Neogaillonia szowitsii (DC.) Lincz. [*Gaillonia szowitsii* DC.]	NE	x	x		x		
Phuopsis stylosa (Trin.) Benth. & Hook. f. ex B. D. Jacks.	NE		x		x	x	
Rubia rigidifolia Pojark.	VU B1ab(iii)+2ab(iii)	x	x				
Rubia transcaucasica Grossh.	DD		x	x			
RUTACEAE							
Dictamnus caucasicus (Fisch. & C. A. Mey.) Grossh.	DD			x		x	
Haplophyllum buhsei Boiss.	NE				x		
Haplophyllum ciscaucasicum (Rupr.) Grossh. & Vved.	EN B2ab(iii)					x	
Haplophyllum kowalevskyi Stschegl.	NE		x		x		
Haplophyllum schelkovnikovii Grossh.	VU D2	x	x				
Haplophyllum villosum (M. Bieb.) G. Don	NE	x	x	x	x	x	x
SALICACEAE							
Populus gracilis Grossh.	DD		x				
Populus hyrcana Grossh.	DD		x	x			
Populus tremula L. subsp. *microtremula* (Grossh.) Kamelin	DD					x	
Salix alba L. subsp. *micans* (Andersson) Rech. f. [*Salix micans* Andersson]	DD			x		x	
Salix kazbekensis A. K. Skvortsov	NE		x	x		x	
Salix kikodseae Goerz	EN B1ab(iii)+2ab(iii)			x		x	
Salix kuznetzowii Laksch. ex Goerz	NT		x	x		x	
Salix micans Andersson [*Salix alba* L. subsp. *micans* (Andersson) Rech. f.]	DD			x		x	
Salix pantosericea Goerz	LC			x		x	
Salix phlomoides M. Bieb.	DD		x				

TAXON	RED LIST ASSESSMENT	AR	AZ	GE	IR	RU	TU
Salix rizeensis Güner & Ziel.	EN B1ab(i,ii,iii)+ 2ab(i,ii,iii)						x
Salix trabzonica A. K. Skvortsov	EN B1ab(i,ii,iii,iv)+ 2ab(i,ii,iii,iv)						x
SANTALACEAE							
Thesium laxiflorum Trautv.	LC			x			
Thesium maritimum C. A. Mey.	VU B1ab(iii)+2ab(iii)		x				
Thesium procumbens C. A. Mey.	NE	x	x	x		x	
Thesium szowitsii A. DC.	NE	x	x				
SAXIFRAGACEAE							
Chrysosplenium albowianum Kuth.	DD			x			
Saxifraga abchasica Oett.	DD			x			
Saxifraga akinfievii Galushko & Kudrjasch.	DD					x	
Saxifraga artvinensis V. A. Matthews	CR B2ab(ii,iii)						x
Saxifraga carinata Oett.	EN B1ab(v); C2a(i)					x	
Saxifraga cartilaginea Willd. [*Saxifraga paniculata* Mill. subsp. *cartilaginea* (Willd. ex Sternb.) D. A. Webb.]	NE	x	x	x	x	x	x
Saxifraga caspica Sipliv.	DD		x			x	
Saxifraga caucasica Sommier & Levier	LC			x		x	
Saxifraga charadzeae Otsch.	NT					x	
Saxifraga colchica Albov	DD			x		x	
Saxifraga columnaris Schmalh.	VU B1ab(i,ii,iii,v)					x	
Saxifraga dinnikii Schmalh. ex Akinf.	VU B1ab(iii); C2a(i)					x	
Saxifraga iranica Bornm.	NE				x		
Saxifraga kolenatiana Regel	DD	x	x	x		x	x
Saxifraga kusnezowiana Oett.	VU D2			x			
Saxifraga mazanderanica Rech. f.	NE				x		
Saxifraga oettingenii Galushko & Kudrjasch.	DD					x	
Saxifraga paniculata Mill. subsp. *cartilaginea* (Willd. ex Sternb.) D. A. Webb. [*Saxifraga cartilaginea* Willd.]	NE	x	x	x	x	x	x

TAXON	RED LIST ASSESSMENT	AR	AZ	GE	IR	RU	TU
Saxifraga pontica Albov	DD		x	x			
Saxifraga pseudolaevis Oett.	NE		x	x		x	
Saxifraga ramsarica Jamzad	NE				x		
Saxifraga ruprechtiana Manden.	VU B1ab(iii)+2ab(iii)			x		x	
Saxifraga scleropoda Sommier & Levier	LC			x		x	
Saxifraga sosnowskyi Manden.	DD			x		x	
Saxifraga subverticillata Boiss.	DD			x		x	
Saxifraga trautvetteri Manden.	VU D2			x			
Saxifraga verticillata Losinsk.	NE	x	x			x	
SCROPHULARIACEAE							
Digitalis ciliata Trautv.	NE			x		x	
Digitalis ferruginea L. subsp. *schischkinii* (Ivanina) K. Werner [*Digitalis schischkinii* Ivanina]	NE			x		x	x
Digitalis schischkinii Ivanina [*Digitalis ferruginea* L. subsp. *schischkinii* (Ivanina) K. Werner]	NE			x		x	x
Euphrasia adenocaulon Juz.	NT			x			
Euphrasia alboffii Chabert	VU B1ab(iii)					x	
Euphrasia amblyodonta Juz.	NE		x	x		x	x
Euphrasia caucasica Juz.	NE		x	x		x	
Euphrasia daghestanica Juz.	DD		x				
Euphrasia georgica Kem.-Nath.	DD		x				
Euphrasia grossheimii Kem.-Nath.	DD			x			
Euphrasia karjaginii Kem.-Nath.	VU D2		x				
Euphrasia kemulariae Juz.	DD			x			
Euphrasia lebardensis Kem.-Nath. [*Euphrasia ossica* Juz.]	NE		x	x		x	
Euphrasia macrodontha Juz.	EN B2ab(ii,iii,v)			x		x	
Euphrasia nisamii Kem.-Nath.	DD		x			x	
Euphrasia ossica Juz. [*Euphrasia lebardensis* Kem.-Nath.]	NE		x	x		x	
Euphrasia petiolaris Wettst.	NE			x		x	x

TAXON	RED LIST ASSESSMENT	AR	AZ	GE	IR	RU	TU
Euphrasia sosnowskyi Kem.-Nath.	NE		x	x			
Euphrasia svanica Kem.-Nath.	DD			x			
Euphrasia woronowii Juz.	DD			x			
Linaria armeniaca Chav.	NE	x	x	x			x
Linaria corrugata Karjagin	CR B2ab(v)		x				
Linaria genistifolia (L.) Mill. subsp. *artvinensis* P. H. Davis	VU B1ab(i,ii,iii)						x
Linaria meyeri Kuprian.	NE			x		x	
Linaria schelkownikowii Schischk.	NE	x	x	x			
Linaria schirvanica Fomin	NE		x	x			
Linaria sintenisii P. H. Davis	DD						x
Linaria zangezura Grossh.	LC	x	x				
Melampyrum alboffianum Beauverd	DD			x			
Melampyrum arvense L. var. *elatius* Boiss. [*Melampyrum elatius* (Boiss.) Soó]	NE			x		x	x
Melampyrum caucasicum Bunge	NE		x	x	x		
Melampyrum chlorostachyum Beauverd	NE	x	x	x		x	
Melampyrum elatius (Boiss.) Soó [*Melampyrum arvense* L. var. *elatius* Boiss.]	NE			x		x	x
Melampyrum mulkijanianii T. N. Popova	NE	x		x		x	
Melampyrum stenophyllum Boiss.	NE			x		x	
Paederotella daghestanica (Trautv.) Kem.-Nath.	VU B1ab(iii)+2ab(iii)					x	
Paederotella pontica (Rupr. ex Boiss.) Kem.-Nath.	VU B2ac(ii)			x		x	
Pedicularis atropurpurea Nordm.	NE			x		x	x
Pedicularis balkharica E. A. Busch	NT					x	
Pedicularis chroorrhyncha Vved.	NT					x	
Pedicularis crassirostris Bunge	NE	x		x		x	x
Pedicularis daghestanica Bonati	LC					x	
Pedicularis elisabethae T. N. Popova	NT			x			
Pedicularis nordmanniana Bunge	NE			x		x	x

TAXON	RED LIST ASSESSMENT	AR	AZ	GE	IR	RU	TU
Pedicularis ochrorrhyncha Galushko & T. N. Popova	DD					x	
Pedicularis panjutinii E. A. Busch	NT			x		x	
Pedicularis subrostrata C. A. Mey.	NT					x	
Pedicularis tatianae Bordz.	NT					x	
Pedicularis wilhelmsiana Fisch. ex M. Bieb.	NE	x		x	x	x	x
Pseudolysimachion spicatum (L.) Opiz subsp. *transcaucasicum* (Bordz.) Gabrieljan	EN B2ab(iii,iv,v)	x					
Rhamphicarpa medwedewii Albov	NE			x			x
Rhinanthus colchicus Vassilcz.	DD			x		x	
Rhinanthus schischkinii Vassilcz.	LC			x		x	
Rhynchocorys orientalis (L.) Benth.	NE	x	x	x		x	x
Rhynchocorys stricta (K. Koch) Albov	NE			x		x	x
Scrophularia atropatana Grossh.	NE	x	x		x		x
Scrophularia capillaris Boiss. & Balansa	CR B2ab(ii,iii)						x
Scrophularia charadzeae Kem.-Nath.	DD					x	
Scrophularia diffusa Sommier & Levier	DD		x	x			
Scrophularia grossheimii Schischk. ex Grossh.	NE	x	x	x		x	x
Scrophularia hyrcana (Grossh.) Grossh.	NE		x				
Scrophularia imerethica Kem.-Nath.	NT			x			
Scrophularia lateriflora Trautv.	NE		x	x		x	
Scrophularia macrobotrys Ledeb.	NE	x	x	x		x	
Scrophularia minima M. Bieb.	NE		x	x		x	
Scrophularia mollis Sommier & Levier	NT					x	
Scrophularia olgae Grossh.	VU B1ab(ii,iii)+2ab(ii,iii)	x					
Scrophularia oxysepala Boiss.	NE				x		
Scrophularia ruprechtii Boiss.	NE		x	x		x	
Scrophularia scopolii Hoppe ex Pers.	NE	x		x			x
Scrophularia sosnowskyi Kem.-Nath.	NE			x			x

TAXON	RED LIST ASSESSMENT	AR	AZ	GE	IR	RU	TU
Scrophularia sprengeriana Sommier & Levier	DD			x		x	
Scrophularia svartiana Gabrieljan	NE	x			x		
Scrophularia takhtajanii Gabrieljan	CR	x					
Scrophularia thesioides Boiss. ex Buhse	NE	x	x				x
Scrophularia zuvandica Grossh.	NE		x				x
Verbascum adzharicum P. P. Gritz.	VU D2			x			
Verbascum alpigenum K. Koch	NE			x			x
Verbascum artvinense E. Wulff	DD						x
Verbascum atroviolaceum (Sommier & Levier) Murb.	NE	x				x	
Verbascum decursivum Hub.-Mor.	CR B2ab(ii)						x
Verbascum drymophilum Hub.-Mor.	EN B2ab(i,ii,iii)						x
Verbascum eriorrhabdon Boiss.	NE			x			x
Verbascum erivanicum E. Wulff	NE	x	x				
Verbascum formosum Fisch. ex Schrank	NE	x	x	x		x	
Verbascum gabrielianae (Hub.-Mor.) Hub.-Mor.	LC	x	x				
Verbascum gracilescens Hub.-Mor.	EN B1ab(i,ii,iii)+ 2ab(i,ii,iii)						x
Verbascum hajastanicum Bordz.	VU D2	x					x
Verbascum horticolum Hub.-Mor.	DD	x					
Verbascum megricum (Tzvelev) Hub.-Mor.	NE	x	x				
Verbascum paniculatum E. Wulff	NE	x	x	x			
Verbascum patris (Bordz.) Bordz.	DD	x					
Verbascum sessiliflorum Murb.	VU D2			x			
Verbascum sevanense Hub.-Mor.	DD	x					
Verbascum stachydiforme Boiss. & Buhse	CR B2ab(ii,iii,iv,v)		x				
Verbascum suworowianum (K. Koch) Kuntze	NE	x			x		x
Verbascum transcaucasicum E. Wulf	CR B2ab(ii); D						x

TAXON	RED LIST ASSESSMENT	AR	AZ	GE	IR	RU	TU
Verbascum varians Freyn & Sint. var. *trapezunticum* Murb.	CR B2ab(ii,iii)						x
Verbascum wilhelmsianum K. Koch	LC			x		x	
Veronica albanica K. Koch	VU D2		x				
Veronica allahuekberensis Öztürk	CR B2ab(ii)						x
Veronica amoena M. Bieb.	NE		x	x			
Veronica bogosensis Tumadz.	EN B1ab(ii,iii,v)+ 2ab(ii,iii,v); C2a(i)					x	
Veronica borissovae Holub	DD					x	
Veronica caucasica M. Bieb.	NE		x	x		x	
Veronica ceratocarpa C. A. Mey.	NE	x	x	x	x	x	x
Veronica charadzeae Kem.-Nath.	DD			x		x	
Veronica colchica Kimer.	NT			x			
Veronica crista-galli Steven	NE		x	x	x	x	
Veronica euphrasiifolia Link	CR				x		
Veronica filifolia Lipsky	EN A2a+c; B1ab(iii,iv)					x	
Veronica galathica Boiss.	NE	x		x			
Veronica imerethica Kem.-Nath.	DD			x		x	
Veronica liwanensis K. Koch	NE	x		x			x
Veronica minuta C. A. Mey.	LC		x			x	
Veronica monticola Trautv.	NE			x		x	x
Veronica oltensis Woronow	EN B1ab(i,ii,iii)+ 2ab(i,ii,iii)						x
Veronica petraea (M. Bieb.) Steven	NE		x	x		x	
Veronica propinqua Boriss.	NE		x	x		x	
Veronica schistosa E. A. Busch	LC			x		x	
Veronica telephiifolia Vahl	NE			x		x	x
Veronica tumadzhanovii Mardal.	NT			x			
SOLANACEAE							
Atropa caucasica Kreyer	NE	x	x	x		x	
Solanum woronowii Pojark.	NT	x		x			x
STAPHYLEACEAE							

TAXON	RED LIST ASSESSMENT	AR	AZ	GE	IR	RU	TU
Staphylea colchica Steven	VU A2c; B1ab(iii)+ 2ab(iii)			x		x	
TAMARICACEAE							
Reaumuria kuznetzovii Sosn. & Manden.	NT			x			
THYMELAEACEAE							
Daphne albowiana Woronow ex Pobed.	EN B2ab(iii)			x		x	
Daphne axiliflora (Keissl.) Pobed.	NE		x	x			
Daphne baksanica Pobed.	CR B2ab(ii,v); C2a(i)					x	
Daphne caucasica Pall.	LC			x		x	
Daphne circassica Woronow ex Pobed.	EN A2ab(i,iv); C2a(i)					x	
Daphne glomerata Lam.	LC	x		x		x	x
Daphne pseudosericea Pobed.	EN B2ab(iii)			x		x	
Stelleropsis caucasica Pobed.	EN B1ab(v)					x	
Stelleropsis magakjanii (Sosn.) Pobed.	DD	x	x				
TILIACEAE							
Tilia begoniifolia Steven [*Tilia caucasica* Rupr.]	LC	x	x	x			
Tilia caucasica Rupr. [*Tilia begoniifolia* Steven]	NE	x	x	x			
TRAPACEAE							
Trapa colchica Albov	CR C2a(i)			x			
Trapa hyrcana Woronow	NE		x	x		x	
Trapa maleevii V. N. Vassil.	EN B1ab(iii,v)+2ab(iii,v)			x			
ULMACEAE							
Ulmus georgica Schchian	DD			x			
URTICACEAE							
Parietaria cryptorum K. Koch	NE	x		x			
Parietaria kemulariae Schchian	VU D2			x			
VALERIANACEAE							
Pseudobetckea caucasica (Boiss.) Lincz.	NT			x		x	
Valeriana cardamines M. Bieb.	NE		x	x		x	
Valeriana colchica Utkin	NE			x		x	

TAXON	RED LIST ASSESSMENT	AR	AZ	GE	IR	RU	TU
Valeriana daghestanica Rupr. ex Boiss.	NE		x	x		x	
Valeriana eriophylla (Ledeb.) Utkin	NE	x	x	x		x	
Valeriana grossheimii Vorosch.	DD					x	
Valeriana jelenevskyi P. A. Smirn.	LC	x		x		x	
Valeriana kassarica Charadze & Kapeller	NT			x		x	
Valeriana leucophaea DC.	NE	x			x		
Valeriana saxicola C. A. Mey.	NE			x		x	x
Valeriana tiliifolia Troitzky	DD		x	x		x	
Valerianella corniculata C. A. Mey.	NT		x				
VIOLACEAE							
Viola caucasica Kolen.	NE	x	x	x		x	
Viola ignobilis Rupr.	DD	x		x		x	
Viola minuta M. Bieb. subsp. *dagestanica* (Rupr.) Vl. V. Nikitin	NE		x			x	
Viola minuta M. Bieb. subsp. *meyeriana* (Rupr.) Vl. V. Nikitin	NT					x	
Viola minuta M. Bieb. subsp. *minuta*	LC			x		x	
Viola orthoceras Ledeb.	VU B1ab(iii)+2ab(iii)			x			
Viola popoviae Vl. V. Nikitin	DD					x	
Viola somchetica K. Koch	NE	x	x	x	x	x	
Viola vespertina Klokov	DD			x			
Viola wilhelmii Vl. V. Nikitin	DD					x	
ZYGOPHYLLACEAE							
Nitraria komarovii Iljin & Lava	NE		x		x		
MONOCOTS							
ALLIACEAE							
Allium albovianum Vved.	NT			x			
Allium aznavense R. M. Fritsch	CR				x		
Allium baytopiorum Kollmann & Özhatay	CR B1ab(iii)+2ab(iii)						x
Allium candolleanum Albov	DD			x		x	
Allium chevsuricum Tscholok.	DD			x			

TAXON	RED LIST ASSESSMENT	AR	AZ	GE	IR	RU	TU
Allium circassicum Kolak.	CR B2ab(ii,iii)					x	
Allium czelghauricum Bordz.	CR B1ab(iii)+2ab(iii)						x
Allium daghestanicum Grossh.	CR B1ab(ii,iii)+2ab(ii,iii); C2a(ii)					x	
Allium egorovae M. V. Agab. & Ogan.	EN B1ab(iii)+2ab(iii)	x	x		x		
Allium esfandiarii Matin	CR				x		
Allium gracilescens Sommier & Levier	DD			x			
Allium gramineum K. Koch	NE	x	x	x			x
Allium grande Lipsky	EN B1ab(iii,v)+2ab(iii,v)					x	
Allium gunibicum Miscz. ex Grossh.	NT					x	
Allium karsianum Fomin	DD	x	x	x	x		
Allium koenigianum Grossh.	DD						x
Allium kossoricum Fomin	CR B2ab(ii,iv,v)						x
Allium kunthianum Vved.	NE	x	x	x	x	x	x
Allium lenkoranicum Miscz. ex Grossh.	CR B2ab(ii,v)		x				
Allium leonidis Grossh.	DD		x				
Allium leucanthum K. Koch	NE	x	x	x			
Allium mariae Bordz.	EN B1ab(i,ii,v)+ 2ab(i,ii,v)		x				
Allium materculae Bordz.	NE	x	x		x		
Allium mirzojevii Tscholok.	CR B1ab(iii)+2ab(iii)					x	
Allium oltense Grossh.	NE	x					x
Allium otschiauriae Tscholok.	DD			x			
Allium ponticum Miscz. ex Grossh.	NE			x		x	x
Allium psebaicum Mikheev	EN					x	
Allium pseudoalbidum N. Friesen & Özhatay	CR B2ab(iii)						x
Allium rollovii Grossh.	DD						x
Allium rubellum M. Bieb. subsp. *syntamanthum* (K. Koch) Ogan. [*Allium syntamanthum* K. Koch]	DD	x	x		x		
Allium samurense Tscholok.	CR A1ab(ii,iii)+2ab(ii,iii); C2a(ii)					x	

TAXON	RED LIST ASSESSMENT	AR	AZ	GE	IR	RU	TU
Allium schchianae Ogan.	DD	x					
Allium sosnowskianum Miscz. ex Grossh.	EN B1ab(i,ii,iii)+ 2ab(i,ii,iii)						x
Allium struzlianum Ogan.	EN B1ab(i,ii,iii,v)+ 2ab(i,ii,iii,v); D	x					
Allium subvineale Wendelbo	NE				x		
Allium syntamanthum K. Koch [*Allium rubellum* M. Bieb. subsp. *syntamanthum* (K. Koch) Ogan.]	DD	x	x		x		
Allium szovitsii Regel	NE	x	x	x	x		
Allium talyschense Miscz. ex Grossh.	NE	x	x		x		
Allium vasilevskajae Ogan.	DD	x					
Allium woronowii Miscz. ex Grossh.	EN B2ab(iii)	x	x				x
AMARYLLIDACEAE							
Galanthus alpinus Sosn. subsp. *caucasicus* Gagnidze [*Galanthus caucasicus* (Baker) Grossh.]	NE		x	x			
Galanthus alpinus Sosn. var. *alpinus*	NE	x		x		x	x
Galanthus alpinus Sosn. var. *bortkewitschianus* (Koss ex Grossh.) A. P. Davis	EN B1ab(i,ii,iii,iv,v)+ 2ab(i,ii,iii,iv,v)					x	
Galanthus angustifolius Koss	VU B1ab(iii,v)c(ii,iii,iv)+ 2ab(iii,v)c(ii,iii,iv)			x		x	
Galanthus artjuschenkoae Gabrieljan	NE	x	x				
Galanthus bortkewitschianus Koss	EN B1ab(i,ii,iii,iv,v)+ 2ab(i,ii,iii,iv,v)					x	
Galanthus cabardensis Koss	VU B1ac(ii,iv)+2ac(ii,iv)					x	
Galanthus caucasicus (Baker) Grossh. [*Galanthus alpinus* Sosn. subsp. *caucasicus* Gagnidze]	NE		x	x			
Galanthus kemulariae Kuth.	VU D2			x			
Galanthus ketzkhovelii Kem.-Nath.	DD			x			
Galanthus koenenianus Lobin, C. D. Brickell & A. P. Davis	CR B2ab(iii)						x
Galanthus krasnovii A. P. Khokhr.	EN B2ab(iii,v)			x			x
Galanthus lagodechianus Kem.-Nath.	NE	x	x	x		x	

TAXON	RED LIST ASSESSMENT	AR	AZ	GE	IR	RU	TU
Galanthus platyphyllus Traub & Moldenke	NT			x		x	
Galanthus rizehensis Stern	NE			x		x	x
Galanthus schaoricus Kem.-Nath.	EN D			x			
Galanthus transcaucasicus Fomin	NE		x		x		
Galanthus woronowii Losinsk.	NE			x		x	x
Sternbergia alexandrae Sosn.	CR B2ab(v)		x				
ASPHODELACEAE							
Asphodeline dendroides (Hoffm.) Woronow ex Grossh. [*Asphodeline prolifera* (M. Bieb.) Kunth]	NE	x	x	x	x		x
Asphodeline prolifera (M. Bieb.) Kunth [*Asphodeline dendroides* (Hoffm.) Woronow ex Grossh.]	NE	x	x	x	x		x
Asphodeline tenuior (Fisch. ex M. Bieb.) Ledeb.	VU B2ab(iii,v)					x	
Eremurus azerbajdzhanicus Charkev.	DD		x				
COLCHICACEAE							
Colchicum gohariae Gabrieljan	EN B1ab(i,ii,iii)+ 2ab(i,ii,iii)	x					
Colchicum leptanthum K. Perss.	CR B2ab(iii,v)						x
Colchicum ninae Sosn.	EN B1ab(i,ii,iii,v)+ 2ab(i,ii,iii,v)	x					
Colchicum woronowii Bokeria	NE		x	x		x	
Colchicum zangezurum Grossh.	NE	x	x				
Merendera candidissima Miscz. ex Grossh.	CR B2ab(v)		x				
Merendera eichleri (Regel) Boiss.	NT		x			x	
Merendera ghalghana Otschiauri	NT			x		x	
Merendera greuteri Gabrieljan	CR B2ab(i,ii,iii)	x					
Merendera mirzoevae Gabrieljan	NT	x					
CONVALLARIACEAE							
Convallaria majalis L. var. *transcaucasia* (Utkin ex Grossh.) Knorring [*Convallaria transcaucasica* Utkin ex Grossh.]	DD		x	x			x

TAXON	RED LIST ASSESSMENT	AR	AZ	GE	IR	RU	TU
Convallaria transcaucasica Utkin ex Grossh. [*Convallaria majalis* L. var. *transcaucasia* (Utkin ex Grossh.) Knorring]	DD		x	x			x
Polygonatum glaberrimum K. Koch	NE	x	x	x	x	x	x
CYPERACEAE							
Bulbostylis tenerrima (Fisch. & C. A. Mey) Palla	NE			x			x
Carex acrifolia V. I. Krecz. [*Carex enervis* C. A. Mey. subsp. *acrifolia* (V. I. Krecz.) T. V. Egorova]	NE	x	x	x	x	x	x
Carex capitellata Boiss. & Balansa	NE	x	x	x		x	x
Carex enervis C. A. Mey. subsp. *acrifolia* (V. I. Krecz.) T. V. Egorova [*Carex acrifolia* V. I. Krecz.]	NE	x	x	x	x	x	x
Carex hypaneura V. I. Krecz.	NE		x	x		x	
Carex latifrons V. I. Krecz. [*Carex sylvatica* Huds. subsp. *latifrons* (V. I. Crecz.) O. Nilsson]	NE			x		x	x
Carex melanorrhyncha Nelmes	NT						x
Carex micropodioides V. I. Krecz. [*Carex pyrenaica* Wahlenb. subsp. *micropodioides* (V. I. Krecz.) Chandjian]	DD	x	x	x		x	
Carex mingrelica Kük.	LC			x		x	
Carex oligantha Steud.	NE	x	x	x	x	x	x
Carex pontica Albov	NE			x		x	x
Carex pseudofoetida Kük.	NE				x		x
Carex pyrenaica Wahlenb. subsp. *micropodioides* (V. I. Krecz.) Chandjian [*Carex micropodioides* V. I. Krecz.]	DD	x	x	x		x	
Carex sylvatica Huds. subsp. *latifrons* (V. I. Crecz.) O. Nilsson [*Carex latifrons* V. I. Krecz.]	NE			x		x	x
Carex transcaucasica T. V. Egorova	NE	x	x	x			
Kobresia macrolepis Meinsh.	NE		x	x		x	
Kobresia schoenoides (C. A. Mey.) Steud.	NE	x	x	x	x	x	
Pycreus colchicus (K. Koch) Schischk.	NE			x			x

TAXON	RED LIST ASSESSMENT	AR	AZ	GE	IR	RU	TU
Rhynchospora caucasica Palla	EN B2ab(iii)			x			
Scirpus colchicus Kimer.	EN B2ab(i,ii,iiiiv)			x		x	
DIOSCOREACEAE							
Dioscorea caucasica Lipsky	EN B2ab(ii,iii)			x		x	
HYACINTHACEAE							
Bellevalia albana Woronow	VU D2		x				
Bellevalia assadii Wendelbo	CR				x		
Bellevalia montana (K. Koch) Boiss. [*Bellevalia wilhelmsii* (Steven) Woronow]	NE		x	x			
Bellevalia paradoxa (Fisch. & C. A. Mey.) Boiss.	NE	x	x	x			x
Bellevalia tabriziana Turrill	NE				x		
Bellevalia wilhelmsii (Steven) Woronow [*Bellevalia montana* (K. Koch) Boiss.]	NE		x	x			
Hyacinthella atropatana (Grossh.) Mordak. & Zakhar.	NE	x	x		x		
Hyacinthella persica (Boiss. & Buhse) Chouard	NE				x		
Muscari alpanicum Schchian	VU D2			x			
Muscari coeruleum Losinsk.	NT			x		x	
Muscari dolichanthum Woronow & Tron	VU B2ab(ii,iii); C2a(i)			x		x	
Muscari elegantulum Schchian	NT		x				
Muscari grossheimii Schchian	NT		x				
Muscari pallens (M. Bieb.) Fisch.	NE	x	x	x		x	
Ornithogalum brachystachys K. Koch	NE	x	x				
Ornithogalum gabrielianiae Agapova	CR B2ab(i,ii,iii)	x					
Ornithogalum hajastanum Agapova	NE	x	x				x
Ornithogalum hyrcanum Grossh.	VU D2		x				
Ornithogalum imereticum Sosn.	VU D2			x			
Ornithogalum magnum Krasch. & Schischk.	VU B2ab(i,ii,iii,iv)			x		x	
Ornithogalum schischkinii Krasch. [*Ornithogalum sintenisii* Freyn]	NE		x		x	x	

TAXON	RED LIST ASSESSMENT	AR	AZ	GE	IR	RU	TU
Ornithogalum schmalhausenii Albov	DD		x	x			
Ornithogalum sintenisii Freyn [*Ornithogalum schischkinii* Krasch.]	NE		x		x	x	
Ornithogalum tempskyanum Freyn & Sint.	DD	x		x			
Ornithogalum transcaucasicum Miscz. ex Grossh.	NE	x	x				
Scilla caucasica Miscz.	NE	x	x	x			
Scilla mischtschenkoana Grossh.	NE	x	x		x		
Scilla otschiauriae Mordak	VU D2			x			
Scilla rosenii K. Koch	NE	x	x	x			x
Scilla winogradowii Sosn.	DD	x	x	x			x
IRIDACEAE							
Crocus aerius Herb.	EN B1ab(iii)+2ab(iii)						x
Crocus autranii Albov	DD			x			
Crocus biflorus Mill. subsp. *artvinensis* (J. Philippow) B. Mathew	CR B1ab(i,ii,iii)+ 2ab(i,ii,iii)						x
Crocus biflorus Mill. subsp. *fibroannulatus* Kernd. & Pasche	CR B2ab(ii,iii,v)						x
Crocus caspius Fisch. & C. A. Mey. ex Hohen.	NE		x		x		
Crocus gilanicus B. Mathew	NE				x		
Crocus kotschyanus K. Koch subsp. *suworowianus* (K. Koch) B. Mathew [*Crocus suworowianus* K. Koch]	NE			x		x	x
Crocus polyanthus Grossh.	CR B2ab(ii,v)		x				
Crocus scharojanii Rupr.	NE			x		x	x
Crocus suworowianus K. Koch [*Crocus kotschyanus* K. Koch subsp. *suworowianus* (K. Koch) B. Mathew]	NE			x		x	x
Crocus vallicola Herb.	NE			x		x	x
Gladiolus caucasicus Herb.	NE	x	x	x			
Gladiolus dzhavakheticus Eristavi	NE	x		x			
Gladiolus hajastanicus Gabrieljan	VU B1ab(i,ii,iii)+ 2ab(i,ii,iii)	x					

TAXON	RED LIST ASSESSMENT	AR	AZ	GE	IR	RU	TU
Gladiolus kotschyanus Boiss. subsp. *distichus* Gabrieljan	EN B2ab(i,ii,iii)	x					
Gladiolus menitzkyi Gabrieljan	NE	x	x				
Gladiolus szovitsii Grossh. subsp. *pseudopersicus* Ogan. & Gabrieljan	NE	x	x				
Iris acutiloba C. A. Mey.	EN B2ab(iii)c(v)		x			x	
Iris alexeenkoi Grossh.	NT		x			x	
Iris annae Grossh.	CR B2b(ii,v)c(ii,iii)		x				
Iris camillae Grossh.	VU D2		x				
Iris carthaliniae Fomin	NE		x	x		x	
Iris caucasica Steven	NE	x	x	x		x	x
Iris colchica Kem.-Nath.	NT			x		x	
Iris demetrii Achv. & Mirzoeva [*Iris prilipkoana* Kem.-Nath.]	VU B2ab(iii,v)	x	x				
Iris furcata M. Bieb.	NE	x	x	x		x	
Iris grossheimii Woronow ex Grossh.	EN B1ab(iii)+2ab(iii)	x	x				
Iris hyrcana Woronow ex Grossh.	VU D2		x				
Iris iberica Hoffm.	NE	x	x	x			
Iris imbricata Lindl. [*Iris sulphurea* K. Koch]	NE	x	x		x		
Iris lazica Albov	NE			x			x
Iris marschalliana Bobrov	EN B1ab(iii,v)+2ab(iii,v)					x	
Iris notha M. Bieb.	VU B2ab(ii,iv,v)					x	
Iris paradoxa Steven	NE	x	x	x	x		
Iris prilipkoana Kem.-Nath. [*Iris demetrii* Achv. & Mirzoeva]	VU B2ab(iii,v)	x	x				
Iris pseudonotha Galushko	LC					x	
Iris schelkownikowii (Fomin) Fomin	EN B2ab(v)		x				
Iris schischkinii Grossh.	DD		x				
Iris sulphurea K. Koch [*Iris imbricata* Lindl.]	NE	x	x		x		
Iris taochia Woronow ex Grossh.	EN B1ab(i,ii,v)+2ab(i,ii,v)						x
Iris timofejewii Woronow	EN B2ab(v)					x	

TAXON	RED LIST ASSESSMENT	AR	AZ	GE	IR	RU	TU
Iris winogradowii Fomin	VU D2			x			
JUNCACEAE							
Juncus elbrusicus Galushko	NT					x	
Luzula abchasica Novikov [*Luzula stenophylla* Steud.]	NE		x	x		x	
Luzula stenophylla Steud. [*Luzula abchasica* Novikov]	NE		x	x		x	
LILIACEAE							
Erythronium caucasicum Woronow	NE			x	x	x	
Fritillaria collina Adams	NE	x	x	x		x	
Fritillaria grandiflora Grossh.	CR B1ab(v)		x				
Fritillaria lagodechiana Charkev.	NT			x		x	
Fritillaria latifolia Willd.	NE			x		x	x
Fritillaria lutea M. Bieb.	DD		x				
Fritillaria orientalis Adams	NT			x		x	
Fritillaria pinardii Boiss. subsp. *hajastanica* Gabrieljan	NE	x	x				
Gagea bezengiensis Levichev	NT					x	
Gagea caroli-kochii Grossh.	NE	x	x	x			
Gagea chanae Grossh. var. *chanae*	NE	x	x	x		x	x
Gagea chanae Grossh. var. *oppressa* Levichev	NE					x	
Gagea charadzeae Davlian.	NE	x	x	x			
Gagea commutata K. Koch [*Gagea sarmentosa* K. Koch]	NE		x	x		x	
Gagea cuneata Levichev & Murtaz.	DD					x	
Gagea daghestanica Levichev & Murtaz.	DD					x	
Gagea eleonorae Levichev	NE	x	x	x		x	
Gagea helenae Grossh.	NE	x	x	x		x	
Gagea improvisa Grossh.	VU D2		x				
Gagea menitskyi Levichev	NE	x					
Gagea reticulata (Pall.) Schult. & Schult. f. var. *pascualis* Levichev	NE	x	x				

TAXON	RED LIST ASSESSMENT	AR	AZ	GE	IR	RU	TU
Gagea sarmentosa K. Koch [*Gagea commutata* K. Koch]	NE		x	x		x	
Gagea sulfurea Miscz.	NE	x	x	x		x	
Gagea tenuissima Miscz.	DD						x
Lilium armenum (Miscz. ex Grossh.) Manden. [*Lilium monadelphum* M. Bieb. subsp. *armenum* (Miscz. ex Grossh.) Kudrjasch.]	LC	x	x				x
Lilium caucasucum (Miscz. ex Grossh.) Grossh. [*Lilium martagon* L. subsp. *caucasicum* Miscz. ex Grossh.]	VU C2a(i)			x		x	
Lilium ciliatum P. H. Davis	EN B1ab(ii,iii,v)+ 2ab(ii,iii,v); C2a(i)						x
Lilium kesselringianum Miscz.	NE			x		x	x
Lilium ledebourii (Baker) Boiss.	NE		x		x		
Lilium martagon L. subsp. *caucasicum* Miscz. ex Grossh. [*Lilium caucasucum* (Miscz. ex Grossh.) Grossh.]	NE			x		x	
Lilium monadelphum M. Bieb. subsp. *armenum* (Miscz. ex Grossh.) Kudrjasch. [*Lilium armenum* (Miscz. ex Grossh.) Manden.]	LC	x	x				x
Lilium monadelphum M. Bieb. subsp. *georgicum* (Manden.) Gagnidze	DD			x			
Lilium monadelphum M. Bieb. var. *szovitsianum* (Fisch. & Avé-Lall.) Elwes [*Lilium szovitsianum* Fisch. & Avé-Lall. subsp. *szovitsianum*]	NE	x		x		x	x
Lilium szovitsianum Fisch. & Avé-Lall. subsp. *szovitsianum* [*Lilium monadelphum* M. Bieb. var. *szovitsianum* (Fisch. & Avé-Lall.) Elwes]	NE	x		x		x	x
Tulipa confusa Gabrieljan	NE	x	x				
Tulipa eichleri Regel	NE		x	x			
Tulipa florenskyi Woronow	NE	x	x		x		
Tulipa gumusanica Terzioğlu	CR B2ab(ii,iii)						x
Tulipa julia K. Koch	LC	x	x				
Tulipa sosnowskyi Achv. & Mirzoeva	NE	x	x				

TAXON	RED LIST ASSESSMENT	AR	AZ	GE	IR	RU	TU
Tulipa urumiensis Stapf	NE				x		
Tulipa wendelboi Matin & Iranshahr	CR				x		
MELANTHIACEAE							
Narthecium balansae Briq.	NE			x		x	x
ORCHIDACEAE							
Cephalanthera caucasica Kraenzl.	VU D2		x		x		
Dactylorhiza amblyoloba (Nevski) Aver.	VU B2ac(iii)	x	x	x		x	
Dactylorhiza armeniaca Herdén	EN B2ac(ii)		x	x		x	x
Dactylorhiza euxina (Nevski) Czerep.	NT	x	x	x		x	x
Dactylorhiza euxina (Nevski) Czerep. var. *markowitchzii* (Soó) Renz & Taubenheim	DD						x
Dactylorhiza incarnata L. subsp. *jugicrucis* Akhalkatsi, R. Lorenz & Mosulishvili	EN D1			x			
Dactylorhiza merovensis (Grossh.) Aver.	NE	x	x				
Dactylorhiza urvilleana (Steud.) H. Baumann & Künkele subsp. *urvilleana*	LC				x		x
Epipactis viridiflora Hoffm. ex Krok. subsp. *kuenkeleana* Akhalkatsi. H. Baumann, R. Lorenz & Mosulishvili	VU D2		x			x	
Himantoglossum formosum (Steven) K. Koch	EN B1ab(ii,iii)		x			x	
Ophrys caucasica Woronow ex Grossh. subsp. *caucasica*	NT			x		x	
Ophrys caucasica Woronow ex Grossh. subsp. *cyclocheila* Aver.	NE	x	x	x		x	
Ophrys oestrifera M. Bieb. subsp. *bremifera* (Steven) K. Richt.	NE	x	x			x	
Orchis collina Banks & Sol. subsp. *chlorotica* (Woronow) Aver.	DD		x			x	
Orchis papilionacea L. subsp. *schirvanica* (Woronow) Soó	NE	x	x			x	
Orchis schelkownikowii Woronow	EN		x				
Traunsteinera sphaerica (M. Bieb.) Schltr.	NE	x	x	x		x	x

TAXON	RED LIST ASSESSMENT	AR	AZ	GE	IR	RU	TU
POACEAE							
Achnatherum ordubadense Tzvelev	DD		x				
Achnatherum roshevitzii Mussajev	VU D2		x				
Agropyron gracillimum Nevski [*Elytrigia gracillima* (Nevski) Nevski]	NE		x	x		x	
Agropyron pinifolium Nevski	VU B1ab(ii,iii)+2ab(ii,iii)					x	
Agrostis balansae (Boiss.) Tzvelev	NE	x		x		x	x
Agrostis marschalliana Seredin [*Agrostis vinealis* Schreb.]	NE	x	x	x		x	
Agrostis trichantha (Schischk.) Tzvelev	NE	x		x			
Agrostis vinealis Schreb. [*Agrostis marschalliana* Seredin]	NE	x	x	x		x	
Alopecurus albovii Tzvelev	NE			x		x	
Alopecurus armenus (K. Koch) Grossh.	DD	x	x				
Alopecurus longifolius Kolak.	VU D2			x			
Alopecurus tuscheticus Trautv.	NE	x	x	x		x	
Avena malzevii Tzvelev	DD		x				
Avenula adzharica (Albov) Holub [*Helictotrichon adzharicum* (Albov) Grossh.]	NE	x	x	x		x	
Avenula armeniaca (Schischk.) Holub [*Helictotrichon armeniacum* (Schischk.) Grossh.]	NE	x	x	x		x	x
Briza marcowiczii Woronow	NE			x		x	x
Bromopsis alexeenkoi (Tzvelev) Czerep.	DD		x			x	
Bromopsis aristata (K. Koch) Holub	NE		x	x		x	
Bromopsis biebersteinii (Roem. & Schult.) Holub subsp. *beckeri* (Tzvelev) Tzvelev	DD		x			x	
Bromopsis biebersteinii (Roem. & Schult.) Holub subsp. *biebersteinii*	NE		x	x		x	
Bromopsis biebersteinii (Roem. & Schult.) Holub subsp. *kazbecki* (Tzvelev) Tzvelev	NE			x		x	
Bromopsis divaricata (Tzvelev) Czerep.	DD			x			

TAXON	RED LIST ASSESSMENT	AR	AZ	GE	IR	RU	TU
Bromopsis gabrielianae Ogan.	CR D	x					
Bromopsis gordjaginii (Tzvelev) Galushko	DD					x	
Bromopsis indurata (Hausskn. & Bornm.) Holub	NE	x	x				
Bromopsis nachiczevanica (Tzvelev) Holub	NE	x	x				
Bromopsis shelkovnikovii (Tzvelev) Holub	NE	x	x	x			
Bromopsis variegata M. Bieb. subsp. *villosula* (Steud.) Tzvelev [*Bromopsis villosula* (Steud.) Holub]	NE	x	x	x	x	x	x
Bromopsis villosula (Steud.) Holub [*Bromopsis variegata* M. Bieb. subsp. *villosula* (Steud.) Tzvelev]	NE	x	x	x	x	x	x
Bromopsis woronowii (Tzvelev) Czerep.	NE	x	x		x		
Bromopsis zangezura Ogan.	NE	x	x				
Bromus tzvelevii Mussajev	CR B1ab(iii)+2ab(iii)		x				
Calamagrostis balkharica P. A. Smirn.	LC					x	
Calamagrostis caucasica Trin.	NE	x	x	x		x	
Calamagrostis czerepanovii Husseinov	DD					x	
Calamagrostis dmitrievae Tzvelev	DD			x			
Calamagrostis iberica (Griseb.) Litv.	NE	x	x	x			
Calamagrostis minarovii Husseinov	DD					x	
Calamagrostis paradoxa Lipsky	NT					x	
Calamagrostis subchalybaea Tzvelev	NT					x	
Calamagrostis tzvelevii Husseinov	DD					x	
Catabrosella calvertii (Boiss.) Czerep.	NE	x	x		x		
Catabrosella variegata (Boiss.) Tzvelev subsp. *leiantha* (Hack.) Tzvelev	NE			x		x	x
Colpodium schelkownikowii Grossh.	DD	x	x				
Dactylis hyrcana (Tzvelev) Mussajev	VU D2		x				
Deschampsia caespitosa (L.) P. Beauv. subsp. *wilhelmsii* (Steud.) Tzvelev	NE	x	x	x		x	

TAXON	RED LIST ASSESSMENT	AR	AZ	GE	IR	RU	TU
Deyeuxia parsana Bor	NE				x		
Digitaria caucasica Henrard	NE			x		x	
Elymus buschianus (Roshev.) Tzvelev	NE			x		x	
Elymus lazicus (Boiss.) Melderis subsp. *lazicus*	EN B1ab(iii)+2ab(iii)						x
Elymus longearistatus (Boiss.) Tzvelev subsp. *sintenisii* Melderis	CR B2ab(iii)						x
Elymus nodosus (Nevski) Melderis subsp. *platyphyllus* Melderis	CR B2ab(ii,iii,v)						x
Elymus prokudinii (Seredin) Tzvelev	NT					x	
Elymus sosnowskyi (Hack.) Melderis [*Elytrigia sosnowskyi* (Hack.) Nevski]	EN B1ab(i,ii)+2ab(i,ii)	x					x
Elymus troctolepis (Nevski) Tzvelev	NE			x		x	
Elytrigia armena (Nevski) Nevski	NE	x	x				x
Elytrigia dshinalica Sablina	DD					x	
Elytrigia gracillima (Nevski) Nevski [*Agropyron gracillimum* Nevski]	NE		x	x		x	
Elytrigia heidemaniae Tzvelev	NE	x	x				
Elytrigia pertenuis (C. A. Mey.) Nevski	VU B1ab(ii)+2ab(ii)		x				
Elytrigia sinuata (Nevski) Nevski	NE	x	x	x			x
Elytrigia sosnowskyi (Hack.) Nevski [*Elymus sosnowskyi* (Hack.) Melderis]	EN B1ab(i,ii)+2ab(i,ii)	x					x
Festuca alexeenkoi E. B. Alexeev	LC		x			x	
Festuca artvinensis Makgr.-Dann.	NE			x			x
Festuca azgarica E. B. Alexeev	DD					x	
Festuca buschiana (St.-Yves) Tzvelev	NE			x		x	x
Festuca caucasica (Boiss.) Hack.	NE			x		x	
Festuca chalcophaea V. I. Krecz. & Bobrov subsp. *euryphylla* (St.-Yves) Makgr.-Dann.	NE						x
Festuca daghestanica (Tzvelev) E. B. Alexeev	LC		x			x	
Festuca djimilensis Boiss. & Balansae	NE			x		x	x
Festuca elbrusica E. B. Alexeev	NT					x	

TAXON	RED LIST ASSESSMENT	AR	AZ	GE	IR	RU	TU
Festuca inguschetica E. B. Alexeev	NT					x	
Festuca karsiana E. B. Alexeev	NE			x			x
Festuca krivotulenkoae E. B. Alexeev	NT					x	
Festuca lazistanica E. B. Alexeev subsp. *giresunica* Markgr.-Dann.	CR B2ab(ii,iii,v)						x
Festuca lazistanica E. B. Alexeev subsp. *lazistanica*	EN B1ab(ii,iii,v)+ 2ab(ii,iii,v)	x					x
Festuca paphlagonica (St.-Yves) Markgr.-Dann. subsp. *villosula* (St.-Yves) Markgr.-Dann.	DD						x
Festuca pontica E. B. Alexeev ex Markgr.-Dann.	EN B1ab(i,ii,iii,v)+ 2ab(i,ii,iii,v)						x
Festuca primae E. B. Alexeev	LC					x	
Festuca sabalanica E. B. Alexeev	EN				x		
Festuca skvortsovii E. B. Alexeev	NE	x	x				
Festuca sommieri Litard.	NE			x		x	x
Festuca transcaucasica (St.-Yves) Tzvelev	CR B1ab(iii)+2ab(iii)		x				
Festuca tzvelevii E. B. Alexeev	NT					x	
Festuca varia Haenke subsp. *woronowii* (Hack.) Tzvelev [*Festuca woronowii* Hack.]	NE	x	x	x		x	x
Festuca woronowii Hack. [*Festuca varia* Haenke subsp. *woronowii* (Hack.) Tzvelev]	NE	x	x	x		x	x
Festuca xenophontis Markgr.-Dann.	EN B1ab(i,ii,iii,v)+ 2ab(i,ii,iii,v)						x
Festuca yaroschenkoi (St.-Yves) E. B. Alexeev	LC		x			x	
Gaudinopsis egorovae Gabrieljan	DD	x					
Glyceria acutiuscula H. Scholz	DD					x	
Glyceria caspia Trin.	NE		x	x	x		
Helictotrichon adzharicum (Albov) Grossh. [*Avenula adzharica* (Albov) Holub]	NE	x	x	x		x	

TAXON	RED LIST ASSESSMENT	AR	AZ	GE	IR	RU	TU
Helictotrichon armeniacum (Schischk.) Grossh. [*Avenula armeniaca* (Schischk.) Holub]	NE	x	x	x		x	x
Hordeum hrazdanicum Gandilyan	LC	x					
Hyalopoa czirahica Husseinov	DD					x	
Hyalopoa hracziana Gabrieljan & Tzvelev	DD	x					
Hyalopoa lakia (Woronow) Tzvelev	CR B2ab(iii)					x	
Hyalopoa pontica (Balansa) Tzvelev	NE			x		x	x
Koeleria albovii Domin subsp. *albovii*	NE	x	x	x			
Koeleria albovii Domin subsp. *fominii* (Domin) Tzvelev	NE	x	x	x			
Koeleria albovii Domin subsp. *loriensis* Nersesian	NT	x					
Koeleria grossheimiana (Tzvelev) Galushko	NT					x	
Koeleria luerssenii (Domin) Domin	NE			x		x	
Melica minor Hack. ex Boiss.	NE		x	x		x	
Milium alexeenkoi (Tzvelev) Tzvelev	VU B2ab(iii)		x			x	
Milium atropatanum Maroofi	NE				x		
Milium schmidtianum K. Koch	NE			x	x	x	x
Milium transcaucasicum Tzvelev	NE	x	x				
Paracolpodium colchicum (Albov) Tzvelev	NE			x		x	x
Paracolpodium tzvelevii Gabrieljan	DD	x					
Phleum tzvelevii Dubovik	NE			x		x	
Poa alexeenkoi (Tzvelev) Czerep.	DD			x			
Poa bulbosa L. subsp. *delicatula* (Tzvelev) Tzvelev	DD		x			x	
Poa caucasica Trin.	NE		x	x		x	x
Poa greuteri Gabrieljan	DD	x					
Poa iberica Fisch. & C. A. Mey.	DD	x	x	x		x	
Poa imeretica Sommier & Levier	NE			x		x	
Poa meyeri Trin. ex Roshev.	NE	x	x				

TAXON	RED LIST ASSESSMENT	AR	AZ	GE	IR	RU	TU
Poa polychroa (Trautv.) Grossh.	DD		x				
Poa primae Tzvelev	DD					x	
Poa pseudosterilis Galkin	DD					x	
Poa rhomboidea Roshev.	NE			x		x	
Poa seredinii Galkin	NE	x	x	x		x	
Psathyrostachys dagestanica (T. Alex.) Nevski	EN B1ab(iii,v)+2ab(iii,v)					x	
Psathyrostachys rupestris (T. Alex.) Nevski	EN B1ab(iii,iv)					x	
Puccinellia bulbosa (Grossh.) Grossh.	NE	x	x				
Puccinellia sevangensis Grossh.	NE	x	x				
Secale kuprijanovii Grossh. [*Secale montanum* Guss. subsp. *kuprijanovii* (Grossh.) Tzvelev]	NE			x		x	
Secale montanum Guss. subsp. *chaldicum* (Fed.) Tzvelev	NE	x	x				
Secale montanum Guss. subsp. *kuprijanovii* (Grossh.) Tzvelev [*Secale kuprijanovii* Grossh.]	NE			x		x	
Sesleria araratica Kit Tan	NT	x					x
Stipa anisotricha P. A. Smirn.	DD	x					
Stipa araxensis Grossh.	NE	x	x				x
Stipa canescens P. A. Smirn. ex Roshev.	NT	x	x				
Stipa daghestanica Grossh.	NE			x		x	
Stipa gaubae Bor	NE		x		x		
Stipa gegarkunii P. A. Smirn.	DD	x					
Stipa karjaginii Mussajev & Sadychov	VU D2		x				
Stipa pennata L. subsp. *lejophylla* (P. A. Smirn.) Tzvelev	DD	x					
Stipa sosnowskyi Seredin	EN B1ab(ii,iii,iv)+ 1ab(ii,iii,iv)					x	
Stipa zuvantica Tzvelev	LC		x				
Trisetum buschianum Seredin	NE			x		x	

TAXON	RED LIST ASSESSMENT	AR	AZ	GE	IR	RU	TU
Trisetum buschianum Seredin subsp. *transcaucasicum* (Seredin) Mosulischvili [*Trisetum transcaucasicum* Seredin]	NE		x	x		x	
Trisetum geghamense Gabrieljan	DD	x					
Trisetum teberdense (Litv.) Charadze	NE			x		x	
Trisetum transcaucasicum Seredin [*Trisetum buschianum* Seredin subsp. *transcaucasicum* (Seredin) Mosulischvili]	NE		x	x		x	
Triticum araraticum Jakubz. [*Triticum timopheevii* (Zhuk.) Zhuk. var. *araraticum* (Jakubz.) Mackey]	NE	x	x				x
Triticum carthlicum Nevski	EW			x			
Triticum durum Desf.	NE		x	x			
Triticum macha Dekapr. & Menabde	EW			x			
Triticum palaeo-colchicum Menabde	EW			x			
Triticum timopheevii (Zhuk.) Zhuk.	EW			x			
Triticum timopheevii (Zhuk.) Zhuk. var. *araraticum* (Jakubz.) Mackey [*Triticum araraticum* Jakubz.]	NE	x	x				x
Triticum vavilovii Jakubz.	NE	x	x				
Triticum zhukovskyi Menabde & Eritzjan	EW			x			
Zingeria kochii (Mez) Tzvelev	VU	x					
RUSCACEAE							
Ruscus colchicus P. F. Yeo	NE			x		x	x
TRILLIACEAE							
Paris incompleta M. Bieb.	NE		x	x		x	x
TYPHACEAE							
Typha alexeevii Mavrodiev	DD		x			x	
Typha azerbaijanensis Hamdi & Assadi	VU				x		

INDEX